T0360660

Direct Modeling for Computational Fluid Dynamics

Construction and Application of Unified Gas-Kinetic Schemes

Advances in Computational Fluid Dynamics

Published

Vol. 2 Adaptive High-Order Methods in Computational Fluid Dynamics
edited by Z. J. Wang (Iowa State University, USA)

Vol. 3 Lattice Boltzmann Method and Its Applications in Engineering
by Zhaoli Guo (Huazhong University of Science and Technology, China) and Chang Shu (National University of Singapore, Singapore)

Vol. 4 Direct Modeling for Computational Fluid Dynamics:
Construction and Application of Unified Gas-Kinetic Schemes
by Kun Xu (Hong Kong University of Science and Technology, Hong Kong)

Forthcoming

Vol. 1 Computational Methods for Two-Phase Flows
by Peter D. M. Spelt (Imperial College London, UK), Stephen J. Shaw (X'ian Jiaotong – University of Liverpool, Suzhou, China) & Hang Ding (University of California, Santa Barbara, USA)

Advances in Computational Fluid Dynamics — Vol. 4

Direct Modeling for Computational Fluid Dynamics

Construction and Application of Unified Gas-Kinetic Schemes

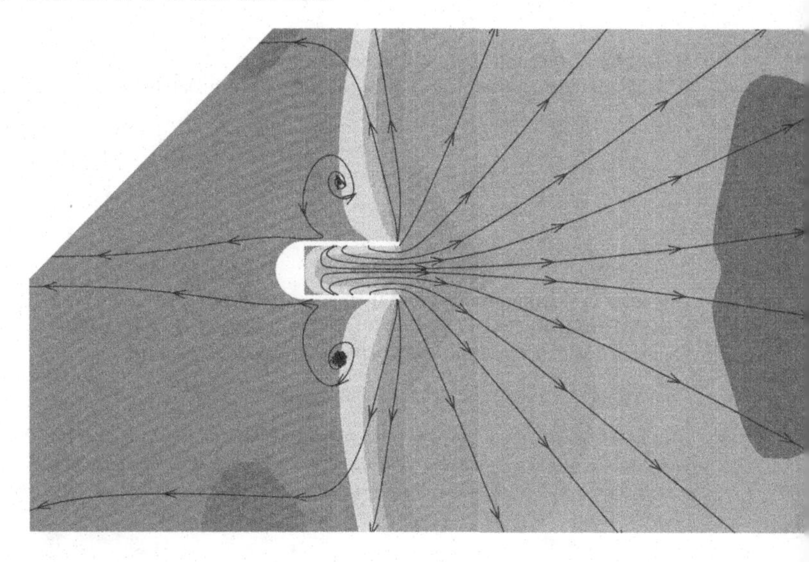

Kun Xu

Hong Kong University of Science and Technology, Hong Kong

World Scientific

NEW JERSEY • LONDON • SINGAPORE • BEIJING • SHANGHAI • HONG KONG • TAIPEI • CHENNAI

Published by

World Scientific Publishing Co. Pte. Ltd.
5 Toh Tuck Link, Singapore 596224
USA office: 27 Warren Street, Suite 401-402, Hackensack, NJ 07601
UK office: 57 Shelton Street, Covent Garden, London WC2H 9HE

British Library Cataloguing-in-Publication Data
A catalogue record for this book is available from the British Library.

Advances in Computational Fluid Dynamics — Vol. 4
DIRECT MODELING FOR COMPUTATIONAL FLUID DYNAMICS
Construction and Application of Unified Gas-Kinetic Schemes

ISBN 978-981-4623-71-1

Typeset by Stallion Press
Email: enquiries@stallionpress.com

Printed in Singapore

The author would like to devote this book to Prof. T.D. Lee of Columbia University who initiated CUSPEA program for many Chinese students to continue their study aboard in 1980s.

Preface

The current Computational Fluid Dynamics (CFD) focuses mainly on the numerical solution of partial differential equations (PDEs). The aim of the CFD is to get the exact solution of the governing equations. Besides introducing the so-called numerical errors, it seems that the numerical mesh size has no any positive dynamic contribution on the solution. As the mesh size and time step go to zero, the CFD algorithm is supposed to converge to the exact solution of the equations, and the limited mesh size is associated with truncation errors only. This CFD principle is based on the belief of the fluid dynamic equations, and makes the equivalence between the fluid dynamics and the equations. Instead of numerical PDE, we propose that the CFD algorithm may be a direct flow modeling in a discretized space, which identifies the flow physics on the scales of mesh size and time step. In other words, the CFD algorithm is to construct discrete numerical governing equations in a space with limited resolution. This monograph will present such a direct modeling principle for CFD algorithm development, and the construction of unified gas-kinetic scheme under such a principle for the flow simulation in all flow regimes.

All fluid dynamic equations have their intrinsic valid modeling scales, such as the mean free path scale of the Boltzmann equation and the hydrodynamic scale of the Navier-Stokes equations. The lost information in the hydrodynamic scale is partially supplied with the modeling of constitutive relationship, which is related to the kinetic scale physics. The current CFD methodology targets on the equations and has no account on the physical modeling scales of these equations anymore. Even with limited mesh size, the CFD is to recover the solution of the PDEs as the mesh size and time step approaching to zero. Under such a CFD practice, the best result is to luckily get the exact solution of the governing equations. But, the flow physics described by the CFD solution is still limited by the modeling scale of the original governing equations. In reality, due to the limited cell resolution, we could never get the exact solution of the original governing equations due to the truncation error. Theoretically, we never know what is the exact underlying governing equation of the CFD algorithm, especially in the cases with unresolved "discontinuities". Therefore, there is NO unique solution when using the approach of numerical PDEs. That is probably the reason why there are so many CFD algorithms for the same PDEs, such as the gigantic amount of approximate Riemann solvers for the Euler equations. The above CFD practice also prevents us from developing multiple scale method if there is no such a governing equation, which is valid in all scales. For example, for the flow around a re-entry air vehicle in near space, the mesh size can vary significantly with respect to the local particle mean free path. There is no such a well-defined governing equation with a continuum variation of modeling scale. Many literatures may claim that the Boltzmann equation is valid in all flow regimes from free molecular to the continuum Navier-Stokes solution. This statement is based on the assumption of fully resolving the mean free path scale physics of the Boltzmann equation everywhere even in the continuum flow regime. It is more or less a statement of brutal force, which cannot be affordable in a real computation. If it were necessary to resolve up to the smallest scale everywhere, there should have no any other scientific discipline except particle physics. In the continuum flow regime, it is unrealistic to set the mesh size to the order of particle mean free path. Instead, we need to construct the governing equations directly in the mesh size scale, and these equations cannot be the Boltzmann equation or the Navier-Stokes equations if the mesh size scale is between the kinetic and hydrodynamic scales. What we are interested in at hydrodynamic scale is the wave propagation and interaction, and at kinetic scale the individual particle transport and collision. Even in the hydrodynamic scale, such as in the unresolved shock region, the macroscopic description seems inadequate to provide necessary mechanism to construct a stable non-equilibrium shock transition. When the shock capturing schemes encounter "carbuncle phenomena", the continuous attempt on different kind of discretization of the PDEs, with the hope of introducing appropriate flow physics which has been ignored in the original Euler equations, can only lead the CFD into a maze. In a discretized space, the CFD should be a multiple scale and multiple physics modeling method.

The aim of CFD is to identify and simulate flow physics in the mesh size scale. This principle of direct modeling is not to solve any specific equation, but to construct flow evolution model. With the variation of the ratio between the mesh size and the local particle mean free path, a direct modeling should be able to capture the flow physics from the kinetic scale particle collision and transport to the hydrodynamic scale wave propagation. The unified gas-kinetic scheme (UGKS) presented in this monograph is mainly about such a direct modeling method, where a continuum spectrum of "governing equations" will be directly obtained through the modeling. The success of the UGKS is due to the adaptation of a time-dependent crossing scale gas evolution solution in the algorithm development, and this local modeling solution recovers the physics from free molecular transport to the macroscopic wave propagation. The specific solution adopted locally in the numerical algorithm depends on the ratio between the numerical time step and the local particle collision time. As a result, with a variation of cell resolution, the UGKS provides a smooth transition of the flow physics of different scales. This methodology is different from other multiscale methods, which target to connect distinctive governing equations.

The author started to work on the gas-kinetic scheme (GKS) more than twenty years ago from the postgraduate period at astronomy department of Columbia university. During the early years, the scheme is mainly to solve the compressible Euler and Navier-Stokes equations through the kinetic formulation. So, in the CFD community, the GKS is mostly regarded as a kind of approximate Riemann solvers, such as the modified flux vector splitting scheme. In the following years, with the further development of kinetic schemes to the non-equilibrium flows and its fully comparison with other CFD algorithms, such as the Godunov method, it is realized that the dynamics in the GKS is rich, which is beyond the Euler and NS equations, especially in the physical modeling of a discontinuous shock layer of a shock capturing gas kinetic scheme. The dissipation in the GKS is provided from the non-equilibrium kinetic particle transport. The dissipative mechanism of the GKS and the Godunov method will be analyzed in Chapter 4. In order to extend the GKS to the rarefied flow computation, much effort has been paid to add more physical ingredients into the GKS construction, such as the generalization of constitutive relationship through the introduction of direction-dependent viscosity coefficient, and the extension of the translational temperature from a scalar to a tensor. But, all these attempts have gotten only partially success in the non-equilibrium flow study. At end, instead of trying different kinds of modification on the macroscopic level, such as the inclusion of Burnett or Super-Burnett terms, a discretized particle velocity is used to capture the peculiarity of the gas distribution function in the nonequilibrium flow regime. The newly developed unified gas-kinetic scheme (UGKS) is an extension of the GKS with the update of both macroscopic flow variables and the microscopic distribution function, and the scheme works very well in all flow regimes. In terms of the algorithm construction, the idea of the UGKS becomes even simpler than GKS, because there is no much kinetic theory needed. With further study, it becomes clear that the gas-kinetic scheme is more or less a direct modeling method. This monograph is basically to present such an understanding. The direct modeling concept may benefit to the CFD community. The current CFD research is mainly about the numerical discretization of well-defined PDEs. The difficulties encountered in the Godunov type shock capturing schemes, such as the shock instability in high Mach number flow computations, may come from the inadequate flow physics in the governing equations in the description of a "discontinuity" in a space with limited resolution. This may be the reason for the non-uniqueness of the CFD solutions as well. Also, the methodology underlying the UGKS is useful for developing multiple scale schemes for other transport process, such as radiative transport and plasma evolution.

The content of this monograph is based on a graduate course taught by the author in the past several years at Hong Kong University of Science and Technology, and short courses at Peking University and National Laboratory of Aerodynamics. This monograph is written for different levels of readers. For students and beginners, the ideas presented here will be useful to give them a wide exposure in CFD study. The mathematics involved in this book is not sophisticated. It can be understood by anyone with basic training on calculus and linear algebra. Some basic knowledge on differential equation and statistical mechanics will be helpful, but not necessary. The book will also benefit to the CFD researchers working on the Euler and Navier-Stokes equations. At least, it presents algorithms which could be a supplement to the existing CFD methods. To understand the similarity and differences between the gas-kinetic scheme and their own in-house CFD method will be a joyful experience. The method presented in this book may be useful in practical engineering applications, especially for vacuum pumps and high speed non-equilibrium flow simulation of near space flight.

Many people have helped and made substantial contributions to the development of the gas-kinetic scheme. I give my sincerely thanks to my collaborators and colleagues: K.H. Prendergast, L. Martinelli, A. Jameson, W.H. Hui, M.D. Su, M. Ghidaoui, T. Ohwada, M. Torrilhon, E. Josyula,

Q.B. Li, T. Tang, H.Z. Tang, J.Q. Li, Z.W. Li, Z.H. Li, M.L. Mao, H. Luo, G.X. Ni, S. Jiang, G.P. Zhao, C.P. Cai, J.C. Huang, J.Y. Yang, Q.H. Sun, G.A. Bird, Z.L. Guo, C.W. Zhong, Q.D. Cai, C.B. Lee, and many others; and supervised students: Y.S. Lian, J.Q. Deng, Y.T. Que, C.Q. Jin, H.W. Liu, J. Luo, S.Z. Chen, P.B. Yu, R.J. Wang, C. Liu, L. Pan, and S. Liu. Without their valuable contributions, the gas-kinetic scheme could never reach the current state of maturity. The development of unified scheme originates from a collaboration with my friend J.C. Huang, to whom I extend my special thanks. Thanks are also due to Z.L. Guo for his comment on the manuscript. I would also like to acknowledge the financial support from the Hong Kong Research Grant Council and State Key Laboratory for Turbulence and Complex Systems at Peking University in the past years.

Finally, I would like to give my thanks to Prof. Ami Harten, who took sabbatical leave in the spring semester of 1993 at Courant Institute, and recommended me to Prof. Antony Jameson as a postdoctoral fellow at Princeton university; to Prof. Jameson, who shifted my research interest from astrophysics to aerospace during the three years at Princeton, and helped me greatly in my professional career; to Dr. Manuel Salas, who gave me the visiting position each summer from 1996 to 2001 at ICASE at NASA Langley, where I got the chance to discuss commonly interesting problems with many world-renowned scientists in CFD community; and to Prof. Bram van Leer who gave me inspiration and encouragement during my difficult times. At end, I would like to thank my wife, Jie Shen, for her love, understanding, and support in the past decades starting from my postgraduate study.

K. Xu

Contents

Chapter 1

Direct Modeling for Computational Fluid Dynamics

Computational fluid dynamics (CFD) is a scientific discipline, which aims to capture fluid motion in a discretized space. The description of the flow behavior depends closely on the scales which are used to identify or "see" it. All theoretical equations, such as the Boltzmann equation or the Navier-Stokes equations, are constructed and valid only on their modeling scales, even though these scales cannot be explicitly observed in these equations. The mechanism of these governing equations depends on the physical modeling, such as the constitutive relationship in the stress and strain of the hydrodynamic equations, and the separation of transport and collision of the kinetic equation. The existence of a few distinct governing equations, such as the Boltzmann equation, the Navier-Stokes equations, and the Euler equations, only presents a partial picture about flow physics in their specific modeling scales. The governing equations between these scales have not been fully explored yet due to the tremendous difficulties in the modeling, even the flow variables to be used for the description of a non-equilibrium flow in the scale between the Navier-Stokes and the Boltzmann equation are not clear. However, the CFD provides us an opportunity to present the flow physics in the mesh size scale. With the variation of the mesh size to resolve the flow physics, the direct modeling of CFD may open a new way for the description and simulation of flow motion. This book is mainly about the construction of numerical algorithms through the principle of direct modeling. The ultimate goal of CFD is to construct the discrete flow dynamic equations, the so-called algorithm, in a discretized space. These equations should be able to cover a continuum spectrum of flow dynamics with the variation of the ratio between the mesh size and the particle mean free path.

Instead of direct discretization of existing fluid dynamic equations, the direct modeling of CFD is to study the corresponding flow behavior in the cell size scale. The direct discretization of a well-defined governing equation may not be an appropriate way for CFD research, because the modeling scale of partial differential equations (PDEs) and mesh size scale may not be matched. Here, besides introducing numerical error the mesh size doesn't actively play any dynamic role in the flow description. For example, the Fourier's law of heat conduction is valid in a scale where the heat flux is proportional to temperature gradient. How could we imagine that such a law is still applicable on a mesh size scale, which can be freely chosen, such as $1cm$, $1m$, or even $1km$? To avoid this difficulty, one may think to resolve everything through the finest scale, such as the molecular dynamics. But, the use of such a resolution numerically in the simulation is not practical due to the overwhelming computational cost, and it is not necessary at all in real engineering applications, since in most times only macroscopic flow distributions are needed, such as the pressure, stress, and heat flux on the surface of a flying air vehicle. Instead of direct discretization of PDEs or resolving the smallest scale of molecular dynamics, a possible way is to model and capture the flow dynamics in the corresponding mesh size scale, and the choice of the mesh size depends on how much information is sufficient to capture the flow evolution in any specific application. Depending on the flow regimes, there is a wide variation between the cell size and the local particle mean free path. Therefore, a multiple scale modeling is needed in the CFD algorithm development, i.e, the construction of the so-called discrete governing equations.

1.1 Physical Modeling and Numerical Solution of Fluid Dynamic Equations

There are different levels in flow modelings. The theoretical fluid mechanics is to apply physical laws in a certain scale with the modeling of the flux and constitutive relationship. Then, based on the construction of discrete physical law, as the control volume shrinking to zero, and with the assumption of smoothness of flow variables in the scale of control volume, the corresponding PDEs are obtained. For the PDEs, even with a continuous variations of space and time, the applicable regime of these equations is on its modeling scale, such as the scale for the validity of constitutive relationship and the fluxes. For the Boltzmann equation, the modeling scale is the particle mean free path and the particle collision time, where the particle collision and transport in such a scale are separated and modeled in an operator splitting way. For the Navier-Stokes (NS) equations, the scale is the dissipative layer thickness, where the Stokes and Fourier's laws are valid. In the NS modeling, the accumulating effect from a gigantic amount of particle collision and transport, instead of individual one, is modeled. It is more or less a mean field approximation. For the Euler equations, the scale becomes even coarse and only the advection wave propagation is followed, where the dissipative wave structures are replaced by discontinuities at contact surface and shock layer. The equations constructed for fluid dynamics can be the potential flow, the Euler equations, the Navier-Stokes equations, and the Boltzmann equations with different modeling scales. After the establishment of these PDEs, research work has been concentrated on the mathematical analysis of these equations, or on their numerical solution. If these equations are obtained from models with distinguishable scale, could we use a continuous variation of scale and get the corresponding governing equations as well? This is the question we will try to answer in this book.

With the above theoretical PDEs, the traditional CFD method is to discretize these PDEs without referring to the underlying modeling scale of these equations. The intrinsic modeling scale of the equation has no direct connection with the definition of numerical mesh size, except resolving the specific layer numerically. The traditional CFD concerns the limiting solution of the scheme as the cell size and time step approaching to zero. The order of accuracy is used to judge the quality of the scheme. In order to improve the accuracy of the solution in the computation, the truncation error and modified equation are studied. Then, according to the numerical errors, the discretization can be adjusted in order to improve the accuracy. Certainly, the numerical stability of the scheme has to be considered as well. This numerical principle is passive and the ultimate goal is to remove the mesh size effect. In this approach, the best result we can get is the exact solution of the governing equations. But, even in this case the solution is still limited by the physical modeling scale of the equations. In practice, the mesh size and time step can never take limiting values, and the exact governing equations of the numerical algorithms become unknown. Usually the PDEs become simplified model of a real physical reality, such as the absence of dissipation in the Euler equations. However, in a discretized space with limited resolution, the inclusion of dissipation becomes necessary for a numerical scheme. Therefore, the artificial one, instead of physical one, is implicitly added in the numerical solution. The vast amount of shock capturing schemes for the Euler equations clearly indicate that the inviscid Euler equations with incomplete physical modeling have been modified with a variety of artificial dissipations, and there is no unique solution at all in the mesh size scale for the Euler solution, even though they may have the same limiting solution with zero mesh size and time step.

In the past decades, the equations to be solved numerically cover the potential flow (60s-70s), the Euler equations (70s-80s), the NS equations (90s- now), and the Boltzmann type equations (90s- now). Even though these equations have been routinely used in a wide range of engineering applications, the acceptability of the numerical schemes depend more or less on the Verification and Validation (V&V) process. Due to the absence of a solid foundation in the numerical PDEs, there is no a precise predicability about the outcome of these schemes. Peter Lax clearly states "the theory of difference schemes is much more sophisticated than the theory of differential equations." Even with great success in the CFD research in the past decades, dynamic flaw in most numerical schemes still exist, such as the numerical shock instability in high Mach number flow simulations. There is still uncertainty about how to design high-order schemes for the capturing of both continuous and discontinuous solutions. Now we are facing an insurmountable difficulty about the numerical principle for the computational fluid dynamics. For any scientific discipline, a fundamental principle is necessary once the subject needs to be promoted from an experience-based study, like the current CFD method, to a principle-based scientific discipline. Now we propose that the principle of CFD is to study flow dynamics in the mesh size scale through the direct physical modeling, the so-called direct construction of discrete governing equations.

1.2 Direct Modeling of Fluid Motion

A gas is composed of molecules, and there are kinematic and dynamic properties for these molecules, such as trajectories of these molecules, molecular mass, mean free path, collision time, and interaction potential between colliding particles. Their representation in a numerical scheme depends on the cell resolution. In a discretized space, in order to identify different flow behavior, a numerical cell size Knudsen number, i.e., $Kn_c = l_{mfp}/\Delta x$, can be defined, which gives a connection between the physical gas property (particle mean free path l_{mfp}) and the cell resolution (Δx). The cell Knudsen number is a parameter to connect the numerical resolution to the molecular reality. In a flow computation, due to the freedom in choosing the cell size and the vast variation of physical condition, such as the flying vehicle at different altitude, the cell's Knudsen number can cover a wide range of values continuously. For example, around a space shuttle flying in near space, the mesh size can take the value of the molecular mean free path in the non-equilibrium shock regions and hundreds of mean free path far away from the shuttle. At different locations, different values of cell Knudsen numbers correspond to different flow physics.

CFD algorithm is to model the flow motion in the cell size scale. Certainly, the modeling can be much simplified by taking $Kn_c \approx 1$ everywhere, where the mesh size is on the same order as the particle mean free path. However from a computational point of view, both efficiency and accuracy of the schemes have to be considered. In a real application, such as the flow around a spacecraft, the particle mean free path can be changed significantly. Besides the difficulties of identifying the local mean free path beforehand, to resolve this scale everywhere will be prohibitively expensive. So, the choosing of a numerical mesh size needs to compromise among the flow information needed for a certain design, the efficiency to achieve such a solution, and the robustness of the scheme if the mesh size is not chosen properly. The purpose of multiple scale modeling is to efficiently capture flow behavior with a variation of scales.

In order to design such a multi-scale modeling scheme, the flow evolution with different Kn_c numbers has to be captured accurately. In the regions of $Kn_c \geq 1$, the same modeling mechanism of deriving the Boltzmann equation needs to be used numerically, such as the modeling of particle transport and collision. In the region of $Kn_c \ll 1$, the modeling in deriving the NS equations should be used in the numerical process, such as the drifting of an equilibrium state in the hydrodynamic limit. Between these two limits, a local flow evolution model has to be constructed as well, even though there is no a valid governing equation yet. Without valid equations, based on numerical PDE approach, it is impossible to design multiscale method, except the artificial brutal connection of different solvers, such as the use of molecular dynamics and the Euler equations in different domain. The aim of the unified scheme is to model the physics in all regimes and construct the algorithm. Different from numerical PDEs, in the direct modeling scheme, the mesh size will actively participate the flow evolution with the changing of the local value Kn_c. In other words, the mesh size and time step are dynamic quantities, where the flow physics is associated explicitly with their resolution, or the accumulation of particle transport and collision in such a scale. The choice of local resolution depends on the specific application.

Many physical problems do need such a unified approach for a valid description of all Kn_c regimes. In the aerospace engineering, we are interested in designing vehicles in the near space, i.e., flying between 20km and 100km. In the flow field, there will have both non-equilibrium flow region with the requirement of the cell size being on the same order of particle mean free path, and the continuum flow region with the cell size being much larger than the local particle mean free path. The challenge is that there is no any distinct boundary between different flow regions. In plasma physics, one needs to recover "macroscopic" modeling in the quasi-neutral region and "kinetic" modeling in non-quasi-neutral region, and there needs a smooth dynamic transition between these models. In radiation and neutron transport, there is optically thick and thin regions, and the dynamics for the photon or neutron transport is different, when a uniform mesh size scale is adopted everywhere. The goal for the development of the direct modeling method is not to use the domain decomposition to construct different flow solvers in different regions, but provides a framework with the automatic capturing of flow physics in different regimes.

The methodology of the unified scheme is to model different flow physics in different regimes in a uniform way, from the particle transport and collision to macroscopic wave propagation. The recipe here is that a time dependent evolution solution from the kinetic to the hydrodynamic scale will be constructed and used in the scheme construction. A valid evolution solution covering different scale physics is critically important here. Therefore, the unified scheme can be considered as an evolution solution-based modeling scheme. Since the kinetic equation itself is valid in the kinetic scale, but the time evolution solution with the account of both transport and collision can go beyond the kinetic scale. The inclusion of intensive particle collision as time step being much larger than the particle collision time will push the solution to the hydrodynamic regime. Since there is no any analytic evolution solution from the Boltzmann equation in general, the key of the unified scheme depends closely on the construction of such an evolution solution. For example, with the integral solution of the kinetic model equation, each term in the solution has to be modeled numerically with the consideration of mesh size scale. The unified scheme provides a general framework for the construction of multiscale method for the transport phenomena with multiple scales physics involved, such as the non-equilibrium flow study and radiative transport. The full Boltzmann collision term can be used as well in its valid scale in the construction of unified scheme.

1.3 Direct Modeling and Multiscale Coarse-graining Models

The macroscopic Navier-Stokes equations can be formally derived from the Boltzmann equation, which in turn can be derived from the Liouville equation via the BBGKY hierarchy[Cercignani (1988)]. Many physical assumptions are built into these derivations and their validations are warranted under certain physical conditions. In the theoretical framework of fluid dynamics, there are only a few distinct successful governing equations, such as the Boltzmann, the Navier-Stokes, and the Euler equations, which can be used to describe the flow physics in the corresponding scales. Between these equations, we may have also the Burnett, or super-Burnett, and many moment equations. But, these equations can be hardly judged from the physical point of view, because they are not directly obtained from first principle of scale-based direct modeling, rather than derived from mathematical manipulation of the kinetic equation, where even the modeling scales of these equations are not clearly defined. The kinetic equation is valid in the kinetic scale, i.e., mean free path and collision time. If the Boltzmann equation is regarded as a valid model in all scales, it still means that the kinetic scale physics is fully resolved and the solution in other scale is obtained through the accumulation of kinetic scale flow evolution. The Boltzmann equation can indeed derive the NS equations which are valid in the hydrodynamic flow regime, but it can also present many other equations through different asymptotic expansions and these equations may not be so successful as the NS ones. Actually, the construction of the NS equations has nothing to do with the Boltzmann equation, and the NS equations were obtained through rational analysis in the macroscopic level earlier. It may not be appropriate to claim that the Boltzmann equation is valid in all scales except they are resolved down to the particle mean free path everywhere.

On the hydrodynamic scale, the macroscopic equations are very efficient, but are not accurate enough to describe the non-equilibrium flow phenomena. Microscopic equation, on the other hand, may have better accuracy, but it is too expensive to be used in the hydrodynamic scale with the required resolution on the kinetic scale. The traditional multiscale modeling is to construct valid technique which combines the efficiency of macroscale models and the accuracy of microscale description. The constitutive relationship is critically important for the macroscopic equations, which has the corresponding physics in microscopic level. For a complex system, the constitutive relations become quite complicated, where too many parameters need to be fitted, and the physical meaning of these parameters becomes obscure. As a result, the original appeal of simplicity and universality of macroscopic model is lost. Central to the multisale coarse-grained modeling is the construction of the constitutive relation which represents the effect of the microscopic processes at the macroscopic level. Many attempts have been done in this kind of approaches [E (2012)].

In the direct modeling, there is no separation between the macroscopic and microscopic description. With a continuous variation of scale, a continuous spectrum of governing equations are recovered computationally. With the changing of modeling scale, a smooth transition from micro to macro gas dynamics will be obtained. In comparison, the generality of the multiscale coarse-graining methodology is limited because the non-equilibrium flow phenomena cannot be fully described through a few constitutive relationship with a limited number of degree of freedom in the macroscopic level only. In the direct modeling method, there is no modification of constitutive relationship in the macroscopic level from microscopic modeling, because the direct modeling is not associated to any specific scale to describe other scale physics. This is also different from the hybrid approach, where micro-macro methods are used in separate regions. In direct modeling method, a continuous scale variation of flow physics is recovered.

1.4 Necessity of Direct Modeling

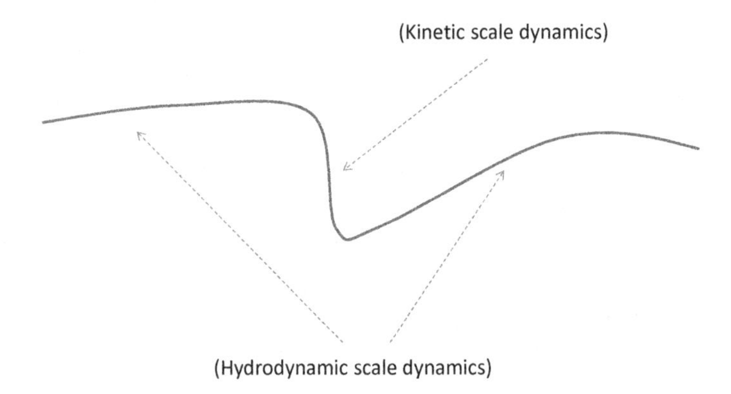

Fig. 1.1 Different physical scales are associated with different dynamics.

In a discretized space, the flow distribution may look like that in Fig. 1.1, where both discontinuous and continuous flow distributions appear.

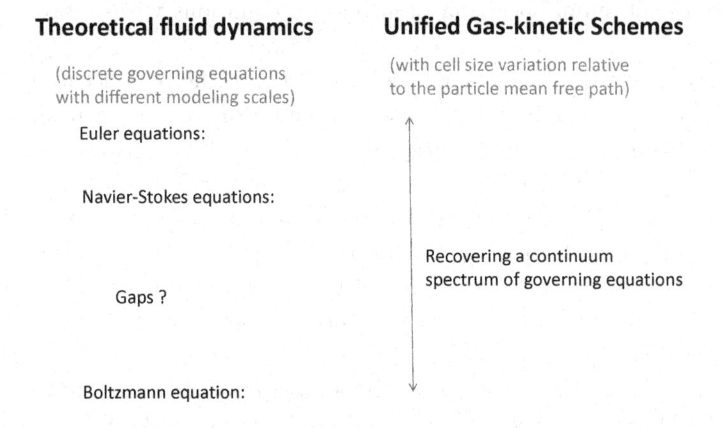

Fig. 1.2 Theoretical and numerical description of flow dynamics.

Theoretically, there is no reason to believe that there is a unique governing equation to describe both discontinuous and continuous flow evolution. In terms of the computation, if the mesh size is not fine enough the non-equilibrium shock region has to be enlarged to the mesh size scale. With the variation of mesh resolution, the non-equilibrium intensity identified in the mesh size scale will be changing as well. Even on the macroscopic level with the cell size being much larger than the shock thickness, the kinetic scale physics is still needed for the capturing of the non-equilibrium flow properties in the enlarged shock layer. On the other hand, in the smooth region, the equilibrium flow evolution can be described by the macroscopic equations. Numerically, all these descriptions are closely related to the mesh size and time step resolution. If both time step and cell size participate actively in the description of flow evolution, the cell size and time step cannot be arbitrarily assigned. In the numerical PDE approach, the cell size and time step are basically separated from the underlying modeling scale of the equation. The sensitivity of the numerical solution on the mesh distribution in the traditional CFD may represent the possible inconsistency between the mesh size scale and modeling scale of the equation. In the direct modeling approach, the choice of the cell size and time step will effect the flow dynamics. Therefore, the direct modeling is not only to develop numerical algorithm, but also to write down governing equations in different scales. As shown in Fig. 1.2, through the direct modeling, the unified scheme is to recover a continuum spectrum of governing equations with different scale mechanism.

The CFD research has made great progress in the past decades. There is no doubt that CFD will become even more important in future. The traditional numerical discretization of PDEs and the independence of flow physics from the numerical mesh resolution limits CFD's further development. Currently, most CFD algorithms are more or less an experience-based research subject, which is guided through the outcome and constraints on the numerical solution in the design of the scheme, such as the total variation diminishing (TVD) and essentially non-oscillatory (ENO) requirements for hyperbolic equation [Harten (1983); Harten *et al.* (1987)]. It lacks a fundamental principle to guide the dynamics. In other words, in the traditional CFD research, the dynamics is fully included in the PDEs, and the numerical computation becomes a tool for the PDEs' solution only. The direct modeling is an attempt to combine the dynamics and computation, which directly models the fluid evolution process in the mesh size scale, and captures its evolution. Certainly, the successful CFD recipes, such as TVD and ENO, are still needed for the discrete data reconstruction, but they are not the fundamental principles. In the unified scheme, there are no specific governing equations to be solved, and it is not expected to make the cell size and time step shrinking to zero in order to legitimate the numerical scheme. We are directly constructing the numerical governing equations in a space with limited cell resolution and following its time evolution.

Chapter 2

Introduction to Gas Kinetic Theory

Kinetic theory studies flow behavior based on the motion of a large number of particles, which is modeled by the kinetic equation for the time evolution of particle distribution function. The unified gas-kinetic scheme is a kinetic evolution solution-based modeling method. The kinetic equation is the Boltzmann equation or other kinetic model equations. Their time evolution solutions with a continuous change of scale will be constructed and used in the algorithm development. In order to correctly use the kinetic equation in the development of unified scheme, a good understanding of the kinetic equation is necessary. In this chapter, we are going to introduce both macroscopic and microscopic flow descriptions with the emphasis on their different modeling scales. In terms of the kinetic equation, the correct understanding of its modeling scale is important. The approaches of a direct numerical discretization of the kinetic equation and an evolution solution-based scheme are fundamentally different. The unified scheme is a numerical algorithm which presents time evolution of both microscopic and macroscopic flow variables in a conservative form. The use of microscopic flow evolution model is necessary in order to properly capture the non-equilibrium transport in kinetic scale, and the use of macroscopic flow variables update is needed as well in order to efficiently capture the hydrodynamic scale flow evolution. These two descriptions are fully coupled, but are not equally important in different regimes. This chapter introduces basic knowledge of macroscopic and microscopic flow descriptions. With the distinguishable modeling scales of hydrodynamic and kinetic equations, it becomes clear that a valid algorithm for simulating flows with both rarefied and continuum regimes can be hardly described by kinetic or hydrodynamic equation alone. A multiscale modeling becomes necessary. In order to construct such a modeling, the full understanding cross-scale gas evolution mechanism is needed.

2.1 Macroscopic Gas Dynamic Equations

There are two ways to describe flow motion with different scales or resolutions. The first one is based on macroscopic quantities, such as mass, momentum and energy densities. The governing equations are the Euler, the Navier-Stokes or higher order moment equations supplied by the equation of state, and the constitutive relationship. These macroscopic equations are derived based on the physical principles and rational analysis. For example, the fluid mechanic equations are about connections between the spatial and temporal derivatives of macroscopic flow variables. In order to validate these equations, we need properly define the flow variables first, which are supposed to be an averaged quantity in a small volume. There is no a precise definition about this volume, which is microscopically large and macroscopically small. Then, the physical conservation law, which is valid independent of the size of the volume, is adopted with the implementation of additional assumptions about the flux function and the constitutive relationship. The validation of the flux function and the constitutive relationship do depend on the scale. They basically use limited number degree of freedom to mimic the core factors for flow evolution. The fluid dynamic equations are used in the study as small as micro-channel flow, and as large as galaxy formation. Based on the dynamic modeling on a control volume with the corresponding assumptions, the connection between temporal and spatial variation of physical quantities is first constructed in a discretized form. Then, as the control volume shrinking to zero, the differential form of the governing equation is obtained. The transformation from a discretized form to a differential form is for the convenience of mathematical analysis and easy representation. Certainly, there is also historic reason, such as the absence of computers at the time in their development. Even with a continuous space and time in differential equations, the validation of these fluid dynamic equations are still on their modeling scales through the use of constitutive relationship. The fluid dynamic equations are phenomenological ones and there is no equivalence between the fluid dynamics and the Navier-Stokes equations, because there are far more degree of freedom in the fluid system than a few variables in the NS equations. At current stage with powerful computers, it may become interesting to think of the foundation of flow computation, i.e., the use of well-defined differential equations or the direct use of a discretized physical modeling.

For the fluid motion, the hydrodynamic equations are the equations for the mass, momentum and energy densities,

$$\mathbf{W}(x_j,t) = \left(\rho(x_j,t), \quad \rho(x_j,t)U_i(x_j,t), \quad \frac{\rho\mathbf{U}^2}{2} + \rho e(x_j,t)\right)^T, \tag{2.1}$$

where ρ is the density, ρe is the internal energy density, U_i is the flow velocity, and $\mathbf{U}^2 = U_1^2 + U_2^2 + U_3^2$ is the square of the macroscopic velocity.

The Navier-Stokes equations merely state the laws of the conservation of mass, momentum and energy, supplied with the constitutive relations, equation of state, and the definition of transport coefficients. The conservation laws for these functions can be written in the following form [Landau and Lifshitz (1959)],

equation of continuity

$$\frac{\partial \rho}{\partial t} + \frac{\partial(\rho U_j)}{\partial x_j} = 0, \tag{2.2}$$

equation of momentum

$$\frac{\partial \rho U_i}{\partial t} + \frac{\partial(\rho U_i U_j)}{\partial x_j} = -\frac{\partial p}{\partial x_i} + \frac{\partial \sigma_{ij}}{\partial x_j} + \rho F_i, \tag{2.3}$$

equation of energy

$$\frac{\partial}{\partial t}\left[\frac{\rho U^2}{2} + \rho e\right] + \frac{\partial}{\partial x_i}\left[U_i\left(\frac{\rho U^2}{2} + \rho e + p\right)\right] = \rho F_i U_i + \frac{\partial}{\partial x_i}(\sigma_{ij}U_j - q_i). \tag{2.4}$$

The closure of the equations (2.2-2.4) is based on two hypotheses, which are

1). The existence of a local thermodynamic equilibrium, which requires the spatial and temporal scales with the inclusion of enough particles and their collisions. This allows us to use the second law of thermodynamics, which holds for quasi-static processes, i.e., the time scale of the variation of these flow variables is much longer than the particle collision time,

$$Tds = de + pd\left(\frac{1}{\rho}\right),$$

and the empirical equation of state,

$$p = p(\rho, T); \qquad e = e(T),$$

where s and T are entropy density and temperature, and p and e are the pressure and internal energy of unit mass.

2). The existence of two linear dissipative relations. Newton's formula for the force of internal friction, and Fick's formula for the vector of thermal flux q_i. Newton's formula is used in generalized form for the viscous stress tensor σ_{ij}. These formulas have the form

$$\sigma_{ij} = \mu \left[\frac{\partial U_i}{\partial x_j} + \frac{\partial U_j}{\partial x_i} - \frac{2}{3}\delta_{ij}\frac{\partial U_k}{\partial x_k} \right] + \zeta\delta_{ij}\frac{\partial U_k}{\partial x_k};$$

$$q_i = -\kappa \frac{\partial T}{\partial x_i}.$$

The first relation expresses the viscous stress tensor in terms of the derivatives of the velocity, and the second links the thermal flux vector with the gradient of the temperature. Theoretically, we don't know the precise scale where the above two relationships can hold. What we can say is that the Navier-Stokes equations are valid in the scale where the above relationships are applicable. As shown in the numerical tests, even at Reynolds number 50 in the cavity flow, the above heat flux definition is incomplete. Besides depending on the temperature gradient, the heat flux can depend on the pressure gradient and other factors as well. Within the framework of phenomenological theory the coefficients of viscosity μ and ζ, and the coefficient of thermal conductivity κ are measured experimentally as functions of ρ and T. As a result, we have a closed set of equations for the "hydrodynamic" variables of ρ, U_i and T. In order to compare the effects from both the viscous term and the heat conduction term, a useful number is defined, which is the Prandtl number,

$$\mathrm{Pr} = \frac{\mu C_p}{\kappa},$$

where C_p is the specific heat at constant pressure, and its physical meaning will be given later. The Prandtl number is practically constant for air, and this value is 0.72 at the common temperature. From a theoretical point of view, the justification of the above Navier-Stokes equations is largely based on the kinetic theory of gases. For example, the second viscosity coefficient ζ is closely related to the internal degree of freedom of a molecule. In this book, the solution of the above NS equations will be recovered from the unified scheme in the hydrodynamic limit, and the transition process, such as from the non-Fourier heat conduction in the low transition regime to the Fourier one, will also be provided by the unified scheme.

The above NS equations can be extended to include high order terms to describe the flow motion, such as third and fourth order tensors, with enlarged number of equations. These extended hydrodynamic equations [Cercignani (1988); Chapman and Cowling (1990); Muller and Ruggeri (1998); Struchtrup (2005); Gu and Emerson (2009)], such as Burnett,

Super-Burnett, Grad 13, R13, and R26, can be hardly constructed from a direct physical modeling, such as the way to construct the NS equations. They are mostly truncated equations derived from the Boltzmann or kinetic model equations with different kind of closure requirements. The aim for the development of extended hydrodynamic equations is for the study of non-equilibrium flows. But, the success of these equations is limited and is mostly validated case by case. The peculiarity of non-equilibrium is associated with distinguishable mechanism, such as shear dominant, heat dominant, or geometric dominant flows, which can be hardly evaluated quantitatively using a few degree of freedom in the solution expansion. Without a clear physical meaning, the equations with higher order derivatives may become problematic in their validation. The use of mathematical requirement, such as the hyperbolicity, in the direct modification of these equations may become even more dangerous and introduce uncertainty. Not only the equations themselves, to incorporate appropriate boundary conditions for these extended hydrodynamic equations also encounters certain difficulties. From a physical point of view, the gas with simple transport and collision will never know what the third or fourth order derivatives are. In next chapter, we are going to further discuss the limitation of extended hydrodynamic equations in the study of non-equilibrium flows, especially the moment methods.

In the real engineering application of the NS equations, the dimensionless formulations are usually adopted. In order to non-dimensionlize the fluid dynamic equations, which include the Boltzmann equation, the characteristic length, time, and mass have to be used, and there are only three independent units. However, with the introduction of temperature as a new unit for the energy, the non-dimensionlization of the equations becomes difficult to understand, because based on the length, mass, and time, we can construct an energy unit already. How could we get another one? What is the relationship between them? Besides the use of mass, length, and time, we may use the Boltzmann constant k, molecular mass m, and time to non-dimensionlize the equations as well. In order to fully understand their intrinsic connections, a detailed analysis of non-dimensionlization of fluid dynamic quantities is summarized in Appendix A.

2.2 Gas Distribution Function of Equilibrium Flow

Another flow description comes from microscopic consideration, i.e., the so-called gas kinetic theory. The modeling scale of the kinetic equation is much smaller than that of hydrodynamic equations, which theoretically can identify or capture more detailed small scale effects, such as the non-equilibrium one in the mean free path scale. The fundamental quantity in the microscopic description is the particle distribution function $f(x_i, u_i, t)$, which gives the number density of molecules in the six-dimensional phase space (x_i, u_i). The evolution equation for the gas distribution function f is the Boltzmann equation which models the flow motion in the particle mean free path scale. Physically, the gas kinetic equation provides more detailed information about the flow dynamics than the macroscopic counterpart, where the macroscopic flow behavior can be considered as a spatial and temporal averaged microscopic gas evolution.

Before introducing the Boltzmann equation for the evolution of distribution function in the next section, here the equilibrium state will be introduced first. Historically, the equilibrium state was obtained before the establishment of the Boltzmann equation. The definition of a gas distribution function is an averaged number density of particles inside a small volume, where the particles motion can be taken into account statistically. For example, the mass density is defined as a collection of individual particles

$$\rho = \sum_i mn_i, \tag{2.5}$$

where m is the molecular mass and n_i is the particle number density at a certain velocity. However, due to the large number of particles in a small volume in common situation, such as $\sum_i n_i = 2.7 \times 10^{19}$ moleculars in 1 cubic centimeter at 1 atmosphere and $T = 273k$, to follow each individual particle is impossible. Instead, a continuous distribution function is used to describe the probability of particles to be located at a certain velocity interval. Statistically, n_i is approximated by a gas distribution function,

$$f(x_i, t, u_i),$$

where (x_i, t) is the location of any point in space and time, $u_i = (u, v, w)$ is particle velocity with three components in the x, y, and z directions, and the relation between n_i and f is

$$mn_i = f(x_i, t, u_i)\Delta u_i,$$

where Δu_i is a small control volume in the particle velocity space. As a result, the summation in Eq. (2.5) can be replaced by an integral

$$\rho = \int\int\int f du dv dw,$$

in the particle velocity space. Both ρ and f become continuous function of space and time. If ρ is defined in the macroscopic equations as an averaged quantity in a scale being larger than the particle mean free path, where f is defined, then the above ρ can be considered as a statistical averaging in the mean free path scale, instead of deterministic one in a macroscopic level.

Theoretically, the Boltzmann equation is a modeling equation on the mean free path scale, which has no information about the variation of macroscopic variables $\mathbf{W}$ beyond the mean free path scale. However, the Chapman-Enskog expansion uses the asymptotic expansion and assumes that f depends on (x_i, t) through macroscopic variables $\mathbf{W}$, such as that $f = f(\mathbf{W}(x_i, t), u_j)$. Since the variation of $\mathbf{W}$ is defined on a hydrodynamic scale, the Chapman-Enskog assumption automatically moves the Boltzmann solution from kinetic to the hydrodynamic scale. As a result, the capability of describing the non-equilibrium state in the Boltzmann equation is lost after using the Chapman-Enskog expansion. This is one of the reasons why the derived macroscopic equations are only valid in the near continuum flow regime, where the hydrodynamic scale flow behavior is studied. Because the equilibrium state is a bridge to connect macroscopic and microscopic flow descriptions, in the following, we are going to introduce the equilibrium gas distribution function first before the Boltzmann equation. Based on the equilibrium state, the classical statistical mechanics and thermodynamic properties can be connected. It provides a good transition from the macroscopic equations to the microscopic description. And a fully understanding of the equilibrium state and the determination of its parameters are essential to understand and to study the gas kinetic theory.

For molecules with internal degree of freedom, such as rotation and vibration, besides the translational motion the distribution function f can take these internal motion into account as well through additional variables ξ_i. The dimension and formulation of ξ_i are defined below. For monotonic gas, the internal degree of freedom N is equal to 0. For diatomic gas, under the normal pressure and temperature, N is equal to 2 which accounts for two independent rotational degrees of freedom. Equipartition principle in equilibrium statistical mechanics shows that each degree of freedom shares an equal amount of energy $\frac{1}{2}kT$, where k is the Boltzmann constant and T is the temperature. For an ideal gas, the total internal energy per unit mass is equal to

$$e = \frac{N+3}{2}\frac{k}{m}T = \frac{N+3}{2}RT,$$

where $R = k/m$ is the gas constant, and the specific heat at constant volume C_v becomes

$$de/dT = C_v = \frac{N+3}{2}R.$$

The pressure p at equilibrium is equal to

$$p = nkT = mn\frac{k}{m}T = \rho RT = \frac{RT}{V},$$

where n is the number density and V is the volume for unit mass. Based on the first law of thermodynamics,

$$\delta Q = de + pdV,$$

at constant pressure with $pdV = d(pV) = d(RT)$, the specific heat at constant pressure C_p is given by

$$C_p = \frac{\delta Q}{\delta T}\mid_p = \frac{(N+3)+2}{2}R. \tag{2.6}$$

From the above equations, we can obtain the ratio of the principal specific heats, which is commonly denoted by γ,

$$\gamma = \frac{C_p}{C_v} = \frac{(N+3)+2}{N+3}. \tag{2.7}$$

So, γ is equal to $5/3$ for monotonic gas ($N = 0$), and $7/5$ for a diatomic gas ($N = 2$).

The thermodynamic aspect of the Navier-Stokes equations is based on the assumption that the departure of the gas distribution function from local equilibrium state is sufficiently small. Although we do not know the real gas distribution function f at this moment, in classical physics we do know the corresponding equilibrium state g locally once we know the mass, momentum and energy densities. The equilibrium state corresponds to the maximum number of microstate under the constraints of fixed mass, momentum, and energy. In the equilibrium state, the flow system can be described using a minimum number degree of freedom. Certainly, the energy can be defined as the combination of translational and the internal energy of molecules. As a result, a single temperature appears in the equilibrium state. If the energy, such as translational, rotational, and vibrational ones, is separated as individual constraint for the construction of equilibrium state, the temperature can appear as a multiple variable, even as a tensor. The derivation of equilibrium state can be found in statistical mechanics books. The equilibrium distribution function was obtained first by Maxwell before the establishment of the Boltzmann equation. In the following, we are going to define the equilibrium distribution and present all its physical properties. In order to understand the internal variable ξ_i in the gas distribution function, let's first write down the Maxwell-Boltzmann distribution g for the equilibrium state,

$$\begin{aligned} g &= \rho \left(\frac{\lambda}{\pi}\right)^{\frac{N+3}{2}} e^{-\lambda[(u_i-U_i)^2+\xi_j^2]} \\ &= \rho \left(\frac{\lambda}{\pi}\right)^{\frac{N+3}{2}} e^{-\lambda[(u-U)^2+(v-V)^2+(w-W)^2+\xi_1^2+...+\xi_N^2]}, \end{aligned} \tag{2.8}$$

where $\xi_j = (\xi_1, \xi_2, ..., \xi_N)$ are the components of the internal particle velocity in N dimensions with an assumption of the same equilibrium temperature, λ is a function of temperature, molecule mass and Boltzmann constant, with the relation $\lambda = m/2kT$, $U_i = (U, V, W)$ is the corresponding macroscopic flow velocity with three components in the x, y, and z directions, and (u, v, w) are the three components of the individual particle velocity. In the above equation, (u_i, ξ_j) are independent variables and the parameters (λ, U_i, ρ) which determine g uniquely are functions of space and time (x_i, t) only. Taking moments of the equilibrium state g, the mass, momentum and energy densities at any point in space and time can be obtained. For example, the macroscopic and microscopic descriptions are related by

$$\begin{pmatrix} \rho \\ \rho U_i \\ \rho E \end{pmatrix} = \int g \begin{pmatrix} 1 \\ u_i \\ \frac{1}{2}(u_i^2 + \xi^2) \end{pmatrix} dudvdwd\xi. \tag{2.9}$$

More specifically,

$$\begin{pmatrix} \rho \\ \rho U \\ \rho V \\ \rho W \\ \rho E \end{pmatrix} = \int_{-\infty}^{\infty} g \begin{pmatrix} 1 \\ u \\ v \\ w \\ \frac{1}{2}(u^2 + v^2 + w^2 + \xi_1^2 + ... + \xi_N^2) \end{pmatrix} dudvdwd\xi_1...d\xi_N, \tag{2.10}$$

from which the total energy density ρE can be expressed as

$$\rho E = \frac{1}{2}\rho \left(U^2 + V^2 + W^2 + \frac{N+3}{2\lambda}\right),$$

which includes both kinetic and thermal energy densities. The detail formulation of the integrations of a Maxwellian distribution function can be found in Appendix C. Note that Eq. (2.8) describes the gas distribution function g in a 3-D space and the value N can be obtained in terms of γ from Eq. (2.7). If we re-define the internal variable ξ_i as a vector in K dimensions, in a 3-dimensional case we have

$$K = N = \frac{-3\gamma + 5}{\gamma - 1}.$$

In 1-D and 2-D flow simulations, the distribution function g has to be modified as follows. For 1-D gas flow, the macroscopic average velocities in y and z directions are equal to zero with $(V, W) = (0, 0)$. So, the random motion of particles in the y and z directions can be included in the internal variable ξ if the translational and rotational temperatures are assumed to be the same, i.e., the so called equipartition principle. As a result, the internal degree of freedom becomes $N + 2$, which is denoted again by K with the relation $K = N + 2$. The distribution function g in the 1-dimensional case goes to

$$\begin{aligned} g &= \rho \left(\frac{\lambda}{\pi}\right)^{\frac{N+3}{2}} e^{-\lambda[(u-U)^2+v^2+w^2+\xi_1^2+\ldots+\xi_N^2]} \\ &= \rho \left(\frac{\lambda}{\pi}\right)^{\frac{N+3}{2}} e^{-\lambda[(u-U)^2+(v^2+w^2+\xi_1^2+\ldots+\xi_N^2)]} \\ &= \rho \left(\frac{\lambda}{\pi}\right)^{\frac{K+1}{2}} e^{-\lambda[(u-U)^2+\xi^2]}, \end{aligned} \tag{2.11}$$

where the dimension of ξ is K. Substituting $N = K - 2$ into Eq. (2.7), we get the relation between K and γ in the 1-D case,

$$K = \frac{3 - \gamma}{\gamma - 1}.$$

For example, for diatomic gas with $N = 2$ and $\gamma = 1.4$, K is equal to 4, and the total energy density goes to

$$\rho E = \frac{1}{2}\rho \left(U^2 + \frac{K+1}{2\lambda}\right).$$

In 2-dimensional case, K is equal to $N+1$, and the equilibrium distribution function is

$$\begin{aligned} g &= \rho \left(\frac{\lambda}{\pi}\right)^{\frac{N+3}{2}} e^{-\lambda[(u-U)^2+(v-V)^2+w^2+\xi_1^2+\ldots+\xi_N^2]} \\ &= \rho \left(\frac{\lambda}{\pi}\right)^{\frac{K+2}{2}} e^{-\lambda[(u-U)^2+(v-V)^2+\xi^2]}. \end{aligned} \tag{2.12}$$

Then, the relation between γ and K becomes

$$K = \frac{4-2\gamma}{\gamma-1}.$$

In 2D case, for diatomic gas, K is equal to 3 and the total energy density becomes

$$\rho E = \frac{1}{2}\rho\left(U^2 + V^2 + \frac{K+2}{2\lambda}\right).$$

In all cases from 1-D to 3-D, the pressure p is related to ρ and λ through the following relation,

$$p = nkT = \frac{\rho}{m} k \frac{m}{2k\lambda} = \frac{\rho}{2\lambda},$$

where n is the particle number density. Note that the pressure is only related to the translational motion of molecules and is independent of the internal degree of freedom N. For flow modeling with different translational and rotational temperature [Xu *et al.* (2007)], the quasi-equilibrium state can be defined as

$$g = \rho\left(\frac{\lambda_t}{\pi}\right)^{\frac{3}{2}}\left(\frac{\lambda_r}{\pi}\right)^{\frac{N}{2}} e^{-\lambda_t(u_i-U_i)^2-\lambda_r\xi^2},$$

where λ_t and λ_r are related to the translational and rotational temperature. In other words, under different constraints on the mass, momentum, translational and rotational energy separately, a quasi-equilibrium state with distinguishable translational and rotational temperature can be obtained.

In the following, we are going to analyze a few limiting cases for the appropriate choice of internal degree of freedom for flow simulations. In order to properly approach to the incompressible flow limit, the internal degree of freedom of the molecular motion has to be reduced as much as possible. For example, in the 2D simulation, the motion of the molecules has to be limited in a 2D plane, and there is no thermal energy in the z-direction and other internal degree of freedom. Therefore, the number of K in the 2D case is equal to zero, and $\gamma = 2$ has the closest match with the incompressible limit [Su *et al.* (1999)]. In the 1D case, the gas for the incompressible flow has $\gamma = 3$, where all molecules move along a string only. In the 3D case, for the incompressible flow computation it is better to use $\gamma = 5/3$ of a monatomic gas. Otherwise, with the inclusion of rotational degree of freedom the bulk viscosity effect will appear, which poisons the incompressible NS solution. In the limit of isothermal flow, theoretically in order to keep the same temperature the internal degree of freedom should go to infinite, where any limited amount of energy is not enough to increase the gas temperature. Therefore, in all 1D, 2D and 3D case, $K = \infty$ corresponds to $\gamma = 1$. A smooth isentropic flow with $p \sim \rho^\gamma$ automatically go to isothermal one.

The unique equilibrium distribution function g corresponds to a state with maximum entropy in classical statistical physics. At any point in space and time, there is a one to one correspondence between g and the macroscopic densities, e.g. mass, momentum and energy. So, from macroscopic flow variables, we can construct a unique equilibrium state. However, in real physical situation, the gas doesn't necessarily stay in a local thermodynamic equilibrium (LTE) state, such as the gas inside a shock or boundary layer, even though we can still construct the corresponding equilibrium one from the macroscopic flow variables there. For the hydrodynamic NS equations, the corresponding gas distribution function is close to the LTE, but with a small amount of deviation. The Boltzmann equation presents an evolution equation for f and the equilibrium state becomes a limiting solution, which requires intensive particle collisions. Theoretically, there is only absolute equilibrium state for a homogeneous flow. The LTE is only an assumption, which cannot be true once there is any gradient of macroscopic flow variables in space.

2.3 Boltzmann Equation

The classical kinetic theory of gases is based on the motion of molecules and the statistical description. Different from molecular dynamics, which is based on the individual particle motion, the gas kinetic theory is a probability theory with mechanical bases. The construction of the Boltzmann equation is the core of kinetic theory. The most successful study of the Boltzmann equation is the derivation of the hydrodynamic equations through the Chapman-Enskog expansion, where the irreversible physical process and the viscosity and thermal conduction coefficients can be connected to the microscopic molecular collisions.

The Boltzmann equation describes the motion of dilute gas molecules by including both transport and collision effects. The underlying assumptions for the derivation of the Boltzmann equation include: particles have only binary collisions for dilute gas; the control volume for modeling the evolution of $f(\vec{x}, \vec{u}, t)$ is on the mean free path scale; any external force will not effect the trajectories of colliding particles; and the incoming colliding particles are uncorrelated. In most of the time, all particles undertake free transport. Since the particle collision must be associated with two particles, the development of the evolution equation for one particle distribution may not be easily understood. The correctness of the equation depends closely on the modeling assumptions, such as the 'molecular chaos hypothesis' and the introduction of inverse collision. In some cases, there may have no associated inverse collision, such as chemical reaction or the collision with the involvement of radiation. Therefore, there is certain underlying limitation of the Boltzmann modeling as well.

The modeling scale in the Boltzmann equation is on the particle mean free path and particle collision time, where the distribution function itself is an averaged statistical quantity on such a scale. Only with this scale, the particle transport and collision can be modeled separately. Therefore, the Boltzmann equation has no reliable information in a scale much less than the particle mean free path, even though the distribution function is still a continuous function of space and time. The scale of effective mechanical particle collision is on the molecular diameter, which is much less than the particle mean free path. Therefore, even with the exact particle trajectory during the collision, only scattering effect is included statistically into the equation. Within a mean free path, it is important to count for collision rate, such as how many collision within unit time, rather than the location where the collision takes place. The detailed collision pairs within the mean free path scale cannot be identified, but their total effect from all collisions within a collision time scale is counted for the re-distribution of particles in the velocity space. Therefore, the validity of the Boltzmann equation for dilute gases relies on these two well-separated scales, where the short time scale is the duration of a collision and longer one is the mean time interval separating two successive collisions of a given molecule.

The modeling from the molecular dynamics, to the Boltzmann equation, and to the hydrodynamics equations, is a coarse graining process of the physical resolution in the spatial and temporal scales. For a gas with number density n, the dilute gas condition is $d << D << l_{mfp}$, where d is the molecular diameter, $D \sim n^{-1/3}$ is the average distance between molecules, and l_{mfp} is the particle mean free path, which can be estimated as $l_{mfp} = 1/\sqrt{2}\pi d^2 n$. Due to the dilute gas condition, the molecules will not interact with each other during the most time of its motion. Therefore, for the description of macroscopic flow variables, a single particle gas distribution function f can be used, which describes the probability in a small control volume in the physical and particle velocity space. The Boltzmann equation models the spatial and time evolution of such a gas distribution function. The scale which can identify this "smooth" variation must be on the scale which include enough particle collision to trigger the smooth variation. The time resolution scale Δt in the Boltzmann modeling is $d/|v| << \Delta t \leq \tau$, which is much longer than the particle impacting time $d/|v|$ and less than the mean collision time $\tau = l_{mfp}/|v|$. But, this time interval Δt is much shorter than the hydrodynamic time scale τ_H for the system approaching to the equilibrium state, which is related to the gradient of macroscopic flow variables. The Boltzmann modeling provides the spatial resolution Δl in the order of $d << \Delta l \leq l_{mfp}$. Therefore, in terms of the physical modeling, the Boltzmann equation captures the dynamics in a smaller scale than that in the hydrodynamic equations, but has less resolution than the molecular dynamics (MD). In the direct discretization of the Boltzmann equation, it is hard to guarantee that the kinetic modeling scale is compatible with the numerical mesh size and time step, especially in the flow covering multiple scale regimes. Therefore, the construction of a unified numerical algorithm based on kinetic equation has to take into account the scales differences. It will be more complicated than the direct discretization of the kinetic equation.

With the definition of the particle distribution function f, which is an averaged quantity in a mean free path scale, and under all above assumptions, the time evolution for f, the so-called the Boltzmann Equation, is

$$f_t + u_i f_{x_i} + a_i f_{u_i} = Q(f, f). \tag{2.13}$$

Here f is the real gas distribution function, a_i is the external force term acting on the particle in i-th direction, and $Q(f, f)$ is the collision operator, which represents the rate of change due to collisions. The left hand side of the Boltzmann equation represents the particle free transport. How far a particle can go is constrained by the right hand side particle collision. From the equation itself, we can easily get the modeling scale of the Boltzmann equation. If there is more than one collision for a individual particle, the transport part of the Boltzmann equation cannot be written in the current way, i.e., the free transport. This limits the upper bound of the scale. On the other hand, the collision term is an averaged quantity at an instant of time t, which has no information of two particle collision trajectory in space and time. This limits the lower bound of the length scale of the Boltzmann equation.

The collision term has gain and loss terms,

$$Q(f, f) = \left(\frac{\partial f}{\partial t}\right)^+_{coll} + \left(\frac{\partial f}{\partial t}\right)^-_{coll}.$$

The total number of molecules scattered out of the element $d\vec{u}d\vec{x}$ at particle velocity $\vec{u}$ and location $\vec{x}$ (averaged in the corresponding scales) as a result of collisions depends on the collision frequency in the gas. More specifically, we need first to consider the collision of a molecule of class $\vec{u}$ with another class $\vec{u}_1$ such that their post-collision velocities are $\vec{u}^*$ and $\vec{u}_1^*$, respectively. This is called forward scattering $\vec{u}, \vec{u}_1 \rightarrow \vec{u}^*, \vec{u}_1^*$. The collision rate for f is about the number of molecules with velocities $\vec{u}$ out of $d\vec{u}d\vec{x}$ in unit time. A molecule of class $\vec{u}$ may be chosen as a test particle moving with relative speed $\vec{g}_r$ among stationary field molecules of class $\vec{u}_1$. The volume swept out in physical space by the cross-section for this class of collision is $g_r \sigma d\Omega$ and the number of class $\vec{u}_1$ molecules per unit volume in physical space is $f_1 d\vec{u}_1$. Therefore, the number of collisions of this class suffered by the test molecule per unit time is, $f_1 g_r \sigma d\Omega d\vec{u}_1$. Since the number of class $\vec{u}$ molecules in the phase space element is $f d\vec{u}d\vec{x}$, the number of class $\vec{u}, \vec{u}_1 \rightarrow \vec{u}^*, \vec{u}_1^*$ collisions per unit time in the element becomes

$$f f_1 g_r \sigma d\Omega d\vec{u}_1 d\vec{u} d\vec{x},$$

where $\sigma d\Omega$ is the differential cross-section for the collision specified. In the above considerations, in the mean free path scale, the specific locations of particles are not identified. The particle distribution in the mean free path scale is averaged homogeneously which likes a mean field approximation.

Note that f denotes the value of the velocity distribution function f at $\vec{u}$ and f_1 the value of f at $\vec{u}_1$. Similarly, f^* and f_1^* are used to denote the values of f at $\vec{u}^*$ and $\vec{u}_1^*$, respectively. The principle of molecular chaos is implicitly used by the expression of the binary collision probability in terms of the product of two single particle distribution functions. In order to obtain gain term, the inverse collision has to be accounted for, $\vec{u}^*, \vec{u}_1^* \rightarrow \vec{u}, \vec{u}_1$, which scatters molecules into class $\vec{u}$. This yields

$$f^* f_1^* g_r (\sigma d\Omega)^* d\vec{u}_1^* d\vec{u}^* d\vec{x},$$

for the collision rate in the phase space element $d\vec{u}^* d\vec{x}$. Due to the symmetry between the direct and inverse collisions, the above equation can be changed to

$$f^* f_1^* g_r \sigma d\Omega d\vec{u}_1 d\vec{u} d\vec{x}.$$

As shown above, the assumption of the molecular chaos is applied to the inverse collision as well. This seems unreasonable because after the forward collision these two molecules at velocities $\vec{u}^*$ and $\vec{u}_1^*$ have correlations. How could they be independent again? Actually, the use of this molecular chaos assumption for the inverse collision is the core for the entropy condition of the Boltzmann equation. It basically assumes that within the mean free path scale all particle collisions are independent. The inverse collision is actually the real collision happened in the control volume and is just categorized as an "inverse" one of another collision pair, because all particles in the velocity space have to be counted in the collisions. This makes mathematical expression simpler. This collision modeling limits the resolution of the Boltzmann equation. That is the reason we continuously emphasize the modeling scale of the Boltzmann equation. The irreversible process in the macroscopic description is mainly due to this coarse graining process, where a macro-state could correspond to gigantic amount of micro-states. The averaging among the microstates in spatial and temporal scales presents a direct of time.

As a result, the rate of change of molecules of class $\vec{u}$ in the phase space element $d\vec{u}d\vec{x}$ due to the combined forward and inverse collisions of class $\vec{u}, \vec{u}_1 \leftrightarrow \vec{u}^*, \vec{u}_1^*$ is obtained by subtracting the loss rate from the gain rate. This gives

$$(f^* f_1^* - f f_1) g_r (\sigma d\Omega) d\vec{u}_1 d\vec{u} d\vec{x}.$$

The total rate of increase of molecules of class $\vec{u}$ in the element as a result of collisions is given by the integration of this expression over the complete cross-section for its collisions with class $\vec{u}_1$ molecules, followed by the integration of the class $\vec{u}_1$ over all velocity space. This required expression becomes

$$\int_{-\infty}^{\infty} \int_0^{4\pi} (f^* f_1^* - f f_1) g_r \sigma d\Omega d\vec{u}_1 d\vec{u} d\vec{x}.$$

Then, the rate of change of velocity distribution function f in a unit control volume at $\vec{u}$ becomes

$$f_t + u_i f_{x_i} + a_i f_{u_i} = \int_{-\infty}^{\infty} \int_0^{4\pi} (f^* f_1^* - f f_1) g_r \sigma d\Omega d\vec{u}_1. \tag{2.14}$$

In the above Boltzmann equation, a collision kernel can be defined as $B(\cos\theta, g_r) = \sigma g_r$, where θ is the angle between pre- and post-collision relative velocities. For hard sphere molecules, the defection angle is determined through $b = d\cos(\theta/2)$, where b is the aiming distance and d is the diameter of the particles. Hence the differential cross-section $\sigma = b|db|/\sin\theta d|\theta|$ is $d^2/4$ and the collision kernel $B = |g_r| d^2/4$. For a general spherically symmetrical interatomic potential $\phi(r)$, the deflection angle is

$$\theta(b, |g_r|) = \pi - 2\int_0^{W_1} \left[1 - W^2 - \frac{4\phi(r)}{m|g_r|^2}\right]^{-1/2} dW,$$

where $W = b/r$ and W_1 is the positive root of the term in the brackets. The $(\eta-1)$-th inverse power-law potential $\phi(r) = C/(\eta-1)r^{\eta-1}$ are called hard and soft potentials when $\eta > 5$ and $\eta < 5$, respectively. Maxwell molecules has the potential with $\eta = 5$.

The detailed derivation of the Boltzmann equation can be found in many text books [Chapman and Cowling (1990); Kogan (1969); Cercignani (1988)]. To understand the gas-kinetic scheme in this book doesn't require us to know the detailed collision cross section under different potential, and all kinds of scattering laws. To understand the collision rate and the modeling scale of the kinetic equation is essential to appreciate the kinetic scheme. Certainly, with the implementation of the full Boltzmann equation collision term, a unified scheme can be developed as well, but the construction of the evolution process to implement the full Boltzmann collision term has to be based on a clear physical picture [Liu *et al.* (2014)]. The construction of unified scheme identifies the scale differences, because the Boltzmann collision term is only valid in the kinetic scale and the unified scheme has a variation of scales. For example, the unified scheme may use a numerical mesh size with 3 or 4 mean free path, with this resolution the Boltzmann equation itself is not applicable and there needs an additional modeling with this specific scale. Therefore, in the design of the unified scheme it more or less likes a modeling process, which is beyond the above Boltzmann modeling. The value of the Boltzmann equation is in its modeling scale. The claim of the superiority of the Boltzmann equation in all scales is groundless physically. The Boltzmann equation just cannot be applied directly in other scales. The solution in other scale must be obtained through the accumulation of kinetic scale physics. For example, starting from the same initial condition, in a cell size with 5 mean free path and a time step of 5 collision time, the solution will become indistinguishable from the full Boltzmann equation and the kinetic model equations with the resolution of time step.

From physical constraints of the conservation of mass, momentum and energy during particle collisions, the following compatibility condition has to be satisfied,

$$\int \psi_\alpha Q(f,f) d\vec{u} = 0, \tag{2.15}$$

where $d\vec{u} = dudvdw$ and $\psi_\alpha = (1, u, v, w, \frac{1}{2}(u^2 + v^2 + w^2))^T$. With the definition of the H-function, $H(\vec{x}, t) = \int f \ln f d\vec{u}$, we have

$$\frac{\partial H}{\partial t} = \int (1 + \ln f) \frac{\partial f}{\partial t} d\vec{u}.$$

Multiplying both sides of the Boltzmann equation by $(1 + \ln f)$ and integrating over $d\vec{u}$, under the condition of external forcing being independent of the particle velocity, the following local balance equation can be obtained

$$\frac{\partial H}{\partial t} + \nabla \cdot \vec{J} = S,$$

where $\vec{J} = \int \vec{u} f \ln f d\vec{u}$ is the entropy flux with the entropy production

$$S = \frac{1}{4} \int_{-\infty}^{\infty} \int_{-\infty}^{\infty} \int_{0}^{4\pi} \ln \left(\frac{f f_1}{f^* f_1^*} \right) (f^* f_1^* - f f_1) g_r \sigma d\Omega d\vec{u}_1 d\vec{u} \leq 0.$$

For an isolated system, the above equation gives the H-theorem established by L. Boltzmann in 1872. A global equilibrium $f = g$ with zero entropy production $S = 0$ requires

$$g^* g_1^* = g g_1,$$

or

$$\ln g^* + \ln g_1^* = \ln g + \ln g_1.$$

Therefore, for an equilibrium state g, $\ln g$ must be proportional to the collision invariants ψ_α,

$$\ln g = \alpha_1 + \alpha_2 u + \alpha_3 v + \alpha_4 w + \alpha_5 (u^2 + v^2 + w^2).$$

Based on the constraints, such as

$$\begin{pmatrix} \rho \\ \rho U \\ \rho V \\ \rho W \\ \rho E \end{pmatrix} = \int_{-\infty}^{\infty} g \begin{pmatrix} 1 \\ u \\ v \\ w \\ \frac{1}{2}(u^2 + v^2 + w^2) \end{pmatrix} dudvdw,$$

for a monatomic gas, the equilibrium state g can be fully determined,

$$g = \rho \left(\frac{\lambda}{\pi} \right)^{\frac{3}{2}} e^{-\lambda[(u-U)^2 + (v-V)^2 + (w-W)^2]}.$$

Note that the above equilibrium state is a global one. There is no such a local equilibrium state derived from the Boltzmann equation. Once there is any flow variable gradient in space, the local equilibrium state becomes an assumption for convenience, which is frequently used in the flow description in the hydrodynamic scale.

2.4 Understanding of Boltzmann Equation

In order to properly apply the Boltzmann equation in a non-equilibrium flow study, especially in the construction of the unified scheme, it is better to have a proper understanding of the modeling in the equation. The most important fact is that the Boltzmann equation represents a physical coarse grained modeling on the mean free path scale. The assumptions for its derivation include

1. Only binary particle collision is included in the particle collision term. The condition for this is that the gas should be dilute enough that the distance between molecules are much larger than the interaction forcing influenced diameter of molecules.

2. The molecules are assumed to be structure-less, which is basically monatomic gas and only elastic collision is included. For diatomic gas with internal degree of freedom, the collision term will become much more complicated [Chapman and Cowling (1990)].

3. The collision is expressed mathematically to be local and instantaneously. It is modeled as a continuous function of space and time. The impact time scale of the collision is much shorter than the mean particle collision time in the Boltzmann modeling. The total number of collisions within a particle collision time and a mean free path scale is averaged and equally distributed in the corresponding space and time, i.e., the collision rate is used.

4. Due to the simple statistics in the modeling of the collision term, such as the collision pair being proportional to ff_1 , the incoming particles have no dynamical co-relations. The molecular chaos is assumed. This is also applied to the inverse collision.

In the Boltzmann equation (2.13), physically the transport term on the left hand side always drives f to non-equilibrium by counting individuality of each particle. For example, the free transport part has a solution,

$$f(\vec{x}, \vec{u}, t) = f_0(\vec{x} - \vec{u}t, \vec{u}, t)$$

and

$$\frac{\partial f}{\partial \vec{u}} = -t\frac{\partial f_0}{\partial \vec{x}} + \frac{\partial f_0}{\partial \vec{u}},$$

can become singular easily once the initial flow distribution is non-homogeneous. On the other hand, the collision term on the right hand side $Q(f, f)$ pushes f back to group behavior by counting the particle's accumulating effect, and reduces the singular behavior of individual particle. Although, $Q(f, f)$ does not change the local mass, momentum and energy, it does re-distribute particles in the phase space and makes the distribution function smooth, and subsequently change the transport mechanism of the particle system, *e.g.* dissipation and heat conduction. The real flow evolution is controlled by the competition between the transport and collision terms. The coupling of the transport and collision in the flux evaluation is the key for the success of the gas-kinetic scheme.

Due to the molecular chaos assumption and statistical nature of the Boltzmann equation, the collision term makes the Boltzmann equation not be time-reversible, $t \rightarrow -t$. In other words, the collision term does not know any time reversing effect in the particle collisions, because in the mean free path scale the Boltzmann equation never takes into account the exact location of individual particle and its trajectory, and the collision rate is determined through a probability theory. Even though the mechanics of a specific particle pair collision is deterministic and time reversible, due to the absence of exact location and time within the modeling resolution it is merged into the Boltzmann equation through averaging. The averaging is a real physical process happening in the scale beyond the impact particle trajectory, where many particle collision effect has been taken into account. The statistical model of the Boltzmann equation is irreversible due to coarse grained averaging in the mean free path scale, which makes the Boltzmann equation satisfy the entropy increasing condition for an isolated system. This is also the main successful aspect of the Boltzmann equation, which provides microscopic description of an irreversible macroscopic thermodynamic system. Basically, the molecular collision is a process of reducing the high quality energy of the system by a tendency of equally distributing the energy to all molecules and in all directions. Even with the resolution of molecular dynamics and time reversible Newton mechanics, any small uncertainty or fluctuation will stop a reversing process and increase the entropy of the system again. The Boltzmann equation describes the most favorable physical process from a non-equilibrium to a local equilibrium one, then to a global equilibrium state for an isolated system. For the collision-less Boltzmann equation, the non-equilibrium property of the system will be remained. A highly non-equilibrium state of a system can be only kept through the supply of high quality energy or low entropy fluxes through the boundary. At the same time, there is inadequate particle collision to equalize the energy among particles. For the same physical reality, the quantitative measurement of the non-equilibrium depends on the scale to describe it, or the scale to include how many particles.

The fundamental property of the Boltzmann equation is to address the time evolution of the system from non-equilibrium to an equilibrium one, from which an equilibrium state can be uniquely determined. For a fluid system, the Boltzmann equation and hydrodynamic equations present different scale description of the flow motion. In comparison with hydrodynamic description, the kinetic equation presents a refined gas evolution model. The development of the unified scheme is basically to use a continuous variation of scale to study the gas system. Since the Boltzmann equation is valid in the kinetic scale, it doesn't provide direct description in other scale. The applicability of the Boltzmann equation to other scale means implicitly that all scales are resolved down to the mean free path one. Practically, it is impossible to use this model in the hydrodynamic flow simulation, especially for the high Reynolds number flow. With the consideration of different kinds of asymptotic expansions, many hydrodynamic equations have been derived from the Boltzmann equation. But, this does not mean that the Boltzmann equation can be directly used in other scale, and a further coarse grained modeling is still needed to extend it to other scale. In the following, only successfully derived hydrodynamic equations from the kinetic theory will be presented. In history, many inappropriate hydrodynamic equations have been obtained from the Boltzmann equation as well through different kinds of asymptotic expansion. The derivation of the governing equations from one scale to other scale needs additional assumption on the space and time evolution of a gas distribution function.

2.5 Relation between Kinetic Theory and Hydrodynamic Equations

The Boltzmann equation describes the time evolution of a gas distribution function over a time scale of molecular mean collision time. The hydrodynamic time scale of macroscopic governing equation has a much larger time scale than the kinetic relaxation time. In the hydrodynamic scale, due to the inclusion of intensive particle collision in scales being much larger than the mean free path and collision time, the averaged or coarse grained gas distribution function over a large amount of particles gets close to the local equilibrium state. So, with the changing scales from the kinetic to hydrodynamic ones, the identified flow motion changes from the particle free transport and collision in a small scale, to the accumulating wave propagation in a large scale. Therefore, in the hydrodynamic scale the evolution of the flow system can be described by a much reduced number of flow variables, because it describes the accumulating effect of gigantic amount of particles. Basically, the macroscopic flow variables, such as the mass $\rho(\vec{x}, t)$, momentum $\rho\vec{U}(\vec{x}, t)$, and energy densities $\rho E(\vec{x}, t)$, are used here. The derivation

of hydrodynamic equations from the kinetic Boltzmann equation is mainly through the Chapman-Enskog expansion [Chapman and Cowling (1990)], where two critical techniques or assumptions used are the following:
(a) the distribution is close to the local equilibrium state, such as $f = g + f^1 + f^{(2)} + ...$ and $f^{(1)}$ and $f^{(2)}$ are proportional to local Knudsen number Kn and Kn^2; and the non-equilibrium parts (term by term) have vanishing contributions to the conservative flow variables,
(b) the normal solution of the distribution function f depends on the space and time $(\vec{x}, t)$ through macroscopic flow variables, such as $f(\vec{x}, t, \vec{u}) = f(\rho(\vec{x}, t), \rho\vec{U}(\vec{x}, t), \rho E(\vec{x}, t), \vec{u})$.

The function of the first assumption is that the deviation from equilibrium state is small, which can be used to linearize the Boltzmann collision term, such as $ff_1 \simeq (g + f^{(1)})(g_1 + f_1^{(1)}) \simeq gg_1 + f^{(1)}g_1 + f_1^{(1)}g$. The second assumption is to separate the hydrodynamic and kinetic scales in the Boltzmann equation. The time and space variations of f, i.e., f_t and f_x on the left hand side of the Boltzmann equation, must go through the macroscopic flow variables, which have variations on the hydrodynamic scale. Therefore, the assumption enforces the left hand side of the Boltzmann equation (transport part) to follow the flow motion in the hydrodynamic scale. Since the collision term in the Boltzmann equation is on the kinetic scale, the separation of the scales makes the Boltzmann equation be solved in a successive approximation. Also, here the characteristic length scale for the hydrodynamic part is defined as the length scale for the variation of macroscopic flow variables, such as $L = \rho/(\partial\rho/\partial x)$. Therefore, the Knudsen number defined by $Kn = l/L$ in the Chapman-Enskog expansion for the equation itself should be different from the Knudsen number defined using a characteristic length of a flying object. The difference becomes more obvious for the microflows, where the spatial and temporal variation of macroscopic variables can be extremely small, but the flow may have a finite Knudsen number, because this Knudsen number is defined by the length scale of the micro-devices, such as the channel height. So, with the same value of Knudsen number, but with different definition for the characteristic length scale, the flow physics can be significantly different. The dynamics of the non-equilibrium hypersonic rarefied flow and non-equilibrium low speed microflow are different, even with the same value of Knudsen number. The length scale used in the Chapman-Enskog expansion is based on the macroscopic flow variation, such as the shock thickness and the boundary layer thickness.

With the above assumptions and considerations of the binary collision $Q(f, f) = I(f|f)$, the lowest order of approximation, i.e. $f = g$, make the collision term vanish $I(g|g) = 0$. To the first order approximation, the Boltzmann equation changes to

$$\frac{\partial}{\partial t}g + \vec{u} \cdot \frac{\partial}{\partial \vec{x}}g = I(g|f^{(1)}) + I(f^{(1)}|g), \tag{2.16}$$

where g is the local equilibrium state. Taking conservative moments on Eq. (2.16), the Euler equations for the inviscid flow can be obtained,

$$\frac{\partial \rho}{\partial t} = -\frac{\partial(\rho U_i)}{\partial x_i},$$

$$\frac{\partial U_k}{\partial t} = -U_i\frac{\partial U_k}{\partial x_i} - \frac{1}{\rho}\frac{\partial(\rho RT)}{\partial x_i},$$

$$\frac{\partial T}{\partial t} = -U_i\frac{\partial T}{\partial x_i} - \frac{2}{3}T\frac{\partial U_j}{\partial x_j}.$$

The above equation (2.16) allows us to determine $f^{(1)}$ with the time and spatial variations of the macroscopic variables of an equilibrium state through the connection in the Euler equations. To the first order approximation with the solution $f_{ns} = g + f^{(1)}$, the Navier-Stokes equations can be recovered by taking the conservative moments of the Boltzmann equation,

$$\int \psi \left(\frac{\partial}{\partial t} f_{ns} + \mathbf{u} \cdot \frac{\partial}{\partial \vec{x}} f_{ns} \right) d\vec{u} = \int \psi I(f,f) d\vec{u} \equiv 0.$$

For the conservative moments, the collision term vanishes for any distribution function, not only for f_{ns}. With the above equation, the Navier-Stokes equations become

$$\int \psi \left(\frac{\partial}{\partial t} g + \frac{\partial}{\partial \vec{x}} g \right) d\vec{u} = -\int \psi \left(\frac{\partial}{\partial t} f^{(1)} + \mathbf{u} \cdot \frac{\partial}{\partial \vec{x}} f^{(1)} \right) d\vec{u},$$

where the dissipative terms are related to the non-equilibrium parts of the gas distribution function. The time variation of $\partial f^{(1)}/\partial t$ can be changed to spatial variation through the Euler equations.

Many references provide detailed derivations of the NS equations through the Chapman-Enskog expansion. Even for the simple flow case, the full solution requires extensive mathematical manipulation, especially in the evaluation of $f^{(1)}$ in Eq. (2.16). In the construction of the unified scheme, theoretically, it is not necessary to have any knowledge about the Chapman-Enskog expansion of the Boltzmann equation, because the NS solution will be recovered instead of constructed by the unified scheme in the continuum limit. The hydrodynamic solution from the NS will be used to validate the the solution from the unified scheme in this limit. In the transition regime, where the flow evolution has a time and spatial scale which is compatible with the particle collision time and mean free path, it is hard to obtain any valid macroscopic equations due to the un-separable scales of macroscopic flow variation in the transport part from the scale of the collision term of the Boltzmann equation. A direct modeling to derive the governing equations between hydrodynamic and kinetic scales becomes difficult due to the lack of specific form of non-equilibrium distribution function. So, in terms of the macroscopic governing equations in different flow regimes through the coarse grained modeling, there is a gap between the NS and the Boltzmann equation. But, numerically the unified scheme presented in this book provides a smooth transition between them and the numerical algorithm becomes the governing equations themselves which fill up the gap. New phenomena, which is different from the NS one, such as heat transport flux from cold to hot region in the transition regime, has been observed from the solution of the unified scheme. Physically, the kinetic scale Boltzmann equation presents particle behavior, and the hydrodynamic NS equations give the wave evolution. In the transition regime, both particle-wave interaction has to be taken into account, which makes the modeling and the construction of valid governing equations difficult, if not impossible.

Based on the full Boltzmann equation, for a monatomic gas the 1st-order approximation of the Chapman-Enskog expansion in hydrodynamic scale gives

$$f_{\text{ns, Boltzmann}} = g\left[1 - \frac{4}{5}\frac{\kappa}{2Rp}\left(\frac{c^2}{2RT} - \frac{5}{2}\right)c_i\partial_{x_i}\ln T \right.$$
$$\left. - \frac{\mu}{RTp}\left(c_ic_j - \frac{1}{3}c^2\delta_{ij}\right)\partial_{x_j}U_i\right], \tag{2.17}$$

where the peculiar velocity $c_i = u_i - U_i$ and $c^2 = \mathbf{c}\cdot\mathbf{c}$. In the above equation, the viscosity coefficient μ and heat conduction coefficient κ are related to the viscous differential cross section σ_μ of the colliding molecules with relative speed g_r through the following relationships [Chapman and Cowling (1990); Bird (1994)],

$$\mu = \frac{\frac{5}{8}\sqrt{\pi mkT}}{(\frac{m}{4kT})^4\int_0^\infty g_r^7\sigma_\mu(g_r)e^{-mg_r^2/4kT}dg_r},$$

and

$$\kappa = \frac{15}{4}\frac{k}{m}\mu.$$

Therefore, for a monatomic gas the Prandtl number from the above Chapman-Enskog expansion is

$$Pr = \frac{C_p\mu}{\kappa} = \frac{5}{2}\frac{k}{m}\frac{\mu}{\kappa} = \frac{2}{3}.$$

With the above distribution function (2.17), and the definition of pressure tensor

$$P_{ij} = p\delta_{ij} - \sigma_{ij} = \int (u_i - U_i)(u_j - U_j) f d\vec{u},$$

the viscous stress σ_{ij} and the heat flux q_i can be obtained,

$$\begin{aligned}\sigma_{ij} &= -\int (u_i - U_i)(u_j - U_j) f_{\text{ns, Boltzmann}} d\vec{u} \\ &= \mu(\partial_j U_i + \partial_i U_j) - 2/3\mu\partial_k U_k \delta_{ij},\end{aligned}$$

and

$$\begin{aligned}q_i &= \frac{1}{2}\int (u_i - U_i)\left((u-U)^2 + (v-V)^2 + (w-W)^2\right) f_{\text{ns, Boltzmann}} d\vec{u} \\ &= -\kappa\partial_i T. \end{aligned} \tag{2.18}$$

Therefore, the NS distribution function (2.17) from the Chapman-Enskog expansion of the full Boltzmann equation with the correct Prandtl number can be written in the following form

$$f_{\text{ns, Boltzmann}} = g + f_{\text{heat, Boltzmann}} + f_{\text{stress, Boltzmann}},$$

where

$$f_{\text{heat,Boltzmann}} = g\frac{1}{5RpT}\left(\frac{c^2}{RT} - 5\right) c_i q_i, \tag{2.19}$$

and

$$f_{\text{stress, Boltzmann}} = -g\frac{\mu}{RpT}\left(c_i c_j - \frac{1}{3}c^2\delta_{ij}\right)\partial_{x_j} U_i. \tag{2.20}$$

The above relationships for recovering of NS equations will be used to validate the simplified gas-kinetic model equations, where one of the important condition for the model equations is to recover the above NS stress and heat flux from the non-equilibrium distribution in the continuum flow limit.

2.6 Kinetic Model Equations

There are many successful Boltzmann model equations which much simplify the particle collision term of the original Boltzmann equation. Here, three model equations, i.e., the BGK [Bhatnagar *et al.* (1954)], the BGK-Shakhov [Shakhov (1968)], and the ES-BGK [Holway (1966)], will be introduced. Besides the conservative properties for the particle collision term, the common requirement for all these models are to recover the Chapman-Enskog expansion for the NS solutions in the continuum flow regime. With the satisfaction of the above requirements, besides the BGK-Shakhov and ES-BGK, these two models can be combined to form new one.

2.6.1 *Bhatnagar-Gross-Krook (BGK) Model*

One of the main functions of the particle collision term is to lead the gas distribution function f back to the equilibrium state g corresponding to the local $\rho, \rho U_i$ and ρE. The particle collision theory assumes that during a time dt, a fraction of dt/τ of molecules in a given small volume undergoes collision, where τ is the average time interval between successive particle collisions for the same particle. The collision term in the BGK model alters the velocity-distribution function from f to g. This is equivalent to assuming that the rate of changes df/dt of f due to collisions is $-(f-g)/\tau$, so the Boltzmann equation without external forcing term becomes [Bhatnagar *et al.* (1954)],

$$\frac{\partial f}{\partial t} + u_i \frac{\partial f}{\partial x_i} = -\frac{f-g}{\tau}. \tag{2.21}$$

At the same time, due to the mass, momentum and energy conservation in particle collisions, the collision term $(g-f)/\tau$ satisfies the compatibility condition,

$$\int \frac{g-f}{\tau} \psi_\alpha d\Xi = 0, \tag{2.22}$$

where $d\Xi = dudvdwd\xi$ and $\psi_\alpha = (1, u_i, \frac{1}{2}(u_i^2+\xi^2))^T$. Eq. (2.21) is a nonlinear integro-differential equation, since the relationship between f and g is connected through a nonlinear way, where $\rho, \rho U_i, \rho E$ for the determination of g are integrals of the function f. The above BGK model coincides in form with the equations in the theory of relaxation process and is therefore sometimes called the relaxation model.

For a gas system, thermodynamic irreversibility is accompanied with dissipation in the system and an increase of entropy. The entropy condition for the BGK model can be easily proved. The Boltzmann H-theorem states that if we define

$$H = \int f \ln f d\Xi,$$

as the entropy density (the entropy in thermodynamics is usually defined as $s = -kH$), and

$$J_i = \int u_i f \ln f d\Xi,$$

as the entropy flux in direction i, the entropy condition implies the following inequality,

$$\frac{\partial H}{\partial t} + \frac{\partial J_i}{\partial x_i} \leq 0. \tag{2.23}$$

In order to prove the above inequality, let's multiply $(1+\ln f)$ on both sides of the BGK model (2.21) and take an integration with respect to $d\Xi$,

$$\int \left(\frac{\partial f}{\partial t} + u_i \frac{\partial f}{\partial x_i}\right)(1+\ln f)d\Xi = \int \frac{g-f}{\tau}(1+\ln f)d\Xi. \qquad (2.24)$$

With the assumption of τ being independent of particle velocity, the above equation gives

$$\frac{\partial}{\partial t}\int f\ln f d\Xi + \frac{\partial}{\partial x_i}\int u_i f \ln f d\Xi = \frac{1}{\tau}\int (g-f)(1+\ln f)d\Xi. \qquad (2.25)$$

From the compatibility condition (2.22), and the fact that $\ln g$ can be expressed as a sum of conservative moments of the collision term, i.e., $\ln g = \sum \alpha_i \psi_i$ we have

$$\int (g-f)\ln g d\Xi = 0.$$

With the definitions of H and J_i, and the relations of $\int (g-f)d\Xi = 0$ and $\int (g-f)\ln g d\Xi = 0$, Eq. (2.25) goes to

$$\begin{aligned}\frac{\partial H}{\partial t} + \frac{\partial J_i}{\partial x_i} &= \frac{1}{\tau}\int (g-f)\ln f d\Xi \\ &= \frac{1}{\tau}\int (g-f)(\ln f - \ln g)d\Xi \\ &\leq 0. \end{aligned} \qquad (2.26)$$

Therefore, the BGK model satisfies the entropy condition, and the particle system will move towards to an equilibrium state due to particle collisions. Boltzmann's H-theorem is of basic requirement because it shows that the Boltzmann equation ensures irreversibility. The entropy condition guarantees the dissipative property of the gas system. But, in recent years it seems that the entropy concept has been over-used in the design of numerical schemes and theoretical analysis. It is certainly true that any realizable scheme should satisfy the entropy condition, but the entropy-satisfying schemes in most times may not be good ones, because the entropy condition only tells us the evolution direction, but not the specific path the real gas system will take. To implement a correct physical path, such as the specific gas evolution model in the construction of interface fluxes in a numerical scheme, is the central point for the algorithm development. Theoretically, there are many entropy-satisfying solutions for a system, but only one is physically realizable.

The detailed derivation from the BGK model to the Navier-Stokes equations can be found in [Vincenti and Kruger (1965)] for monatomic gas, which is also given in Appendix B for polyatomic gas. The further derivations to the Burnett and Super-Burnett orders can be found in [Ohwada and Xu (2004)]. Based on the Chapman-Enskog expansion of the BGK model, the explicit expressions for the coefficients μ and κ in the Navier-Stokes equations can be obtained. For a state close to equilibrium, Eq. (2.21) confirms the fact that the rate to get to equilibrium is proportional to the deviation from equilibrium. The validity of this assertion has been confirmed by comparing the solution with that of the full Boltzmann equation [Cercignani (1988)]. To the Navier-Stokes order of the Chapman-Enskog expansion, the distribution function from the BGK equation is

$$f_{\text{ns, bgk}} = g - \tau\frac{\partial g}{\partial t} - \tau u_i\frac{\partial g}{\partial x_i}, \tag{2.27}$$

with the help of the Euler equations, the above distribution function goes to

$$\begin{aligned} f_{\text{ns, bgk}} &= g\left[1 - \tau\left(\frac{c^2}{2RT} - \frac{5}{2}\right)c_i\partial_{x_i}\ln T \right. \\ &\quad \left. -\tau\frac{m}{kT}\left(c_ic_j - \frac{1}{3}c^2\delta_{ij}\right)\partial_{x_j}U_i\right] \\ &= g + f_{\text{heat,bgk}} + f_{\text{stress,bgk}}, \end{aligned} \tag{2.28}$$

where

$$f_{\text{heat,bgk}} = -\tau g\left(\frac{c^2}{2RT} - \frac{5}{2}\right)c_i\partial_i \ln T,$$

and

$$f_{\text{stress,bgk}} = -\tau g\frac{m}{kT}\left(c_ic_j - \frac{1}{3}c^2\delta_{ij}\right)\partial_j U_i.$$

Theoretically, the collision time τ can be chosen to make one of the above terms of $f_{\text{stress,bgk}}$ or $f_{\text{heat,bgk}}$ be identical to that of Eq. (2.20) and (2.19) of the full Boltzmann equation.

With the above BGK distribution function (2.28), the viscous stress σ_{ij} and the heat flux q_i become,

$$\sigma_{ij} = \tau p(\partial_j U_i + \partial_i U_j) - 2/3\tau p\partial_k U_k\delta_{ij},$$

and

$$q_i = -\frac{5}{2}\frac{k}{m}\tau p\partial_i T, \tag{2.29}$$

which have the same functional dependence on the macroscopic flow variables as that of the full Boltzmann equation, but with different dissipative coefficients. For the BGK model, only one coefficient, i.e., viscosity or heat conduction, can be properly determined by selecting a proper definition of τ. With the definition of dynamic viscosity coefficient $\mu = \tau p$ and $C_p = 5k/2m$ for a monatomic gas, the heat conduction coefficient in the above equations becomes $\kappa = 5k\mu/2m$, and the Prandtl number for the BGK flow system in the NS order becomes fixed with the value of

$$\mathrm{Pr_{bgk}} = \mu C_p/\kappa = 1.$$

It should be noted that the relaxation rate $\nu = 1/\tau$ in the BGK model is not the same as the collision rate ν_c of the Boltzmann equation. The collision rate can be evaluated using kinetic theory [Bird (1994)]. It is velocity-independent for Maxwell molecule $\mu \sim T$, and has the general expression,

$$\nu_c = \frac{(\alpha+1)(\alpha+2)}{3\alpha}\frac{p}{\mu},$$

where μ is the dynamic viscosity coefficient and p is the pressure. For the BGK model, $p/\mu = \nu = 1/\tau$. Here α is the angular scattering exponent. For isotropic scattering ($\alpha = 1.0$), $\nu_c = 2\nu$. For monatomic gas ($\alpha = 2.14$), $\nu_c \simeq 2\nu$. Clearly, the collision rate is generally larger than the relaxation rate with slight modification from different molecular model. The above relationship is needed for a comparison in the solutions from the relaxation model and the full Boltzmann equation or DSMC.

In the following, we present some specific examples derived from the BGK model. For the 1D flow with two component flow velocity (U, V), the Chapman-Enskog expansion gives $f = g - \tau(g_t + ug_x)$. Taking moments of ψ again to the BGK equation with the new f, we get

$$\int \psi(g_t + ug_x)d\Xi = \tau \int \psi(g_{tt} + 2ug_{xt} + u^2 g_{xx})d\Xi.$$

By using the Euler equations to change the time derivatives to spatial derivatives on the right hand side of the above equation, the Navier-Stokes equations with a dynamic viscous coefficient $\mu = \tau p$ can be obtained,

$$\begin{pmatrix} \rho \\ \rho U \\ \rho V \\ \rho E \end{pmatrix}_t + \begin{pmatrix} \rho U \\ \rho U^2 + p \\ \rho UV \\ (\rho E + p)U \end{pmatrix}_x = \begin{pmatrix} 0 \\ s_{1x} \\ s_{2x} \\ s_{3x} \end{pmatrix}_x, \tag{2.30}$$

where

$$s_{1x} = \tau p \left[2\frac{\partial U}{\partial x} - \frac{2}{K+2}\frac{\partial U}{\partial x} \right],$$

$$s_{2x} = \tau p \frac{\partial V}{\partial x},$$

$$s_{3x} = \tau p \left[2U\frac{\partial U}{\partial x} + V\frac{\partial V}{\partial x} - \frac{2}{K+2}U\frac{\partial U}{\partial x} + \frac{K+4}{4}\frac{\partial}{\partial x}\left(\frac{1}{\lambda}\right) \right].$$

For a polyatomic gas, besides the dynamical viscosity there also appears a second viscosity or bulk viscosity coefficient, which is equal to $(2/3)\tau p N/(N+3)$, where N represents the degrees of freedom in the molecule rotation and vibration.

For the 1D flow with only explicit U-velocity only, the above viscous governing equations go to

$$\begin{pmatrix} \rho \\ \rho U \\ \rho E \end{pmatrix}_t + \begin{pmatrix} \rho U \\ \rho U^2 + p \\ (\rho E + p)U \end{pmatrix}_x = \begin{pmatrix} 0 \\ \frac{2K}{K+1}\tau p U_x \\ \frac{K+3}{4}\tau p (\frac{1}{\lambda})_x + \frac{2K}{K+1}\tau p U U_x \end{pmatrix}_x,$$

where in this case K is defined by $K = (3-\gamma)/(\gamma-1)$ and $\rho E = \frac{1}{2}\rho(U^2 + (K+1)/2\lambda)$. Notice that for the 1D flow the molecular motion in both y and z-direction is included as internal variable, and the above expression K is different from the K defined previously for the 1D flow with two velocity components (U, V). For example, for a monatomic gas in 1D case K is equal to 2, which accounts for the molecular motion in the y and z-directions. In this case, the above Navier-Stokes equations go to the standard form,

$$\begin{pmatrix} \rho \\ \rho U \\ \rho E \end{pmatrix}_t + \begin{pmatrix} \rho U \\ \rho U^2 + p \\ (\rho E + p)U \end{pmatrix}_x = \begin{pmatrix} 0 \\ \frac{4}{3}\mu U_x \\ \frac{5}{4}\mu (\frac{1}{\lambda})_x + \frac{4}{3}\mu U U_x \end{pmatrix}_x.$$

Based on the BGK model, many hyperbolic system can be recovered through the Chapman-Enskog expansion, such as the linear advection diffusion equation, the Burgers' equation, the shallow water equations, and the multiple temperature gas dynamic equations. The corresponding schemes can be constructed as well [Xu *et al.* (1996); Xu and He (2003); Xu (1999, 2002b); Xu *et al.* (2008)].

The BGK model presents unit Prandtl number in the continuum limit. For a numerical scheme, the Prandtl number can be corrected by direct modifying the interface heat flux in a finite volume scheme, such as adding a correction part [Chae *et al.* (2000); Xu (2001)] or directly changing the slopes of the temperature gradient [Woods (1993); May *et al.* (2007)]. These numerical treatments can effectively get correct Prandtl number in the continuum flow limit for the NS solutions. Besides the above numerical treatments, a correct Prandtl number can be obtained as well through the Shakhov and Ellipsoid-Statistical models, where the heat flux and stress of the BGK model are corrected separately. The validation of these modifications is based on the Chapman-Enskog expansion up to the Navier-Stokes order. In the following, we are going to introduce the Shakhov and ES-BGK models, and present their derivations based on the Chapman-Enskog expansion.

2.6.2 *BGK-Shakhov Model*

The gas-kinetic BGK-Shakhov model [Shakhov (1968)] replaces the BGK collision term by

$$\frac{f^+ - f}{\tau},$$

where f^+ is a modified "equilibrium" state. In order to understand the explicit form of f^+ for the capturing of a correct Prandtl number, let's get the Chapman-Enskog expansion of the BGK model first,

$$f_{\text{ns, bgk}} = g - \tau(g_t + u_i g_{x_i}) = g + f_{\text{heat,bgk}} + f_{\text{stress,bgk}}.$$

By adjusting the particle collision time τ, such as $\mu = \tau p$, the stress term in the above equation can be identical to that of the full Boltzmann one (2.20),

$$f_{\text{stress,bgk}} = f_{\text{stress, Boltzmann}}.$$

As a result, the heat part of the Chapman-Enskog expansion of the BGK-model becomes

$$f_{\text{heat,bgk}} = \text{Pr} f_{\text{heat, Boltzmann}},$$

where $f_{\text{heat, Boltzmann}}$ is the heat term (2.19) of the full Boltzmann equation. The inclusion of Pr in the above expression is to cancel the correct Prandtl number in the full Boltzmann heat term and makes the unit Prandtl number in the above BGK heat related distribution function. Here $\text{Pr} = 2/3$ can be used for monatomic gas.

Then, the NS distribution function of the full Boltzmann equation can be written as

$$f_{\text{ns, Boltzmann}} = g + f_{\text{heat,bgk}} + f_{\text{stress, Boltzmann}} + (f_{\text{heat, Boltzmann}} - f_{\text{heat,bgk}}).$$

The above correct NS distribution function can be expressed as

$$f_{\text{ns, Boltzmann}} = f_{\text{ns, bgk}} + (f_{\text{heat, Boltzmann}} - f_{\text{heat,bgk}}). \tag{2.31}$$

As a result, the term in the parentheses of the above equation is

$$g^+ = (1 - \text{Pr})g\mathbf{c} \cdot \mathbf{q} \left(\frac{c^2}{RT} - 5 \right) \Big/ (5pRT),$$

where $\mathbf{q}$ is the NS heat flux with correct Prandtl number.

In order to make the Chapman-Enskog expansion of the BGK-Shakhov model has a correct Prandtl number, based on the formulation (2.31) we only need to replace the equilibrium state g in the BGK model which will recover $f_{\text{ns, bgk}}$ in the NS order by f^+, which is defined by the combination of g and g^+

$$f^+ = g \left[1 + (1 - \text{Pr})\mathbf{c} \cdot \mathbf{q} \left(\frac{c^2}{RT} - 5 \right) \Big/ (5pRT) \right] = g + g^+.$$

Therefore, the Chapman-Enskog expansion of the Shakhov model up to NS order becomes

$$\begin{aligned} f_{\text{ns, Shakhov}} &= f^+ - \tau(f_t^+ + u_i f_{x_i}^+) \\ &= f^+ - \tau(g_t + u_i g_{x_i}) + O(\tau^2) \\ &= f_{\text{ns, bgk}} + g^+, \end{aligned}$$

which recovers the Chapman-Enskog expansion of the full Boltzmann equation and has correct values for both viscosity and heat conduction coefficients. The Shakhov model can choose any Prandtl number Pr in its expression of f^+. Both f and f^+ are functions of space x_i, time t, particle velocity u_i. The particle collision time τ is related to the viscosity and heat conduction coefficients, i.e., $\tau = \mu/p$, where μ is the dynamic viscosity coefficient and p is the pressure. Same as the BGK collision term, the mass, momentum, and energy are conserved during particle collisions,

$$\int (f^+ - f)\psi_\alpha dudvdw = 0, \qquad \alpha = 1, 2, 3, 4, 5, \tag{2.32}$$

at any point in space and time. Due to the specific formulation of g^+ which depends on the heat flux, the distribution function f^+ cannot be guaranteed to be positive. This is also one of the main reason why the Shakhov model has not been used widely in the rarefied gas community, especially in the mathematical society. Based on the computational results, we can clearly observe that the BGK-Shakhov model seems have better performance in most rarefied high-speed flow applications than the ES-BGK model, which will be introduced next.

2.6.3 *ES-BGK Model*

Ellipsoidal statistical model (ES-BGK) was proposed in [Holway (1966)]. Same as the Shakhov model, in the Navier-Stokes limit a correct Prandtl number can be obtained from the ES-BGK model through the Chapman-Enskog expansion. The entropy condition of this model has been proved [Andries *et al.* (2000)]. With the definition of a pressure tensor,

$$\bar{P}_{ij} = \frac{1}{\rho}\int (u_i - U_i)(u_j - U_j) f dudvdw,$$

and the corresponding translational temperature $T = Tr(\bar{P}_{ij})/3$, a corrected pressure tensor can be defined as

$$\tilde{T} = (1-\nu)TI + \nu\bar{P}_{i,j} = TI + \nu(\bar{P}_{i,j} - TI), \tag{2.33}$$

which is a linear combination of the initial stress tensor $\bar{P}_{i,j}$ and of the isotropic stress tensor TI. The terms in the parentheses can be considered as a modification of the shear stress which is used to recover the stress term of the full Boltzmann expansion (2.20),

$$(\bar{P}_{i,j} - TI) = -\sigma_{ij} = \frac{1}{\rho}\int (u_i - U_i)(u_j - U_j) f_{\text{stress, Boltzmann}} dudvdw. \tag{2.34}$$

The ES-BGK model replaces the Maxwellian in the BGK model by a Gaussian $G[f]$ defined by the above corrected pressure term in (2.33),

$$G[f] = \frac{\rho}{\det(\sqrt{2\pi\tilde{T}})} \exp\left(-\frac{(u_i - U_i)\tilde{T}_{ij}^{-1}(u_j - U_j)}{2}\right).$$

Thus, the collision term of the ES-BGK model is

$$\frac{G[f] - f}{\tau}.$$

Due to the modification of the shear stress in the Gaussian, to the leading order $G[f]$ can be approximated as

$$G[f] = g + \nu f_{\text{stress, Boltzmann}}.$$

In the following, we are going to present the derivation of the ES-BGK model, especially in the determination of the parameter ν in $G[f]$. Different from the Shakhov model, where the heat flux has been directly modified, the ES-BGK model modifies the shear stress. As mentioned earlier, with a proper choice of collision time τ, here we can make the heat term from the BGK model be identical to that of the full Boltzmann equation,

$$f_{\text{heat,bgk}} = f_{\text{heat, Boltzmann}},$$

and therefore, the stress term of the BGK model becomes,

$$f_{\text{stress,bgk}} = \frac{1}{\text{Pr}} f_{\text{stress, Boltzmann}},$$

where the unit Prandtl number effect of the BGK model has been recovered by modifying the viscosity coefficient of the full Boltzmann stress term. Based on the Chapman-Enskog expansion and Eq. (2.34), the NS distribution function of the ES-BGK model is,

$$\begin{aligned}
f_{\text{ns,es-bgk}} &= G[f] - \tau(G_t + u_i G_{x_i}) = G[f] - \tau(g_t + u_i g_{x_i}) + O(\tau^2) \\
&= g - \tau(g_t + u_i g_{x_i}) + \nu f_{\text{stress, Boltzmann}} \\
&= f_{\text{ns, bgk}} + \nu f_{\text{stress, Boltzmann}} \\
&= f_{\text{ns, Boltzmann}} + (f_{\text{stress,bgk}} - f_{\text{stress, Boltzmann}}) \\
&\quad + \nu f_{\text{stress, Boltzmann}} \\
&= f_{\text{ns, Boltzmann}} + \left(\frac{1}{\text{Pr}} - 1 + \nu \right) f_{\text{stress, Boltzmann}}. \qquad (2.35)
\end{aligned}$$

In order to exactly recover the NS distribution of the full Boltzmann equation from the ES-BGK model, the following condition has to be satisfied

$$\frac{1}{\text{Pr}} - 1 + \nu = 0.$$

Therefore, in order to recover a correct Prandtl number, the parameter ν can be determined from

$$\text{Pr} = \frac{1}{1-\nu},$$

where a monatomic gas has $\text{Pr} = 2/3$ and $\nu = -1/2$.

Up to the NS order, the Chapman-Enskog expansion presents identical distribution functions from the Shakhov, ES-BGK, and the full Boltzmann equation. But, the higher order expansions beyond the NS will be different. Besides the Shakhov and ES-BGK models, the development of other kinetic model equations is based on the Chapman-Enskog expansion to the NS order as well, such as Liu's model [Liu (1990)]. Even though the ES-BGK model has mathematically preferred properties, such as positivity in $G[f]$ and entropy condition, in comparison with the BGK-Shakhov model, many numerical tests in the transitional flow regime show that the Shakhov model may present more accurate solutions than that from the ES-BGK model. The tests include shock structure and Couettee flows. Even with the same NS order distribution function in the continuum regime, the ignored higher order terms from different kinetic models at finite Knudsen number will take effect in the non-equilibrium flow simulations. Also, for a monatomic gas with Pr $= 2/3$, the value of ν is $\nu = -1/2$, which seems present an anti-diffusive term to the BGK stress. This is equivalent to taking an over-relaxation effect in the ES-BGK system. As we know, an anti-diffusive term is easily to make the system unstable. Besides the entropy condition, it may need to prove the stability of the model as well. So, even with the positive gas distribution function in $G[f]$, the dynamics of the ES-BGK model in some cases may not be physically founded.

2.6.4 *Combined Model*

In order to get correct Prandtl number, the BGK-Shakhov and ES-BGK modify the heat flux and stress separately. Recently, these two models have been combined to form a new model [Chen *et al.* (2013)], which has the correct Prandtl number, but provides one more free parameter to take into account other non-equilibrium flow property. The combined model can be written as

$$\frac{\partial f}{\partial t} + u_i \frac{\partial f}{\partial x_i} = \frac{M[f] - f}{\tau}, \tag{2.36}$$

where

$$M[f] = G[f] + g^+$$

with

$$G[f] = \frac{\rho}{\det(\sqrt{2\pi \tilde{T}})} \exp\left(-\frac{(u_i - U_i)\tilde{T}_{ij}^{-1}(u_j - U_j)}{2} \right),$$

$$g^+ = (1 - C_s) g \mathbf{c} \cdot \mathbf{q} \left(\frac{c^2}{RT} - 5 \right) \Big/ (5pRT),$$

and

$$\tilde{T} = TI + C_e(\bar{P}_{i,j} - TI),$$

$$\bar{P}_{ij} = \frac{1}{\rho} \int (u_i - U_i)(u_j - U_j) f dudvdw,$$

$$\bar{P}_{ij} - TI = \frac{1}{\rho} \int (u_i - U_i)(u_j - U_j) f_{\text{stress,Boltzmann}} dudvdw,$$

and

$$T = \text{Tr}(\bar{P}_{ij})/3.$$

In the above equation, C_s and C_e are two new free parameters. As a result, to the NS order, the combined model has the Chapman-Enskog expansion

$$
\begin{aligned}
f_{\text{combined}} &= G[f] + g^+ - \tau(g_t + u_i g_{x_i}) + O(\tau^2) + \ldots \\
&= g - \tau(g_t + u_i g_{x_i}) + C_e f_{\text{stress, Boltzmann}} \\
&\quad + (1 - C_s) f_{\text{heat, Boltzmann}} \\
&= f_{\text{ns, bgk}} + C_e f_{\text{stress, Boltzmann}} + (1 - C_s) f_{\text{heat, Boltzmann}} \\
&= g + \frac{\tau p}{\mu} f_{\text{stress, Boltzmann}} + \frac{\mu C_p}{\kappa} f_{\text{heat, Boltzmann}} \\
&\quad + C_e f_{\text{stress, Boltzmann}} + (1 - C_s) f_{\text{heat, Boltzmann}} \\
&= g + \left(\frac{\tau p}{\mu} + C_e \right) f_{\text{stress, Boltzmann}} \\
&\quad + \left(\frac{\mu C_p}{\kappa} + 1 - C_s \right) f_{\text{heat, Boltzmann}},
\end{aligned}
$$

where μ and κ are the real physical viscosity and heat conduction coefficients from the full Boltzmann equation. As a result, based on the combined model the viscosity and heat conduction coefficients can be determined from

$$
\frac{\tau p}{\mu} + C_e = 1 \quad \text{and} \quad \frac{\mu C_p}{\kappa} + 1 - C_s = 1,
$$

which give

$$
\mu = \frac{\tau p}{1 - C_e},
$$

$$
\kappa = \frac{\mu C_p}{C_s}.
$$

With a fixed Prandtl number, the relationship among free parameters C_e and C_s is

$$
\mathrm{Pr} = \frac{C_p \mu}{\kappa} = \frac{C_s \tau p}{(1 - C_e)\mu}.
$$

If $\mu = \tau p$ is further used for the determination of the collision time τ in the combined model, the C_e and C_s are connected by

$$
\mathrm{Pr} = \frac{C_s}{1 - C_e},
$$

where the BGK-Shakhov and ES-BGK are two limiting cases with ($C_s = \mathrm{Pr}, C_e = 0$) and ($C_s = 1, C_e = 1 - \frac{1}{\mathrm{Pr}}$). Theoretically, a variation of C_s and C_e may be used for the capturing of high-order terms in the Chapman-Enskog expansion or high-order moments of the Boltzmann equation.

2.7 Summary

In this chapter, basic kinetic theory, the Boltzmann equation, and several kinetic model equations are introduced. The derivations from the kinetic equations to the hydrodynamic NS equations have been briefly presented. Different modeling scales have been emphasized. All these are composed of the basic knowledge needed to understand and develop gas-kinetic schemes. Even for the flow solver targeting to the macroscopic governing equations, the kinetic theory based scheme provides a rich mechanism in the capturing of numerical shock structure through the non-equilibrium modeling.

All above kinetic model equations are basically the relaxation models. For the BGK model, it is fortunate that an evolution solution can be obtained. If τ is a local constant, Eq. (2.21) has an integral solution [Kogan (1969)],

$$f(x_i, t, u_i, \xi) = \frac{1}{\tau}\int_{t_0}^{t} g(x_i - u_i(t-t'), t', u_i, \xi)e^{-(t-t')/\tau}dt' \\ +e^{-(t-t_0)/\tau}f_0(x_i - u_i(t-t_0), t_0, u_i, \xi), \tag{2.37}$$

where f_0 is the gas distribution function f at t_0, and g is the equilibrium state in (x_i, t). As a special case, we examine a gas with homogeneous distribution function whose state at time t_0 is given by $f(t_0)$, which does not depend on the spatial coordinates. In this case, it follows from the conservation laws that the mass ρ, momentum ρU_i and energy ρE are constant in space and time, and nothing will be changed macroscopically. The corresponding equilibrium state g is a constant in space and time as well. From Eq. (2.37), we have

$$f(t) = (1 - e^{-(t-t_0)/\tau})g + e^{-(t-t_0)/\tau}f(t_0). \tag{2.38}$$

The distribution function tends to an equilibrium state g exponentially, with a characteristic relaxation time τ which is equal to the time interval between particle collisions. For example, the denser the gas is, the faster the equilibrium is attained. From this example, we can observe that the gas-kinetic description provides more information than the macroscopic counterparts. Although, all macroscopic flow quantities are homogeneous and time independent, the particle distribution is actually a function of time. Consequently, the dissipative property of the gas system is time dependent. The evolution from f to g is a process with an increase of entropy. So, the dissipative character of this gas system is captured in the kinetic level. A detailed comparison between the DSMC and BGK model equation in terms of the time evolution of gas distribution functions under different initial conditions in the above homogeneous case have been conducted recently [Sun *et al.* (2014)].

At end, it is emphasized again that the Boltzmann equation is a model for the flow physics in the mean free path and mean collision time scale. It doesn't provide a direct description for the flow physics in other scales, such as hydrodynamic one. The integral solution of the BGK model in Eq. (2.37) goes beyond the kinetic scale physically due to the accumulation of the dynamic effect of particle collisions and transport. The integral of the equilibrium state, i.e., $(1/\tau)\int_{t_0}^{t} g(x_i - u_i(t-t'), t', u_i, \xi)e^{-(t-t')/\tau}dt'$, provides a hydrodynamic evolution model and recovers the NS gas distribution function directly when $(t-t_0) \gg \tau$ [Xu (1993)]. In other words, the integration of the equilibrium state is basically a hydrodynamic evolution model in space and time, which is beyond the kinetic one. It is fortunate that the integral solution of the BGK model presents an evolution process from the kinetic to hydrodynamic scale and this solution can be used as a modeling to construct a multiscale numerical algorithm. The Boltzmann equation itself does not indicate any explicit hydrodynamic evolution process, because it is a kinetic scale model only. Certainly, its time evolution solution of the distribution function in space and time will be valid in other scales as well. But, there is no such a general solution due to the complexity of the collision term for the full Boltzmann equation. A few special solutions, such as the equilibrium state and the Krook-Wu solution [Krook (1977)], can be hardly used to construct a multiscale numerical method. However, the Boltzmann collision term can be still used in the unified scheme whenever the local flow condition is in the kinetic regime, such as the local time step is less than the particle collision time.

Chapter 3

Introduction to Nonequilibrium Flow Simulations

In this chapter, we will briefly introduce numerical methods which are used in the non-equilibrium flow studies. At the same time, the basic framework for the construction of unified gas-kinetic scheme will be presented.

3.1 Nonequilibrium Flow Study

The study of rarefied gas flow becomes important due to its wide applications in engineering design and analysis, including hypersonic flows on re-entry space vehicle, micro-eletro-mechanical systems (MEMS) flows, chemical reactive flows, and even turbulent flows. The field of rarefied gas dynamics is mainly concerned with flow that cannot be completely described by the Navier-Stokes equations with small deviation from equilibrium. The non-equilibrium appears with the scale changes from the hydrodynamic to kinetic one. A large value of the Knudsen number, which is defined as the ratio of the molecular mean free path to the characteristic length, indicates the necessary condition for the emergence of the non-equilibrium, but not necessarily sufficient. Traditional gas dynamics equations model the flow motion in the hydrodynamic scale with the advection and diffusive process of macroscopic flow variables. However, the highly non-equilibrium mechanism can be modeled mostly in the kinetic level. The modeling deficiency of the continuum hydrodynamics has been observed in the regions with high gradient and close to boundary, where the flow motion is significantly effected by the non-equilibrium mechanism.

If the Knudsen number is used to characterize the degree of rarefaction, the rarefied flow results from either a large mean free path or small characteristic length scales. Gas flow problems involving large mean free paths are often encountered in high altitude atmospheric flow, such as those around reentry vehicles and spacecraft in low Earth orbit; vacuum systems as used in industrial processes including chemical vapor deposition and freeze drying; and in low density plume flows from spacecraft or rockets. Flow problems for which rarefaction effects occur due to small length scales include hypersonic flows, where local characteristic lengths may be very small, such as the thickness of shock wave and boundary layer around the leading edge of a flat plat; low speed flows through or around MEMS; and other mechanical devices, such as computer hard drives, which involve small air gaps.

Practical engineering flow problems may involve the regions with the coexistence of hydrodynamic equilibrium and kinetic non-equilibrium ones, where the characteristic length scale or the particle mean free path may vary significantly throughout the flow field between fully rarefied and continuum ones. With a reasonable number of mesh points, it is very hard in a practical computation to have the mesh size to be on the same order as the particle mean free path everywhere, which is required by the direct Bolztmann solver and direct simulation Monte Carlo (DSMC) method. Although there are techniques which extend the continuum approach by the use of slip boundary conditions to rarefied regime or the use of more moment equations to enlarge the hydrodynamic system, these methods cannot fill up the entire gap between the continuum and DSMC method. The extension of an entirely continuum or rarefied simulation methods to the domain covering the whole transition regime cannot be physically accurate and numerically efficient. In these cases, a multi-scale flow modeling with a smooth transition from the particle behavior in the kinetic scale to the wave propagation in the hydrodynamic scale is needed. The unified gas-kinetic scheme targets to fill up the gap between kinetic and hydrodynamic descriptions, where a transition is provided through the direct modeling in the algorithm development.

3.2 Numerical Methods for Non-equilibrium Flows

For rarefied flow computations, the most successful numerical method is the direct simulation Monte Carlo (DSMC) method. Other emerging methods in recent years are the direct Boltzmann solvers and the moments methods. The DSMC and direct Boltzmann solver are basically the single scale methods, where their modeling requires the cell resolution on the order of particle mean free path. In these methods, the hydrodynamic flow behavior can be obtained only through the long time integration of kinetic scale evolution. This requires a gigantic amount of computational resources. Therefore, the hybrid methods which adapt kinetic and hydrodynamic solvers in different flow regimes become popular in engineering applications, but with difficulties in its connection due to the imbalance of the information provided from different methods. For the moment and extended hydrodynamic equations, due to the requirement of moment closure and the basic assumption on the distribution function expansion, they are mainly used for the continuum or near continuum flows. Only in recent years, much attention has been paid to the development of multiscale method, such as the asymptotic preserving (AP) schemes for the capturing of both continuum and kinetic flow physics in a uniform numerical treatment. For other transport equations, such as the radiative transport, the development of AP scheme has been under intensive investigation. However, the accuracy of most AP schemes has never been seriously tested, especially in the capturing of diffusive solution in the continuum regime. The unified gas-kinetic scheme developed for the whole flow regimes has been extensively tested and validated. In this chapter, these methods will be introduced briefly.

3.2.1 *DSMC and Direct Boltzmann Solver*

The development of accurate numerical methods for the whole flow regime is challenging. At the current stage, the DSMC method is the most effective and dominant numerical method for molecular simulation of dilute gases [Bird (1994)]. For monatomic simple gas, the equivalence between the DSMC and the Boltzmann equation has been proved [Wagner (1992)]. In practice, the DSMC is a direct physical modeling, which can be extended to wider applications than that modeled through the Boltzmann equation, even for the flows where no corresponding Boltzmann equation has been constructed, such as the chemical reactive flow. The consistency between the DSMC and the direct Boltzmann solver helps us to understand both methods. Based on the Boltzmann equation,

$$f_t + \mathbf{u} \cdot \nabla f = Q(f, f),$$

the main feature of the DSMC method and direct Boltzmann solver is to split the above equation into two sub-processes:
(1) relaxation in accordance to the collisional operator of the Boltzmann equation

$$\frac{\partial f}{\partial t} = Q(f, f),$$

(2) free-molecular transport

$$\frac{\partial f}{\partial t} = -\mathbf{u} \cdot \nabla f.$$

The Boltzmann solver directly discretizes the above equations; and the DSMC uses particles to mimic the above physical processes. A valid physical process which is consistent with the above numerical splitting treatment is that the cell size and time step used have to be less than particle mean free path and collision time. This should not be surprising because the modeling scale in the construction of the Boltzmann equation is indeed the kinetic one, and the separation of collision and transport is explicitly used. Only under this condition, the pair of particles chosen for collision in the modeling scale or in the computational cell is independent of the locations of these particles. Actually, the limitation on the cell size in DSMC has been proved by [Alexander *et al.* (1998)], who showed the strong dependence of the viscosity and heat conduction coefficients on the cell size. For the hard spheres, when the collision cells are one mean free path wide the errors in the viscosity and thermal conductivity are 7.5% and 4.5%, which cannot be neglected. And the error can go to 20% when using cells of the size of two mean free path. So, the DSMC and direct Boltzmann solver are flow solvers on the kinetic scale only. For the continuum flow simulation, such as that for the NS flow, a cell size can get to ten to hundred of mean free path. The modeling in DSMC and Boltzmann solver constraints their extension to the continuum flow simulations, because they need to resolve down to the mean free path scale even for the continuum flow computation.

Most current research related to the further development of the DSMC method is on the modeling of collision procedure for complicated gas viscosity laws and the reduction of statistical noise due to limited number of particles. On the other hand, due to the particle nature and direct statistical modeling in the DSMC method, the lack of a direct connection with the kinetic equation may evoke certain mistrust of its solution and may lead to certain difficulties in a systematic approach to the increase of method's accuracy and efficiency. The necessity to construct a close connection between DSMC solution and the solution of kinetic equation is inevitable due to a number of reasons [Belotserkovskii and Khlopkov (2010)]. Numerous solutions have been obtained by DSMC method, but most of them were not repeated with the help of other methods. The connection between DSMC and kinetic solution may help the analysis of both methods and improvement of their effectiveness. Such a connection may give hint to formulate a general approach to the construction of methods, and may perhaps exclude any false modification of these methods. Unfortunately, at the current stage, there is no such a method based on the kinetic equation which is as trustable as DSMC for the rarefied flow computation. The unified scheme presented in this book is targeting to provide such an alternative. Based on the current progress on the unified scheme, we have not found any significant difference between the DSMC and unified solution in a gigantic amount of test cases studied so far.

Since the DSMC closely follows the physical process of particle transport and collision, there is no any space for DSMC to make fundamental mistakes. The only concern is to know the exact flow the DSMC is really simulating because the real particle number used in the simulation is significantly different from the physical one, how to implement a general equation of state and viscosity law in the particle method through the modeling of particle scattering in the collision. The accuracy and efficiency of the DSMC are also related to the governing equations the DSMC is trying to solve. Even though it is a direct modeling, the kinetic theory is still needed in the construction and understanding of the method.

The Boltzmann equation describes the time evolution of the velocity distribution function of a monatomic dilute gas with binary elastic collisions. Many Boltzmann equation solvers have been developed [Aristov (2001); Filbet *et al.* (2006); Wu *et al.* (2013)]. Since the Boltzmann equation is a modeling equation on the kinetic scale, its validation in the whole flow regime is under the assumption with the resolution up to the mean free path. Even though hydrodynamic equations are obtained after mathematical manipulation of the Boltzmann equation through asymptotic expansion, it does not mean that a similar technique is applicable in a numerical scheme. A direct discretization of the Boltzmann equation in the hope to get a valid scheme in all Knudsen number regime is difficult. Actually, it is not necessary at all because the hydrodynamic equations are accurate enough for the continuum flows. The importance for the unified scheme is that it can make a smooth transition from a kinetic to a hydrodynamic model in its algorithm discretization. So, even starting from the kinetic Boltzmann equation, new modeling is still needed to capture the flow evolution in other scales. The claim of the Boltzmann equation being applicable to the entire flow regime is theoretically valid, but practically meaningless.

Similar claim, such as the MD method being applicable to the entire flow regime, is valid as well theoretically.

In the framework of deterministic kinetic approximation, the most popular class of methods is based on the so-called discrete velocity method (DVM) or Discrete Ordinate Method (DOM) of the Boltzmann equation [Chu (1965); Yang and Huang (1995); Li and Zhang (2009); Mieussens (2000); Aristov (2001); Kolobov *et al.* (2007)]. These methods use regular discretization of particle velocity space. Numerically, they use the same operator splitting method as DSMC to solve the Boltzmann equation. Therefore, the same constraint on the cell size and time step is applied. Most of these methods can give accurate numerical solution for high Knudsen number flow computation, such as those from the upper transition to the free molecule regime. However, in the low transition and continuum flow regime, their solutions have not yet been well validated. In the continuum flow regime, similar to the DSMC method, they have great difficulty in the capturing of the Navier-Stokes solutions, especially for the high Reynolds number flows. The requirement of the time step being less than the particle collision time makes these methods be prohibitively expensive in the continuum flow application, but this constraint is fully required for an operator splitting method. Numerically, we have the freedom to determine the cell size, which is equivalent to the physical resolution to identify the gas system. However, without additional modeling, the direct adaptation of DOM with a large mesh size will be problematic. In order to get a stable scheme with large time step, it is natural to use implicit or semi-implicit method for the collision part. However, for the near continuum flow simulation, once the numerical cell size is much larger than the particle mean free path, and with the above DOM method the numerical discretization error due to free transport in the flux evaluation becomes dominant, such as the large numerical dissipation being proportional to the cell size. Even though an implicit scheme could overcome the stability restriction and use of a large time step, the accuracy barrier can be hardly removed. These schemes have the same dissipative mechanism as the Flux Vector Splitting (FVS) methods, where the numerical cell size is pretended to be the particle mean free path [Xu and Li (2001)], subsequently deteriorates the NS solution.

In recent years, attention has been paid to the further improvement of DOM methods [Mieussens (2000); Pieraccini and Puppo (2007); Filbet and Jin (2010); Bennoune *et al.* (2008); Degond *et al.* (2006)]. In order to develop kinetic schemes for all flow regime, much effort has been paid on the development of the so-called asymptotic preserving (AP) scheme. As defined in [Filbet and Jin (2010)], a kinetic scheme is AP if (a). it preserves the discrete analogy of the Chapman-Enskog expansion when the Knudsen numbers go to zero. (b). in the continuum regime, the time step is not restricted by the particle collision time. Besides the above two conditions, we may need a third one as well, (c). the scheme has at least second-order accuracy in both continuum and free molecule regimes, and the scheme can effectively get the NS solution. Therefore, the AP approach is an ideal choice to develop kinetic solvers. For example, in the application of AP scheme to the nozzle simulation, with a uniform time step in the whole domain, this time step can be much larger than the particle collision time in the inner high density region, and be on the same order or less in the outer rarefied and free molecular regions. Many attempts have been devoted to the construction of AP methods. One of the AP schemes with the NS asymptotic solution in the continuum regime is the method developed in [Bennoune *et al.* (2008)]. Their deterministic method is based on a decomposition of the Boltzmann equation into a system which couples a kinetic equation (non-equilibrium) with a fluid one (equilibrium). The fluid part of this system degenerates, for small particle collision time, into the NS equations. However, the physical basis of separating a distribution function into an equilibrium and a non-equilibrium part is questionable. How could a particle know that it should behave partially as a continuum flow with correlation with others, and partially as rarefied with its individual transport?

Even though individual particle movement has distinct transport and collision process, once this process is described by a statistical model, such as the Boltzmann equation, the transport and collision processes are coupled everywhere in space and time. To separate them numerically, such as the operator splitting method, is just to get back the original modeling. In this way, the advantage of constructing a time evolution solution and extending its solution to other scale is lost. Many existing AP methods use a simple upwind discretization for the particle transport of the Botltzmann equation. This upwind treatment is perfectly valid in the rarefied regime, but is associated with large numerical dissipation in the continuum flow regime. These methods can be only claimed to have AP property for the inviscid Euler solutions, because numerical dissipation due to free transport will definitely deteriorate the NS solutions, especially in the high Reynolds number flow computations, where the cell size is much larger than the particle mean free path. On the other hand, the Euler limit doesn't show any legitimate for the AP scheme development at all because the direct use of conservation with ad hoc flux can easily get the Euler solution. This Euler solution is not sensitive to the flow transport mechanism, such as the appearance of many approximate Riemann solvers [Toro (2009)], which can be equally used for the Euler solution. In engineering application now, the method with the Euler solution in the continuum regime is basically useless for the air-vehicle design. The neglect of the particle collision effect in the flux transport may not cause any problem in rarefied regime, since the numerical error which is proportional to time step or mesh size can be much less than the physical one which is proportional to the particle collision time and mean free path. For example, in the shock structure calculation with 20 or 30 points, the cell size is much less than the particle mean free path. But, this kind of numerical dissipation cannot be tolerated in the continuum flow regime. The upwind discretization in the flux vector splitting and flux difference splitting has different mechanism [Hirsch (2007); Toro (2009)]. It cannot be blindly used in a numerical discretization for any transport term even though it is extremely successful in the construction of shock capturing schemes for the inviscid Euler equations.

3.2.2 *Hybrid Scheme*

There are many flow problems where both rarefied and continuum flow regimes can co-exist. An example is a hypersonic flow around a blunt body, where the shock wave and boundary layer with large spatial gradients settle to the non-equilibrium flow regime, and other smooth regions may be close to the equilibrium one. For the continuum flow, the Navier-Stokes equations are well-defined and validated. In order to design a scheme which can be applied to both continuum and rarefied flow computations, a combination of NS and DSMC or NS and direct Boltzmann solver becomes a natural choice. Since the Boltzmann solver is much more expensive than NS solver, the NS one should take as large as possible in the computational domain. In recent years, the hybrid methods which combine NS and kinetic approaches have been used to model flows which have both continuum and rarefied regimes [Schwartzentruber and Boyd (2006); Schwartzentruber *et al.* (2007); Degond *et al.* (2006); Bourgat *et al.* (1996); Tiwari (1998); Coron and Perthame (1991); Wijesinghe *et al.* (2004)]. A buffer zone is designed to couple different approaches with the assumption of correctness of both methods in this zone. These approaches may sensitively depend on the location of the interface between different methods. Certainly, it is required that both methods are reliable in the buffer zone. But, in reality it may become that neither method can be applied there. For example, in many hybrid methods, it is assumed that in the buffer zone the flow can be correctly described by the NS equations, which means that the extension of the kinetic approach to the buffer zone should have the correct NS limit as well. But, this is just the difficult part for the kinetic solvers. Two popular hybrid methods which are used in aerospace industry is the CFD/DSMC and CFD/Boltzmann solver [Schwartzentruber and Boyd (2006); Kolobov *et al.* (2007); Burt *et al.* (2011)].

For any hybrid scheme, a boundary between different flow regimes has to be clearly defined, and a valid data exchange between different domains needs to be established. The continuum breakdown parameters are usually defined based on the CFD simulation data, where the flow variable gradients or the NS stress can be directly used to set up a threshold between different domains [Tiwari (1998)]. For the DSMC/CFD hybrid method, another difficulty is about how to overcome the statistical fluctuation in the DSMC simulation. A CFD solver may be sensitive to the noise introduced through the boundary. However, for the direct Boltzmann solver there is no noise generated at the interface. A smooth transition between CFD and Boltzmann solver may be obtained, especially when a gas-kinetic solver is used for the CFD part [Kolobov *et al.* (2007)]. The weakness of the CFD/Boltzmann approach is that the direct Boltzmann solver is much more expensive than the DSMC method in high Mach number case. The imbalance of the amount of information contained in different methods intrinsically brings the questions about the reliability of the hybrid method as well. A recent study based on the molecular dynamics simulation for the compressible shock wave and low speed channel flow identifies the difficulty to set up a valid criterion between the rarefied and continuum regimes [Dongari (2012)], where the level of non-equilibrium is a function of both Knudsen number and Mach number, and there are great varieties in the definition of Knuden number.

3.2.3 *Extended Hydrodynamics and Moment Methods*

Besides the above hybrid methods, for rarefied flow study there are also many other approaches, such as moments method and the extended hydrodynamic equations. The DSMC and direct Boltzmann solver present the kinetic scale physics, and attempt to go to continuum flow description. For the extended hydrodynamics and moment equations, their foundation is on the hydrodynamic flow evolution, and extends to the rarefied flow regime. Based on the Boltzmann equation, Hilbert discussed the existence and uniqueness of the Boltzmann equation [Hilbert (1912)]. It shows that there exists a one-to-one mapping between the gas distribution function which is a solution of the Boltzmann equation and the first five moments. These moments represent density, bulk velocity components and temperature. Hilbert's approach can be used for obtaining the Euler equations. In order to derive both Euler and Navier-Stokes equations from the Boltzmann equation, an approach developed independently by Chapman and Enskog can be used. In the Chapman-Enskog approach, the distribution function is expanded as a power series about the equilibrium distribution function [Chapman and Cowling (1990)]. In this approach, it is assumed that time is not an explicit argument of the distribution function and the time derivative of the first five macroscopic moments, but is an implicit argument by virtue of the dependence of the distribution function on the macroscopic moments and their spatial derivatives. A set of moment equations can be derived successively, such as the Euler equations, the Navier-Stokes equations, the Burnett equations, and the super-Burnett equations. The use of Burnett and super-Burnett equations can be challenging, as these equations are ill-posed and subject to numerical instabilities, which need to be regularized [Zhong *et al.* (1993)].

Another interesting approach is the moment method proposed in [Grad (1949)], where the distribution function is expanded in terms of a complete set of orthogonal polynomials (e.g. Hermite polynomials), whose coefficients correspond to velocity moments. This expansion is truncated by retaining terms up to a certain selected order, resulting in a closed set of moment equations up to this order. Although Grads moment approach has been widely used for solution of the Boltzmann equation, it is best suited to problems for which the velocity distribution function can be expressed as a perturbation about the equilibrium distribution, i.e., the near continuum flow. In other words, Grads method is not very efficient for treating problems with general non-equilibrium distributions, as Grads moment method assumes that the zeroth order term of the expansion recovers the equilibrium distribution by virtue of the properties of Hermite polynomial (basis) functions. Recently, the Chapman-Enskog expansion and the Grad moments methods have been combined in the developing of regularized methods [Levermore (1996)], such as R13 and R26 [Struchtrup (2005); Struchtrup and Torrihon (2003); Gu and Emerson (2009)]. The success of these methods is mainly in the low speed microflows at modest Knudsen number. Based on kinetic models, many other approaches for obtaining generalized hydrodynamic equations and development of the corresponding schemes have been conducted as well, but with limited success [Muller and Ruggeri (1998); Jou *et al.* (1996); Levermore (1996); Struchtrup (2005); Gorth and McDonald (2009); Xu (2002a); Xu and Josyula (2006); Xu *et al.* (2007); Xu and Liu (2008); Xu and Guo (2011)].

Theoretically, merely implementing higher-order terms in continuum model equations may not always result in good predictions of transitional flow. The difficulties may come from the boundary conditions and the non-linear constitutive relationship. More importantly, once the Knudsen number is used, any continuum model equation obtained through the expansion solution of the Boltzmann equation leads the macroscopic equations in the hydrodynamic scale. However, the non-equilibrium flow physics happens in the kinetic scale. So, it will not be surprising to observe the difficulties in the moment methods, because the barrier due to the different physical modeling scale limits their applicable regime.

To construct macroscopic governing equations through the truncated Chapman-Enskog expansion encounters tremendous difficulties, it introduces more difficulties when solving these equations numerically. Even though deriving macroscopic descriptions from the kinetic equation is the central topic of kinetic theory during its development, with the expansion of the gas distribution function on the basis of the Hermite orthogonal polynomials in velocity space, the expansion coefficients are exactly the velocity moments of the distribution function. The truncation of higher-order terms in a Hermite expansion do not directly alter the velocity moments of the distribution function. Therefore, the resulting equations for the Hermite coefficients can be directly constructed out of projecting the Boltzmann equation onto a truncated Hermite polynomial basis using the standard Galerkin procedure. Thus, instead of constructing extended hydrodynamic equations, a purely kinetic description of fluid system which is equivalent to that obtained through Grad's expansion procedure can be obtained [Shan *et al.* (2009b)]. As a result, a much simpler set of governing equations with kinetic nature can be constructed and solved numerically [Cai and Li (2010); Cai *et al.* (2012); Meng *et al.* (2013a,b)]. In this representation, higher approximations to the Boltzmann equation beyond the Navier-Stokes level can be constructed easily by merely expanding the equilibrium distribution to higher orders and adopting quadratures of a sufficiently high degree of precision. The level of accuracy is increased as higher-order terms in the truncated expansion are retained and quadratures of sufficient degree of precision employed. Theoretically, the hydrodynamic behavior of the expansion system can be made to correspond to any defined level in the Chapman-Enskog expansion to recover all kinds of high-order hydrodynamic equations. Practically, the real application of the distribution function expansion method in realistic flow problems is not so encouraging.

In summary, the moment equations or the distribution function expansion methods are based on the expansion of the equilibrium state in functional space and the possible use of the moment closure. Since the central expansion point is an equilibrium state, all these methods are suitable to describe accumulating particle behavior, such as these in the continuum and near continuum flow regime. The intensive particle collisions there merge individual particle behavior and form "wave" structure which can be described through a few degree of freedom, such as the limited number of moments or expansion coefficients. As the system approaches to the rarefied regime, the individual particle character will appear and become more and more distinguishable from each other. In the limiting free molecular motion, the gas distribution function becomes a gigantic amount of delta functions, which can be hardly recovered through Hermit expansion. Also, due to the independency of individual particle in the free molecular limit, their moments can hardly have any physical or dynamic connection. So, any moment closure used here, such as a higher order moment can be expressed as a summation of low-order ones, must be inappropriate. Therefore, in principle the moment method is not suitable for rarefied flow simulation, even though they can provide some kind of qualitative flow behavior in the low transition regime, but their results cannot be elevated to a quantitative level. For a highly non-equilibrium flow, the mathematical convergence of the high-order Chapman-Enskog expansion and high-order moment expansion are also questionable.

3.3 Direct Modeling in Unified Gas Kinetic Scheme

Physically, different flow regimes are defined through the Knudsen number, which is the ratio of the particle mean free path to the characteristic length scale, such as the continuum ($Kn \leq 10^{-3}$), transition ($10^{-3} < Kn < 10$), and free molecular ones ($Kn \geq 10$). Numerically, the computation takes place in a discretized space. With the current computer power, a three dimensional engineering application can be done with a mesh size on the order of $10^3 \times 10^3 \times 10^3$ in the physical space. With such a mesh distribution, such as the computation around a flying space vehicle, the flow regime to be identified depends on the local mesh size and the local particle mean free path. In terms of the computation, it is hard to use the traditional Knudsen number to define different flow regimes. What we need to model in the computation is the flow physics when the ratio of cell size to the particle mean free path varies significantly. A unified scheme targets to capture the corresponding flow physics with respect to the cell's Knudsen number, where the mechanism from particle free transport to the hydrodynamic wave propagation will be captured numerically.

3.3.1 *General Methodology*

The unified scheme is based on the direct physical modeling of the flow motion in a discretized space. This process of constructing a numerical scheme is similar to the modeling process in deriving theoretical fluid dynamic equations, but without shrinking the control volume to zero. Different mesh size scale in terms of local particle mean free path will notify different transport phenomena. The direct modeling is to simulate the physical phenomena which can be observed in the mesh size scale, such as the construction of the governing equation with cell size resolution. Certainly, the accuracy of the physical reality to be modeled depends on the requirement of practical engineering design, and the availability of computational resources. If the cell size comes to a scale of particle mean free path, then the Boltzmann modeling physics, such as transport and collision, can be used to construct the local gas evolution [Bird (1994); Aristov (2001)]. If the cell size is much larger than the particle mean free path, the accumulating transport and collision effect up to the NS solutions need to be recovered. Here, we need to clearly distinguish between traditional numerical discretization of the Boltzmann equation, the so-called numerical PDEs, and the direct modeling method, such as the unified scheme shown in this book. In the numerical PDE approach, if f represents a particle distribution function and u is the particle velocity in the x-direction, a direct discretization will approximate uf_x as

$$uf_x = \begin{cases} u(f_j - f_{j-1})/\Delta x, & u \geq 0, \\ u(f_{j+1} - f_j)/\Delta x, & u < 0, \end{cases} \tag{3.1}$$

where f_{j-1} and f_{j+1} are the particle distribution functions at the neighboring cells of cell j. This is the so-called upwind approximation according to particle velocity. In order to validate the above discretization, there are underlying assumptions. The cell size Δx should be on a scale which allows particle free transport. If Δx is on the scale of hundreds particle mean free path, the above approximate is inappropriate because a particle will not take free transport across such a long distance without encountering other particles. So, the direct use of upwind here will introduce physical inconsistency. Due to the success of the upwind concept in the numerical solution of hyperbolic equations, this concept has been over-used to other flow systems without considering the validating condition for upwind. A free transport between numerical cells in the discretization will automatically pretend the cell size as the particle mean free path. Hence, a simple discretization of PDE without considering the real physical process may introduce dynamic deficiency and distort the physical picture.

On the contrary, we emphasize the concept of direct modeling in CFD. The direct modeling is to construct an evolution model for the particle transport across a cell interface. If the cell size is on a scale which is much larger than the particle mean free path, the evolution solution will present the equilibrium wave propagation in the hydrodynamic scale. If the cell size is on the scale less than the particle mean free path, this solution will go to particle free transport and the above upwind discretization gets recovered. If the cell size has a scale between the kinetic and the hydrodynamic one, the direct modeling should give a time dependent non-equilibrium gas evolution which accounts for the competition between the particle collision and transport. To properly construct such an evolution solution through direct modeling is the underlying principle for the development of the unified scheme. The unified scheme is a direct flow modeling of physical process rather than a purely numerical discretization of PDEs. Both continuum and rarefied flow evolution should be obtained under the same numerical recipe, because the time-dependent modeling solution includes the evolution process from a non-equilibrium to an equilibrium one. The weights between these two limiting solutions depend on the ratio of time step to the particle collision time. Therefore, the unified method has advantages in simulating flow with both continuum and rarefied regions.

The direct modeling in a discretized space is the starting point in the development of the unified scheme. The corresponding fluid dynamic equations simulated by the unified scheme will depend on the local variation of mesh size and time steps. The numerical governing equation should change according to the space and time resolution. The mesh size and time step become dynamic quantities for the flow description. The quality of the scheme cannot be evaluated using the traditional truncation error analysis because there is no a fixed governing equation to be solved. Actually, there is no such a valid governing equation in the transition regime.

In order to construct such a scheme, the physical mechanism from kinetic to hydrodynamic one needs to be used. Even though the Boltzmann and other kinetic model equations are valid in the kinetic scale only, but their evolution solutions could go beyond the kinetic level. The use of the kinetic equation's evolution solution is the key in the algorithm development. This methodology distinguishes it from the single scale modeling scheme, such as DSMC and direct Boltzmann solver. It also distinguishes from the scheme which solves numerically a specific governing equation.

3.3.2 *Modeling in Discretized Space*

The unified scheme is a direct physical modeling in a discretized space. The "governing" equation is the numerical algorithm itself. Since the cell size and time step never go to zero, there is no purpose to "get" the so-called PDEs. The discretized space is divided into control volume, i.e., $\Omega_{i,j,k}(\vec{x}) = \Delta x \Delta y \Delta z$ with the cell sizes $(\Delta x) = x_{i+1/2,j,k} - x_{i-1/2,j,k}$, $\Delta y = y_{i,j+1/2,k} - y_{i,j-1/2,k}$, and $\Delta z = z_{i,j,k+1/2} - z_{i,j,k-1/2}$ in a physical space. The temporal discretization is denoted by t^n for the n_{th} time step. The particle velocity space is discretized by Cartesian mesh points with velocity spacing Δu, Δv, and Δw with a volume $\Omega_{l,m,n}(\vec{u})$, around the center of the (l, m, n)-velocity interval at (u_l, v_m, w_n). The fundamental flow variable in a discretized space is the averaged gas distribution function in a control volume (i, j, k), at time step t^n, and around particle velocity (u_l, v_m, w_n),

$$\begin{aligned} f(x_i, y_j, z_k, t^n, u_l, v_m, w_n) &= f^n_{i,j,k,l,m,n} \\ &= \frac{1}{\Omega_{i,j,k}(\vec{x})\Omega_{l,m,n}(\vec{u})} \int_{\Omega_{i,j,k}} \int_{\Delta u \Delta v \Delta w} f(x, y, z, t^n, u; v, w) d\vec{x} d\vec{u}. \end{aligned} \tag{3.2}$$

The time evolution of a gas distribution function in the computational space is due to the particle transport through cell interface and the particle collisions inside each cell, which re-distributes particles in the velocity space. The direct modeling in a discretized space gives

$$\begin{aligned} f^{n+1}_{i,j,k,l,n,m} = f^n_{i,j,k,l,m,n} &+ \frac{1}{\Omega_{i,j,k}} \int_{t^n}^{t^{n+1}} \sum_{r=1} u_r \hat{f}_r(t) \Delta S_r dt \\ &+ \frac{1}{\Omega_{i,j,k}} \int_{t^n}^{t^{n+1}} \int_{\Omega_{i,j,k}} Q(f) d\vec{x} dt, \end{aligned} \tag{3.3}$$

where $\hat{f}_r$ is the time-dependent gas distribution function at a cell boundary, which is integrated along the surfaces of the control volume $\Omega_{i,j,k}$, u_r is the particle velocity normal to the cell interface, ΔS_r is the r_{th} cell interface area, and $Q(f)$ is the particle collision term, which redistributes the particle in the velocity space. The above equation is the fundamental governing equation through the direct modeling in a discretized space. It presents explicitly an evolution equation or algorithm on the scale of the control volume. The size of the control volume depends on the required resolution to describe the flow physics. Someone may say that Eq. (3.3) is an integral form of the Boltzmann equation, which is basically the same as the Boltzmann equation. But, the truth is that the Boltzmann equation can be derived from the modeling in (3.3). This modeling is more fundamental than the Boltzmann equation in the following aspects. (a) it doesn't require a continuous f function in space and time. (b) the choices of cell size and time step are not limited by the particle mean free path and collision time. (c) if the above equation is considered to be derived from the Boltzmann equation, a common mistake is to use upwind to approximate the interface distribution function. Here, due to the large variation of the cell size to the local particle mean free path, the collision effect has to be explicitly included in the time evolution of the interface gas distribution function.

If we take conservative moments ψ_α on Eq. (3.3), i.e.,

$$\psi = (\psi_1, \psi_2, \psi_3, \psi_4, \psi_5)^\top = \left(1, u, v, w, \frac{1}{2}(u^2 + v^2 + w^2)\right)^\top,$$

where $d\vec{u} = dudvdw$ is the volume element in the phase space, due to the conservation of conservative variables during particle collisions, the update of conservative variables becomes

$$\mathbf{W}^{n+1}_{i,j,k} = \mathbf{W}^n_{i,j,k} + \frac{1}{\Omega_{i,j,k}} \int_{t^n}^{t^{n+1}} \sum_{r=1} \Delta\mathbf{S}_r \cdot \vec{\mathbf{F}}_r(t) dt, \tag{3.4}$$

where $\mathbf{W}$ is the volume averaged conservative mass, momentum, and energy densities inside each control volume, and $\vec{\mathbf{F}}$ is the flux for the corresponding macroscopic flow variables across the cell interface. The flux can be obtained from a time-dependent gas distribution function as well. Now the discrete governing equations of the unified scheme are the equations (3.3) and (3.4). These equations are the fundamental governing equations for the description of flow motion in all flow regimes where the flow physics solely depends on the time evolution of the gas distribution function at a cell interface and the particle collision inside each control volume. Theoretically, Eqs. (3.3) and (3.4) are the direct modeling of the physics, and there is no any inaccuracy introduced yet. But, the quality of the scheme depends on the modeling of interface flux and particle collision term in the cell size and time step scales.

3.3.3 *Multiple Scale Gas Evolution Model*

In the above unified scheme, the engine is the time-dependent gas distribution function at a cell interface and the particle collision inside each cell. The time dependent gas distribution function at a cell interface needs to mimic the physical process in the mesh size scale, where the ratio between the particle mean free path and the numerical cell size can be varied significantly. In the unified scheme, in order to model the gas evolution process, the gas-kinetic BGK model, the Shakhov model, the ES-BGK model, the Rykov model for diatomic gases, and even the full Boltzmann equation, can be used. Basically, a local time evolution solution of the gas distribution function at a cell interface and the particle collision inside each cell are needed.

In order to capture the flow physics from the kinetic to the hydrodynamic scales, the basic idea for the evolution solution f at a cell interface is the following,

$$f = (1-\omega) f_{\text{hydrodynamic}} + \omega f_{\text{kinetic}}, \tag{3.5}$$

where ω is a time dependent function, which may depend on the ratio of the time t and the particle collision time τ, such as $\omega = \exp(-t/\tau)$. The time step Δt with $t \in [0, \Delta t]$ depends on the cell size. More sophisticated formulation will be presented for the unified scheme in Chapter 5. But, the idea is the same. The above formulation presents a gas evolution process from the kinetic scale ($\omega = 1$), which takes into account the particle free transport, to the hydrodynamical scale ($\omega = 0$), which describes a nearly equilibrium solution. In the hydrodynamic limit, the Navier-Stokes or multi-temperature hydrodynamic equations should be recovered from $f_{\text{hydrodynamic}}$. The physics in the transition regime between the hydrodynamic and kinetic scales is mainly determined through the construction of the function ω. For the time dependent collision term inside each cell, it can be written as

$$Q(f) = \Gamma Q_{models} + (1-\Gamma) Q_{Boltzmann},$$

where Γ depends on the time step Δt and the particle collision time τ, Q_{models} is approximate collision model, such as BGK and Shakhov, and $Q_{Boltzmann}$ is the full Boltzmann collision model. The unified scheme doesn't require the time step being less than the particle collision time. Many experiments show that any initial non-equilibrium gas distribution function will eventually get to the equilibrium one due to the particle collision. After 3 or 4 particles collision, the gas distribution functions from the full Boltzmann collision term and the approximate collision models basically have no difference. Therefore, we can simply design $\Gamma = \mathrm{H}(\Delta t/\tau - n_c)$, where H is the Heaviside function and n_c is a critical threshold. Practically, the unified scheme does require a scale dependent collision term. The full Boltzmann collision term plays a role only in a small subset of the collision process. Even with the choice of $\Gamma = 1$, the unified scheme presented in the later Chapters can present reasonable and accurate results in the whole flow regime.

In a highly non-equilibrium region, such as inside a high Mach number dissipative shock layer, the cell size must be less than the particle mean free path in order to resolve the non-equilibrium shock structure. In these regions, the kinetic part in Eq. (3.5) will contribute mostly in the final gas distribution function f. In the continuum flow region, due to the sufficient number of particle collisions and with the condition of time step being much larger than the local particle collision time, the contribution from the hydrodynamical part in Eq. (3.5) will be dominant in the solution of the distribution function. The limiting hydrodynamical distribution function corresponds to the Chapman-Enskog expansion for the NS solution. A naive approach may target to the Euler solution in the hydrodynamical scale. This target will much deteriorate the quality of the scheme, because for the Euler limit there is no any dissipative wave structure needs to be resolved, and any conservative scheme with the update of mass, momentum, and energy can easily get to the target. Actually, the equilibrium state inside each control volume can be simply reconstructed instead of updated. In other words, we know precisely the equilibrium state which can be numerically used to get the Euler solutions. The importance of the evolution model (3.5) is the time-dependent solution, where the physical process from the non-equilibrium to the equilibrium is modeled.

In order to discretize the collision term in Eq. (3.3) efficiently, the unified scheme will update the macroscopic variables first. Then, the updated macroscopic flow variables will be used to model the collision term in Eq. (3.3) implicitly. The implicit treatment of the particle collision term releases the difficulty due to the stiff collision effect in the near continuum flow regime. In the near continuum flow regime, the cell size can be much larger than the particle mean free path. The distribution function is close to the equilibrium state and this solution is not sensitive to the collision term at all. As presented in previous sections, all collision models have the same NS distribution function in the hydrodynamic limit.

For the continuum flow simulation, the NS gas distribution function is well defined through the Chapman-Enskog expansion. It is unnecessary to use Eq. (3.3) to update f anymore if only NS solution is our concern, because the NS distribution function can be reconstructed from macroscopic variables. Therefore, the numerical scheme here can be directly developed using Eq. (3.4) with a modeled time-dependent gas distribution function at a cell interface for the fluxes. The gas-kinetic scheme (GKS) for the continuum flow simulation [Xu (2001); Xu *et al.* (2005)], which will be introduced in the next chapter, has been developed along this line. In GKS, a continuous particle velocity space space is used and there is no need to discretize the particle velocity space at all.

3.3.4 *Discretization and Integration in Particle Velocity Space*

In the unified gas-kinetic scheme (UGKS), a continuous gas distribution function f is discretized with a set of discrete velocity points and its moment is evaluated using quadrature rule. Then, the update of the gas distribution function is done at each discrete point [Yang and Huang (1995); Li and Zhang (2009)]. The selections of the discrete points and the range of the velocity space in the discrete ordinate method depend on the specific problems.

In a finite particle velocity domain $\Omega(\mathbf{u}) \in [\mathbf{u}_{min}, \mathbf{u}_{max}]$, a set of points $\mathbf{u}_\alpha = (u_i, v_j, w_k)$ are used to discretize the space, where $i = 1, \dots, nx$, $j = 1, \dots, ny$ and $k = 1, \dots, nz$ are index to represent the velocity points (i, j, k).

Once the value of f at discrete velocity point $f_{i,j,k}$ is given, the macroscopic flow variables in the physical space can be obtained with the numerical quadrature rule for the integration in the particle velocity space. For example, the macroscopic variables for monatomic gas can be evaluated as

$$W = \begin{pmatrix} \rho \\ \rho U \\ \rho V \\ \rho W \\ \rho E \end{pmatrix} = \sum_{i=1}^{nx} \sum_{j=1}^{ny} \sum_{k=1}^{nz} \begin{pmatrix} 1 \\ u_i \\ v_j \\ w_k \\ \frac{1}{2}(u_i^2 + v_j^2 + w_k^2) \end{pmatrix} W_i W_j W_k f_{i,j,k},$$

$$\begin{pmatrix} P_{xx} & P_{xy} & P_{xz} \\ P_{yx} & P_{yy} & P_{yz} \\ P_{zx} & P_{zy} & P_{zz} \end{pmatrix} = \sum_{i=1}^{nx} \sum_{j=1}^{ny} \sum_{k=1}^{nz} \begin{pmatrix} c_i c_i & c_i c_j & c_i c_k \\ c_j c_i & c_j c_j & c_j c_k \\ c_k c_i & c_k c_j & c_k c_k \end{pmatrix} W_i W_j W_k f_{i,j,k},$$

$$\begin{pmatrix} q_x \\ q_y \\ q_z \end{pmatrix} = \frac{1}{2}\sum_{i=1}^{nx}\sum_{j=1}^{ny}\sum_{k=1}^{nz}\begin{pmatrix} c_i \\ c_j \\ c_k \end{pmatrix}(c_i^2 + c_j^2 + c_k^2)W_iW_jW_kf_{i,j,k} \tag{3.6}$$

where W_i, W_j and W_k represent the quadrature weights in integrations, respectively, $c_i = u_i - U$, $c_j = v_j - V$ and $c_k = w_k - W$ represent the peculiar velocity.

In comparison with the CFD methods for the Navier-Stokes equations, the kinetic solver with discrete particle velocity space is very expensive in terms of computer memory requirement and computational efficiency. The efficiency mainly depends on the number of discrete points needed to update a gas distribution function. The use of the quadrature rule is important to evaluate the moments of a discrete distribution function. The order of quadrature rule will affect the accuracy of macroscopic variables and the satisfaction of conservative constraint for the mass, momentum and energy. In general, the high-order quadrature rules are recommended to evaluate the moments in Eq. (3.6).

With the discretization of particle velocity space, a finite domain in the velocity space has been used. The size of the domain depends on the flow velocity and temperature. Theoretically, the domain can be estimated within $\left[-\alpha\sqrt{\frac{1}{2}RT}, \alpha\sqrt{\frac{1}{2}RT}\right]$ and $\alpha \geq 4$. For a given problem, the domain should be large enough to contain all possible distribution functions in the whole physical space,

$$\mathbf{u}_{min} = \min_{x\in\Omega}\left\{\mathbf{U} - \alpha\sqrt{\frac{1}{2}RT}\right\}, \quad \mathbf{u}_{max} = \max_{x\in\Omega}\left\{\mathbf{U} + \alpha\sqrt{\frac{1}{2}RT}\right\}, \tag{3.7}$$

where the flow velocity $\mathbf{U}$ and temperature T have to be estimated first.

The general numerical quadrature for a function $f(x)$ in a given domain $[a, b]$ can be expressed as

$$\int_a^b f(x)dx = \sum_{i=1}^{N} w_i f(x_i), \tag{3.8}$$

where x_i are the quadrature points and w_i are the corresponding quadrature weights of the integration rule. The selection of quadrature rule depends on the problems solved. Two kinds of quadrature rules are adopted in our research: the 4th-order composite Newton-Cotes rule, and the Gauss quadrature rule.

For the 4th-order Newton-Cotes rule, the domain $[a, b]$ is divided into N sections with equal size $h = (b - a)/N$, and the the quadrature rule reads

$$\begin{aligned}\int_a^b f(x)dx \approx & \frac{2h}{45}[7f(x_0) + 32f(x_1) + 12f(x_2) + 32f(x_3) + 7f(x_4) \\ & +7f(x_4) + \cdots + 7f(x_{N-4}) + 7f(x_{N-4}) \\ & +32f(x_{N-3}) + 12f(x_{N-2}) + 32f(x_{N-1}) + 7f(x_N)] \\ = & \sum_{i=0}^{N} w_i f_i,\end{aligned} \tag{3.9}$$

and the quadrature weights w_i have the following form

$$w_i = \begin{cases} \frac{14h}{45}, & i = 0 \quad or \quad N; \\ \frac{28h}{45}, & mod(i, 4) = 0; \\ \frac{14h}{45}, & mod(i, 4) = 2; \\ \frac{64h}{45}, & \text{others.} \end{cases} \tag{3.10}$$

The Newton-Cotes rule can be used to any kind of flow simulation, including the hypersonic or highly non-equilibrium flows. However, the disadvantage of this rule is the enormous memory requirements due to the equally spaced velocity grid points.

For low speed flows with small temperature variation, where the distribution function concentrates on the zero velocity and may be close to the equilibrium state, the high-order Gauss-Hermite rule is the best choice,

$$\int_{-\infty}^{\infty} f(x)dx = \int_{-\infty}^{\infty} e^{-x^2}[e^{x^2} f(x)]dx \approx \sum_{i=1}^{N} w_i e^{x_i^2} f(x_i), \tag{3.11}$$

where e^{-x^2} is the weighting function, $x_i (i = 1, \ldots, N)$ are the positive roots of the Hermite polynomial of degree N, and w_i are the corresponding weights of Gauss-Hermite quadratures, which can be evaluated through

$$w_i = \frac{2^{N-1} N! \sqrt{\pi}}{N^2 [H_{N-1}(x_i)]^2}, \tag{3.12}$$

where H_i is the ith Gauss-Hermite polynomial [Shizgal (1981); Steen *et al.* (1969)]. The advantage of the Gauss-Hermite quadrature rule is its relatively high accuracy. However, due to the limited number of quadrature points, this rule cannot be used in the high temperature or high Mach number flows. It should be noted that, for highly non-equilibrium flows, even the temperature or fluid velocity is low, the velocity distribution function may become bizarre, and there maybe exist singular points. So the limited number of points may not be enough to represent a real distribution, and plausible results may be obtained in high Knudsen number flow simulations. So, in this case an adaptive mesh in the velocity space is an appropriate choice.

3.4 Summary

This chapter presents general numerical methods for the non-equilibrium flow computations. Even with many successful methods, such as the DSMC, the direct Boltzmann solvers, AP schemes, moment methods, extended hydrodynamic equations, and hybrid methods, there is still room for the development of a unified scheme, which covers different flow regimes smoothly.

The basic idea of the modeling and the construction of a unified gas-kinetic scheme (UGKS) is presented. The UGKS provides a general framework for the construction of a scheme for all Knudsen number flow through direct modeling. Even though the kinetic BGK, Shakhov, ES-BGK, and even the full Boltzmann equation, can be used to construct the local evolution solution at a cell interface, and to model the collision term inside each cell, the UGKS is not targeting to solve these kinetic equations, but to use their solution to do the modeling in the construction of the numerical algorithm. In the discretized space, the real governing equation of the UGKS changes with the variation of the cell size in terms of particle mean free path. Actually, the numerical algorithm itself is the discrete governing equation, where the flow physics and cell resolution are closely coupled. The numerical cell size and the time step do contribute dynamically to the gas evolution. The scheme is basically a physical description of the flow motion in the mesh size and time step scale. This fact distinguishes it from any other numerical PDE approach, where the cell size and time step need to approach to zero in order to properly recover the PDE's solution. The dynamics provided in the UGKS covers all scale physics from the kinetic to the hydrodynamic, and a continuum spectrum of governing equations of different scales are recovered and used in the computation.

The unified scheme updates the gas distribution function in both physical and particle velocity space. In the rarefied flow regime, the distribution function is usually smooth in the physical space, but becomes peculiar in the velocity space. In the continuum flow regime, the distribution function changes dramatically in the physical space, but is smooth in the velocity space. The unified scheme takes this into account, where the interface flux and inside cell collision play different roles in different regimes. In the continuum flow regime, the interface flux takes a dominant role for the quality of the scheme, and in the rarefied flow regime the collision term inside each cell has important contribution for the accuracy of the scheme.

Chapter 4

Gas Kinetic Scheme

4.1 Introduction

In the past decades, the gas-kinetic scheme (GKS) for the Navier-Stokes solutions has been developed, and successfully applied for the continuum flow simulations from nearly incompressible to hypersonic viscous and heat conducting flows [Xu (1993); Prendergast and Xu (1993); Xu and Prendergast (1994); Xu *et al.* (1995); Xu (2001); Su *et al.* (1999); Xu *et al.* (2005, 2008); May *et al.* (2007); Kumar *et al.* (2013); Jiang and Qian (2012); Li *et al.* (2011); Righi (2014)]. The GKS is a finite volume method for the update of macroscopic conservative flow variables with the construction of a time dependent gas distribution function at a cell interface for the flux evaluation. Here a continuous particle velocity space is used for the gas distribution function, because from macroscopic flow variables the NS distribution function can be constructed through the Chapman-Enskog expansion, see Chapter 2. Therefore, the efficiency of the GKS is similar to the Riemann solution-based NS flow solver, where the same CFL condition is used for the determination of time step. The GKS is able to present accurate NS solution in the smooth flow regime and have favorable shock capturing capability in the shock region. Even for the continuum flow simulation, it seems that the non-equilibrium flow physics is still needed for providing reliable numerical dissipation in the discontinuous shock region.

The reason we introduce the GKS for the continuum flow computation in this chapter is that the unified gas-kinetic scheme (UGKS) for all flow regime is a natural extension of GKS. To fully understand GKS will be helpful for the easy acceptance of the unified method. In terms of numerical modeling the UGKS is simpler than the GKS, because there is no any Chapman-Enskog theory needed there. The gas distribution function is updated in UGKS, instead of reconstructed in GKS. More importantly, based on the GKS methodology, we can get a fully comparison of the mechanism between

GKS and the Godunov method, where the advantages and weakness of these methods can be analyzed through a physical way [Li *et al.* (2011); Ohwada *et al.* (2013)], instead of traditional numerical analysis.

For the continuum flow computation, a finite volume scheme is a discretized conservation law for the update of conservative flow variables inside each control volume. Since there is one to one correspondence between the macroscopic flow variables and the NS gas distribution function, there is no reason to update the gas distribution function anymore. Therefore, we only need to update macroscopic flow variables through Eq. (3.4). Also, we need to understand that the NS equations are phenomenological model, which is not equivalent to the fluid dynamics. Even in the continuum flow regime there may have cases where the NS cannot be applied or present incorrect solutions, such as these heat related ghost effect [Sone (2007)]. For the continuum flow computation, it is not guaranteed that the NS is still a valid modeling in the discontinuous region where the mesh size is much larger than the NS shock thickness. If we would like to study the smooth transition from the rarefied to the continuum flow and check the validity of the NS equations, the UGKS can be faithfully used. Even though there is no source term for the conservative macroscopic variable update inside each control volume, instead of using free transport the construction of a time dependent gas distribution function with the inclusion of particle collision is critically important for the evaluation of interface flux. The inclusion of particle collision term in the interface flux evaluation is the core for the success of the GKS and is the main difference between GKS and many other kinetic solvers [Pullin (1980); Reitz (1981); Deshpande (1986); Perthame (1992); Mandal and Deshpande (1994); Chou and Baganoff (1997)]. The recipes inside GKS can be used to design other accurate kinetic schemes [Ohwada (2002); Ohwada and Kobayashi (2004); Tang (2012); Yang *et al.* (2013, 2014)].

In the past decades, for the shock capturing scheme, the most important achievement is the introduction of nonlinear limiter for the oscillation free initial data reconstruction [Boris and Book (1973); van Leer (1977)]. As a result, an interface discontinuity for the flow variable is introduced, where the kinematic numerical dissipation is implicitly added through the preparation of such an initial data, where the kinetic energy has been converted into the thermal one under the constraint of total energy conservation in the reconstruction [Xu (2001)]. Due to the interface discontinuity, in order to evaluate the fluxes across a cell interface, the Riemann solution of the Euler equations is commonly adopted and becomes a back-bone for most modern shock capturing schemes. The reason for the wide use of Riemann solution in CFD algorithm development may be the following. First, the inviscid Euler equations are dynamically close to the Navier-Stokes equations, where the physical dissipative terms in NS equations can be simply added into the Euler system through operator splitting approach. Second, the viscous and heat conduction terms cannot cope with the discontinuous initial condition, such as the immediate divergences of dissipative term at a cell interface. So, the use of hyperbolic equations is a natural choice to capture the flow evolution from a discontinuity, and the Euler equations become idealized well-defined equations. Third, the Riemann problem of the Euler equations with two constant states has an exact solution, which can be used for the flux construction [Godunov (1959)]. Also, the use of Riemann solution introduces dynamics into the CFD algorithm, which is more or less beyond the traditional numerical PDE methodology. Fourth, since the main numerical dissipation is implicitly added through the initial condition with discontinuity, the scheme will not be so sensitive to the Riemann fluxes and the physical dissipation in the Riemann solver doesn't play an important role in the capturing of shock. As a result, it attracts people to design all kinds of approximate Riemann solvers. This is similar to the Lattice Boltzmann Method (LBM). Due to its dramatic simplification, to recover a simple boundary condition could involve tens, even hundreds papers. Even with the wide spread of Riemann solvers, from a physical point of view, the NS mechanism should be more appropriate to describe the flow evolution. The absence of a direct NS solver with a discontinuous initial data is due to the non-existence of the exact solution and the inconsistency of the dissipative term with a discontinuity. But, this doesn't mean that the use of the Euler equations is legitimate here. In a space with limited resolution, the approximate Riemann solvers, which may not solve the Euler equations at all, may be more physically founded than the exact Riemann solution. If we consider the interface discontinuity as a physical reality, the physical solution always exists. The numerical algorithm needs to recover such a gas evolution process from a discontinuity, instead of sticking on what kinds of equations may have solution. Fundamentally, the zero thickness initial discontinuity triggers flow physics in a different scale from the hydrodynamic NS and Euler equations. The physical process analysis will be presented in this chapter. Following the numerical PDE methodology, except designing specific remedies when meeting problems, there is no fundamental principle to guide the CFD moving forward.

The Riemann solver based CFD methods for the compressible flow computations have achieved great success in the past decades for the capturing discontinuous solutions. The Riemann solutions, including many approximate Riemann solvers [Toro (2009)], have been widely studied and used. But, the weakness of the schemes starts to emerge as well. For example, for the nonlinear system it is hard to figure out its precise dissipative mechanism in comparison with the central difference method. The solution of the upwind scheme may depend on the limiter sensitively in the smooth region. The shock instability in the high speed applications is another unsolved problem. The solution depends closely on the mesh distribution. Even for the exact Riemann solution, the carbuncle phenomena cannot be avoided [Li *et al.* (2011)]. So, at the current stage, the real numerical challenge for the continuum flow computation is about the accuracy and robustness of the shock capturing scheme in hypersonic flow computation, and the capability of capturing both discontinuous and smooth viscous solution in multidimensional cases. It seems hard to develop higher-order schemes without breaking the barrier of the first order dynamics in the Riemann solution.

The algorithm structure of the gas-kinetic scheme for the continuum flow computation is similar to the shock capturing scheme based on the Riemann solution. However, instead of using the physics of the Riemann solution in the flux calculation, a time dependent multiscale modeling is used for the construction of a gas distribution function at a cell interface. Due to the kinetic evolution part in the distribution function, a non-equilibrium dissipative mechanism, which is consistent with the physical shock structure formation, is included explicitly in the flux function. Based on the comparison between the GKS and Riemann solution-based shock capturing schemes, it shows the importance of including non-equilibrium dissipative mechanism in the flux function inside a discontinuity, instead of using the equilibrium Euler solution there. For a discontinuous solution, even though the physical shock structure cannot be fully resolved numerically in the mesh size scale, a physically consistent numerical dissipation is still needed to keep a stable and oscillation-free shock transition. Also, the multi-dimensionality is important for the shock capturing scheme in order to get accurate solution in the smooth viscous flow computation. The Riemann solution does not provide a multi-dimensional flow evolution, and its wave propagating direction is solely determined by the mesh orientation. In CFD community, the upwind and central difference schemes belong to two different categories, i.e., one for the discontinuous initial data and the other for the continuous one. Actually, the discontinuous and continuous flow distributions depend mainly on the cell resolution and the physical layer thickness. The choosing of the upwind or central difference should depend on the mesh resolution, instead of a pre-assigned numerical technique. The mesh size should actively participate the flow evolution, rather than a passive parameter associated with numerical error only, because the same physical solution may appear differently with different cell resolution, and the evolution "dynamics" should be closely associated with the mesh size. The GKS provides such a scheme with a smooth transition between upwind and central difference modeling.

In this chapter, we will introduce GKS, analyze the dissipative mechanism in it, and compare it with the Riemann-solver-based CFD methods. The CFD principle is the direct modeling of flow physics in the mesh size scale instead of targeting any governing equation with a specific modeling scale. The mesh size and time step should actively participate in the gas evolution. The CFD algorithm itself is a governing equation. Through the current study and analysis, hopefully the CFD research can move forward, and go beyond the numerical PDE methodology.

4.2 Gas Kinetic Scheme

Similar to the MUSCL (Monotone Upwind Scheme for Conservation Laws) type approach [van Leer (1979)], the first step for the GKS is to interpolate the macroscopic flow variables inside each computational cell. For a second order GKS, the van Leer limiter is used for the initial data reconstruction of conservative variables [van Leer (1977)]. The fundamental task in the construction of a finite-volume gas-kinetic scheme is to evaluate a time-dependent gas distribution function f at a cell interface, from which the numerical flux can be obtained. Besides incorporating the slopes in the normal direction along a cell interface, the slopes in the tangential direction on both sides of a cell interface will also participate in the time evolution of a gas distribution function in the multidimensional GKS method. In what follows, we take 2D case as an example. Denote $x = 0, y \in [-\Delta y/2, \Delta y/2]$ as a cell interface, on both sides of this interface the interpolated macroscopic variables have gradients $(\nabla\rho, \nabla(\rho U), \nabla(\rho V), \nabla(\rho E))_{left,right}$, where $\rho, \rho U, \rho V$, and ρE are the densities of mass, momentum, and energy in 2D case. The gradients can be in any direction around the cell interface. For the continuum flow computation, the GKS updates the conservative flow variables inside each control volume ΔV in the following way,

$$\mathbf{W}_{i,j}^{n+1} = \mathbf{W}_{i,j}^{n} + \frac{1}{\Delta V}\sum_{l}\int_{t^n}^{t^{n+1}} \vec{\mathbf{F}}_l \cdot \vec{\Delta} S_l dt, \tag{4.1}$$

where $\vec{\mathbf{F}}_l$ is the flux at the center of the cell interface. This flux is evaluated from a time dependent gas distribution function. The extension of the GKS to an arbitrary control volume is presented in the Appendix D.

Based on the general framework for the construction of gas kinetic scheme, a time dependent gas distribution function f in GKS at a cell interface $(x_{i+1/2}, y_j)$ at time t is modeled as, which is the same as the integral solution of BGK model,

$$f(x_{i+1/2}, y_j, t, u, v, \xi) = \frac{1}{\tau}\int_0^t g(x', y', t', u, v, \xi) e^{-(t-t')/\tau} dt' + e^{-t/\tau} f_0(x_{i+1/2} - ut, y_j - vt), \tag{4.2}$$

where $x' = x_{i+1/2} - u(t-t'), y' = y_j - v(t-t')$ are the trajectory of a particle motion and f_0 is the initial gas distribution function f at the beginning of each time step ($t = 0$). The terms f_0 and g are related to the kinetic and hydrodynamic scale flow transport. Two unknowns g and f_0 must be specified in Eq. (4.2) in order to obtain the solution f. Their constructions are based on the modeling. The non-equilibrium distribution function $f(x_{i+1/2}, y_j, t, u, v, \xi)$ and the equilibrium one $g(x_{i+1/2}, y_j, t, u, v, \xi)$ correspond to the same mass, momentum, and energy at the same location in the physical space, which satisfy the compatibility condition,

$$\int f\psi_\alpha d\Xi = \int g\psi_\alpha d\Xi, \quad \alpha = 1, 2, 3, 4, \tag{4.3}$$

and $\psi = (1, u, v, \frac{1}{2}(u^2 + v^2 + \xi^2))$, and $d\Xi = dudvd\xi$. In order to simplify the notation, $(x_{i+1/2} = 0, y_j = 0)$ will be used.

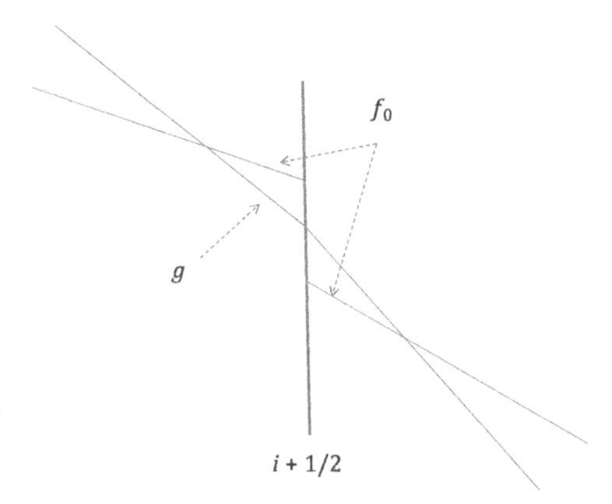

Fig. 4.1 Schematic distributions of initial non-equilibrium and equilibrium states at $t = 0$.

For a multidimensional GKS, the initial gas distribution function f_0 is constructed as,

$$f_0 = \begin{cases} g^l\left(1 + a^l x + b^l y - \tau(a^l u + b^l v + A^l)\right) & x \leq 0 \\ g^r\left(1 + a^r x + b^r y - \tau(a^r u + b^r v + A^r)\right), & x > 0 \end{cases} \tag{4.4}$$

where g^l and g^r are Maxwellian distribution functions on the left and right hand sides of a cell interface, a^l and a^r are related to the slopes in the normal direction, b^l and b^r in the tangential direction. The τ related terms correspond to the non-equilibrium part of the Chapman-Enskog expansion for the the NS solution. Note that the non-equilibrium parts have no net contribution to conservative flow variables, i.e.,

$$\begin{aligned} &\int (a^l u + b^l v + A^l)\psi_\alpha g^l d\Xi = 0, \\ &\int (a^r u + b^r v + A^r)\psi_\alpha g^r d\Xi = 0. \end{aligned} \tag{4.5}$$

All terms in f_0, g^l, g^r, a^l, a^r, b^l, and b^r will be defined later.

After constructing f_0, the equilibrium state g around $(x = 0, y = 0, t = 0)$ is modeled as

$$g = g_0 \left(1 + (1 - H[x])\bar{a}^l x + H[x]\bar{a}^r x + \bar{b}y + \bar{A}t\right), \tag{4.6}$$

where $\bar{b}$ is a term related to the flow variation in the tangential direction, and $H[x]$ is the Heaviside function defined as

$$H[x] = \begin{cases} 0, & x < 0, \\ 1, & x \geq 0. \end{cases}$$

Here g_0 is a local Maxwellian distribution function located at the center of a cell interface $(x = 0, y = 0)$. In both f_0 and g, $a^l, A^l, a^r, A^r, \bar{a}^l, \bar{a}^r, \bar{b}$, and $\bar{A}$ are related to the derivatives of a Maxwellian in space and time. The schematic distributions of f_0 and g around a cell interface are shown in Fig. 4.1. The initial discontinuous macroscopic flow distribution is similar to those used in a Generalized Riemann problem (GRP) [Ben-Artzi and Falcovitz (2003)], which evolves to a continuous one in the GKS. As analyzed in [Ohwada and Kobayashi (2004)], the contributions from f_0 and g in Eq. (4.2) have the upwind and central difference properties, which make the GKS have advantages in capturing both discontinuous and continuous flows in different regions.

The dependence of $a^l, a^r, ..., \bar{A}$ on the particle velocity can be obtained from a Taylor expansion of a Maxwellian and have the following form,

$$a^l = a_1^l + a_2^l u + a_3^l v + a_4^l \frac{1}{2}(u^2 + v^2 + \xi^2) = a_\alpha^l \psi_\alpha,$$

$$A^l = A_1^l + A_2^l u + A_3^l v + A_4^l \frac{1}{2}(u^2 + v^2 + \xi^2) = A_\alpha^l \psi_\alpha,$$

$$\cdots$$

$$\bar{A} = \bar{A}_1 + \bar{A}_2 u + \bar{A}_3 v + \bar{A}_4 \frac{1}{2}(u^2 + v^2 + \xi^2) = \bar{A}_\alpha \psi_\alpha,$$

where $\alpha = 1, 2, 3, 4$ and all coefficients $a_1^l, a_2^l, \ldots, \bar{A}_4$ are local constants.

With the initial data reconstruction, the distributions

$$\bar{\mathbf{W}} = \left(\bar{\rho}_i, \rho\bar{U}_i, \rho\bar{V}_i, \rho\bar{E}_i\right)^T,$$

inside each cell are obtained. At the center of a cell interface $(x_{i+1/2}, y_j)$, the left and right macroscopic states are denoted by

$$\bar{\mathbf{W}}_i(x_{i+1/2}, y_j) = \begin{pmatrix} \bar{\rho}_i(x_{i+1/2}, y_j) \\ \rho\bar{U}_i(x_{i+1/2}, y_j) \\ \rho\bar{V}_i(x_{i+1/2}, y_j) \\ \rho\bar{E}_i(x_{i+1/2}, y_j) \end{pmatrix}$$

and

$$\bar{\mathbf{W}}_{i+1}(x_{i+1/2}, y_j) = \begin{pmatrix} \bar{\rho}_{i+1}(x_{i+1/2}, y_j) \\ \rho\bar{U}_{i+1}(x_{i+1/2}, y_j) \\ \rho\bar{V}_{i+1}(x_{i+1/2}, y_j) \\ \rho\bar{E}_{i+1}(x_{i+1/2}, y_j) \end{pmatrix}.$$

By using the relation between the gas distribution function f and the macroscopic variables, we get

$$\int g^l \psi_\alpha d\Xi = \bar{\mathbf{W}}_i(x_{i+1/2}, y_j); \quad \int g^l a^l \psi_\alpha d\Xi = \vec{n} \cdot \nabla \mathbf{W}^l \tag{4.7}$$

$$\int g^r \psi_\alpha d\Xi = \bar{\mathbf{W}}_{i+1}(x_{i+1/2}, y_j); \quad \int g^r a^r \psi_\alpha d\Xi = \vec{n} \cdot \nabla \mathbf{W}^r \tag{4.8}$$

where $\nabla \mathbf{W}^l$ and $\nabla \mathbf{W}^r$ are the gradients of macroscopic variables on the left and right hand sides of a cell interface, and $\vec{n}$ is the normal direction. Similarly, in the tangential direction b^l and b^r can be obtained from

$$\int g^l b^l \psi_\alpha d\Xi = \vec{t} \cdot \nabla \mathbf{W}^l; \quad \int g^r b^r \psi_\alpha d\Xi = \vec{t} \cdot \nabla \mathbf{W}^r, \tag{4.9}$$

where $\vec{t}$ is the unit vector in the tangential direction along the cell interface.

With the definition of Maxwellian distributions in 2D,

$$g^l = \rho^l \left(\frac{\lambda^l}{\pi}\right)^{\frac{K+2}{2}} e^{-\lambda^l((u-U^l)^2 + (v-V^l)^2 + \xi^2)}$$

and

$$g^r = \rho^r \left(\frac{\lambda^r}{\pi}\right)^{\frac{K+2}{2}} e^{-\lambda^r((u-U^r)^2 + (v-V^r)^2 + \xi^2)},$$

and from Eqs. (4.7) and (4.8), the equilibrium states g^l and g^r are fully determined by density, velocity, and temperature, which can be obtained from the local conservative macroscopic variables at the same location. The parameters a^l and a^r are related to the local slopes of macroscopic variables. For example, the solution for the coefficients in the parameters a^l and a^r can be found from

$$M_{\alpha\beta}a_\beta = M(a_1, a_2, a_3, a_4)^T = \frac{1}{\rho}\vec{n} \cdot \nabla \mathbf{W},$$

on the left and right hand sides separately, where the matrix M is given by $M_{\alpha,\beta} = (1/\rho)\int g\psi_\alpha\psi_\beta d\Xi$. The solutions for the above equations are given in Appendix C. Similarly, the parameters b^l and b^r can be determined from Eq. (4.9).

After determining the terms a^l, b^l, a^r, and b^r, A^l and A^r in f_0 can be found from Eq. (4.5), which are

$$\begin{aligned} M^l_{\alpha\beta}A^l_\beta &= -\frac{1}{\rho^l}\int \psi_\alpha(a^l u + b^l v)g^l d\Xi, \\ M^r_{\alpha\beta}A^r_\beta &= -\frac{1}{\rho^r}\int \psi_\alpha(a^r u + b^r v)g^r d\Xi, \end{aligned} \tag{4.10}$$

where $M^{l,r}_{\alpha\beta} = \int g^{l,r}\psi_\alpha\psi_\beta d\Xi/\rho^{l,r}$. Again, $M_{\alpha\beta}$ has the same formulation and it depends on the specific equilibrium states used in the integration. The solutions of A^l and A^r can be obtained using the same methods in Appendix C.

For the equilibrium state g in Eq. (4.6), the corresponding values of ρ_0, U_0, V_0 and λ_0 in g_0,

$$g_0 = \rho_0\left(\frac{\lambda_0}{\pi}\right)^{\frac{K+2}{2}} e^{-\lambda_0((u-U_0)^2+(v-V_0)^2+\xi^2)},$$

can be found as follows. Taking the limit $t \to 0$ in Eq. (4.2) and substituting its solution into Eq. (4.3), the conservation constraint at $(x_{i+1/2}, y_j, t = 0)$ gives

$$\int g_0\psi_\alpha d\Xi = \mathbf{W}_0 = \int_{u>0}\int g^l\psi_\alpha d\Xi + \int_{u<0}\int g^r\psi_\alpha d\Xi, \tag{4.11}$$

where $\mathbf{W}_0 = (\rho_0, (\rho U)_0, (\rho V)_0, (\rho E)_0)^T$. Since g^l and g^r are obtained earlier, the above moments can be evaluated explicitly, see the Appendix C for the moments evaluation. Then, g_0 is uniquely determined. For example, λ_0 in g_0 can be found from

$$\lambda_0 = (K+2)\rho_0 \Big/ \left(4\left((\rho E)_0 - \frac{1}{2}((\rho U)_0{}^2 + (\rho V)_0^2)/\rho_0\right)\right).$$

Then, $\bar{a}^l$ and $\bar{a}^r$ in the expansion of g, see Eq. (4.6), can be obtained through the relation of

$$\frac{\bar{\mathbf{W}}_{i+1}(x_{i+1}, y_j) - \mathbf{W}_0}{\rho_0 \Delta x^+} = \bar{M}^0_{\alpha\beta} \begin{pmatrix} \bar{a}^r_1 \\ \bar{a}^r_2 \\ \bar{a}^r_3 \\ \bar{a}^r_4 \end{pmatrix} = \bar{M}^0_{\alpha\beta} \bar{a}^r_\beta, \tag{4.12}$$

and

$$\frac{\mathbf{W}_0 - \bar{\mathbf{W}}_i(x_i, y_j)}{\rho_0 \Delta x^-} = \bar{M}^0_{\alpha\beta} \begin{pmatrix} \bar{a}^l_1 \\ \bar{a}^l_2 \\ \bar{a}^l_3 \\ \bar{a}^l_4 \end{pmatrix} = \bar{M}^0_{\alpha\beta} \bar{a}^l_\beta, \tag{4.13}$$

where $\Delta x^- = x_{i+1/2} - x_i$ and $\Delta x^+ = x_{i+1} - x_{i+1/2}$ are the distances between the center of cell interface to cell centers. Since the matrix $\bar{M}^0_{\alpha\beta} = \int g_0 \psi_\alpha \psi_\beta d\Xi / \rho_0$ is known, $(\bar{a}^r_1, \bar{a}^r_2, \bar{a}^r_3, \bar{a}^r_4)^T$ and $(\bar{a}^l_1, \bar{a}^l_2, \bar{a}^l_3, \bar{a}^l_4)^T$ can be evaluated accordingly. The term $\bar{b}$ in Eq. (4.6) can be evaluated by

$$\int \bar{b} g_0 \psi d\Xi = \int_{u>0} b^l g^l \psi d\Xi + \int_{u<0} b^r g^r \psi d\Xi, \tag{4.14}$$

or from reconstructed macroscopic flow variables along the cell interface.

Up to this point, we have determined all parameters in the initial gas distribution function f_0 and the equilibrium state g at the beginning of each time step $t = 0$. After substituting Eq. (4.4) and Eq. (4.6) into Eq. (4.2), the gas distribution function f at a cell interface can be expressed as

$$\begin{aligned} f(x_{i+1/2}, y_j, t, u, v, \xi) = {} & (1 - e^{-t/\tau}) g_0 \\ & + \left(\tau(-1 + e^{-t/\tau}) + t e^{-t/\tau}\right) \\ & \left(\bar{a}^l u H[u] + \bar{a}^r u (1 - H[u]) + \bar{b} v\right) g_0 \\ & + \tau(t/\tau - 1 + e^{-t/\tau}) \bar{A} g_0 \\ & + e^{-t/\tau} \left((1 - (t + \tau)(u a^l + v b^l)) H[u] g^l \right. \\ & \left. + (1 - (t + \tau)(u a^r + v b^r))(1 - H[u]) g^r\right) \\ & + e^{-t/\tau} \left(-\tau A^l H[u] g^l - \tau A^r (1 - H[u]) g^r\right). \end{aligned} \tag{4.15}$$

The only unknown left in the above expression is $\bar{A}$. Since both f (Eq. (4.15)) and g (Eq. (4.6)) contain $\bar{A}$, the integration of the conservation constraint Eq. (4.3) at $(x_{i+1/2}, y_j)$ over the whole time step Δt gives

$$\int_0^{\Delta t} \int (g - f) \psi_\alpha dt d\Xi = 0,$$

which reduces to

$$\begin{aligned}
\bar{M}^0_{\alpha\beta}\bar{A}_\beta &\equiv \frac{1}{\rho_0}(\partial\rho/\partial t, \partial(\rho U)/\partial t, \partial(\rho V)/\partial t, \partial(\rho E)/\partial t)^T \\
&= \frac{1}{\rho_0}\int \left[\gamma_1 g_0 + \gamma_2\left(\bar{a}^l u H[u] + \bar{a}^r u(1 - H[u]) + \bar{b}v\right) g_0\right. \\
&\quad + \gamma_3\left(H[u]g^l + (1 - H[u])g^r\right) \\
&\quad + \gamma_4\left((a^l u + b^l v)H[u]g^l + (a^r u + b^r v)(1 - H[u])g^r\right) \\
&\quad + \gamma_5\left((a^l u + b^l v + A^l)H[u]g^l\right. \\
&\quad \left.\left. + (a^r u + b^r v + A^r)(1 - H[u])g^r\right)\right]\psi_\alpha d\Xi,
\end{aligned} \tag{4.16}$$

for the determination of $\bar{A}$, where

$$\begin{aligned}
\gamma_0 &= \Delta t - \tau(1 - e^{-\Delta t/\tau}), \\
\gamma_1 &= -(1 - e^{-\Delta t/\tau})/\gamma_0, \\
\gamma_2 &= \left(-\Delta t + 2\tau(1 - e^{-\Delta t/\tau}) - \Delta t e^{-\Delta t/\tau}\right)/\gamma_0, \\
\gamma_3 &= (1 - e^{-\Delta t/\tau})/\gamma_0, \\
\gamma_4 &= \left(\Delta t e^{-\Delta t/\tau} - \tau(1 - e^{-\Delta t/\tau})\right)/\gamma_0, \\
\gamma_5 &= -\tau(1 - e^{-\Delta t/\tau})/\gamma_0.
\end{aligned}$$

The above multidimensional GKS is similar to the directional splitting GKS in [Xu (2001)], except additional terms b^l, b^r, and $\bar{b}$ related to the flow variations in the tangential direction.

Finally, the time-dependent numerical fluxes in the normal-direction across the cell interface can be computed by

$$\begin{pmatrix} \mathcal{F}_\rho \\ \mathcal{F}_m \\ \mathcal{F}_n \\ \mathcal{F}_E \end{pmatrix}_{i+1/2} = \int_{-\Delta y/2}^{\Delta y/2}\int u \begin{pmatrix} 1 \\ u \\ v \\ \frac{1}{2}(u^2 + v^2 + \xi^2) \end{pmatrix} f(x_{i+1/2}, y_j, t, u, v, \xi) d\Xi dy, \tag{4.17}$$

where $f(x_{i+1/2}, y_j, t, u, v, \xi)$ is given by Eq. (4.15). By integrating the above equation to the whole time step, we can get the total mass, momentum and energy transport in a finite volume GKS. The above flux can be used in a finite volume scheme with arbitrary mesh orientation, even moving cell interface, see Appendix D.

The above modeling of the solution f at a cell interface is based on the integral solution of the BGK model, where a unit Prandtl number is recovered in the hydrodynamic limit. In order to simulate the flow with any value of Prandtl number, the heat transport part in the energy flux can be modified according to the Prandtl number. With a gas distribution function f at the cell interface Eq. (4.15), the time-dependent heat flux can be evaluated precisely,

$$q = \frac{1}{2}\int (u-U)\left((u-U)^2 + (v-V)^2 + \xi^2\right) f d\Xi, \tag{4.18}$$

where the average velocities U and V are given by

$$U = \int uf d\Xi \Big/ \int f d\Xi, \quad V = \int vf d\Xi \Big/ \int f d\Xi.$$

Then, the Prandtl number in GKS can be modified by changing the heat flux (4.18) in the energy flux transport,

$$\mathcal{F}_E^{new} = \mathcal{F}_E + \left(\frac{1}{\mathrm{Pr}} - 1\right) q, \tag{4.19}$$

where $\mathcal{F}_E$ is the energy flux in Eq. (4.17). Besides the above heat flux modification, in order to get correct Prandtl number the Shakhov and ES-BGK models can be directly used in the integral solution of the kinetic model equation as well. The use of Shakhov and ES-BGK models have been implemented in the unified gas-kinetic scheme [Chen *et al.* (2013)]. For the continuum flow computation without updating the gas distribution function, the current fix is accurate enough to get the NS solution.

In the above second order scheme, the cell interface gas distribution function $f(x_{i+1/2}, y_j, t, u, v, \xi)$ doesn't depend on y explicitly. For a higher order GKS, such as the 3rd-order GKS [Li *et al.* (2010); Luo (2012); Luo and Xu (2013)], f at the cell interface will depend explicitly on y, which can be integrated in terms of y without using Gaussian point integration.

4.3 Analysis of Gas Kinetic Scheme

In this section, we are going to analyze the GKS. Many issues related to the kinetic limits, collision time, and high-order extension will be addressed. As emphasized earlier, the direct modeling for CFD is not targeting to solve any specific PDE. The NS solution obtained from GKS is only in cases where the mesh size is fine enough to resolve the NS wave structure. In the unresolved cases, it is meaningless to talk about the NS solution. But the analysis of the physical mechanism and modeling in unresolved case is still needed.

Collision time

In a well-resolved dissipative region, such as the cell size Δx is smaller than the scale of the dissipative layer determined by the physical viscosity, the initial reconstruction theoretically should present an almost continuous flow distribution across a cell interface. The collision time τ in this case can be naturally determined by the relation

$$\tau = \mu/p,$$

where μ is the dynamical viscosity coefficient and p is the pressure. This is a well-known result from the Chapman-Enskog expansion of the kinetic BGK model [Vincenti and Kruger (1965)], see Chapter 2. For the viscosity coefficient, μ can take any reasonable value in the determination of τ, such as the Sutherland's law,

$$\mu = \mu_\infty \left(\frac{T}{T_\infty}\right)^{3/2} \frac{T_\infty + S}{T + S},$$

where T_∞ and S are the temperatures with the values $T_\infty = 285K$ and $S = 110.4K$. The above particle collision time is implemented in the GKS in the following way. The collision time in the initial distribution function f_0 is determined from the reconstructed macroscopic flow variables at the left and right hand sides of a cell interface. For the collision time in the integration of the equilibrium state, based on $\mathbf{W}_0$ in Eq. (4.11), we can evaluate $\mu(\mathbf{W}_0)$ according to the above formulation, where T is the temperature evaluated from $\mathbf{W}_0$. Then, the local collision time τ is defined with the value $\tau = \mu(\mathbf{W}_0)/p$, where p is the pressure and is a function of $\mathbf{W}_0$ too.

Theoretically, the dissipative structure, such as the shock thickness, is solely determined by the physical viscosity. The structure should be independent of the cell size and time step used in a numerical scheme. The Navier-Stokes equations can be accurately solved if the cell size is fine enough to resolve the wave structure. Otherwise, the physical solution has to be replaced by a numerical one. For example, the physical shock thickness may be replaced by the numerical cell size in a shock capturing scheme. In such a situation, there is no meaning to talk about the solution of the original NS equations in this region. If we prefer to avoid the implicit artificial dissipation, which have been used in many shock capturing schemes, it is better to honestly admit that in this situation the NS solution with the original viscosity cannot be properly maintained. An effective viscosity is needed to enlarge the shock thickness to the numerical cell size. Therefore, besides the appearance of discontinuous initial reconstruction, the collision time has to include both physical and numerical effect. In the un-resolved discontinuity region, an additional numerical dissipation is required. But, the question is what kind of dissipative mechanism can be used to construct such a mesh size related shock structure. Since the magnitude of the jump at the cell interface indicates the un-resolvedness of a flow structure, the collision time τ used in GKS takes this into account, and a generalized collision time is defined with the inclusion of pressure jump,

$$\tau = \frac{\mu(\mathbf{W}_0)}{p(\mathbf{W}_0)} + \alpha_n \frac{|\rho_l/\lambda_l - \rho_r/\lambda_r|}{|\rho_l/\lambda_l + \rho_r/\lambda_r|}\Delta t, \tag{4.20}$$

where Δt is the CFL time step, which is related to the cell size by CFL condition. The second part corresponds to the numerical viscosity with α_n being an adjustable parameter from 1 to 5. In the smooth flow region or around the slip line, the additional collision time is very small or diminishes due to the continuous pressure distribution. Based on the numerical tests, the enhancement of the particle collision time does not deteriorate the boundary layer calculations once the layer is well resolved, but enhances the robustness of the GKS in shock capturing capability. The associated dissipative mechanism in GKS will be analyzed later when it is compared with the Godunov method.

The modification of particle collision time in the GKS is not purely increasing the viscosity coefficient of the NS solution, but introduces the kinetic scale mechanism into the scheme, such as particle free transport. Even though the shock jump with a width of 2 or 3 cell size can be captured nicely by the Godunov method, the dissipation there for the construction of such a shock structure is not very clear, but is surely from the numerics only, such as the preparation of the initial data at the beginning of each time step [Xu and Li (2001)]. The use of an enlarged particle collision time here is to enhance the dynamic dissipation from the kinetic scale-related mechanism, i.e., the so-called free transport of f_0 in the evolution solution. Physically, a particle really encounters a limited number of collisions across a shock layer. A non-equilibrium discontinuous gas distribution function of f_0 with free transport mechanism mimics the mechanism inside a real physical shock layer. Consequently, the shock thickness on the mean free path scale is enlarged to the mesh size thickness as the numerical particle mean free path. In other words, the construction of numerical shock transition in a shock capturing GKS borrows the real physical mechanism, and the particle mean free path is artifically increased to the mesh size scale. Therefore, even for the continuum flow computation the GKS is still a multiple scale modeling method with both kinetic (discontinuous region) and hydrodynamic (smooth region) gas evolution. More analysis will be presented later when comparing GKS with Riemann solver.

With a further consideration of different dynamic effect of the collision time in the time-dependent gas distribution function (4.15), two collision times can be defined. One is the physical one

$$\tau_p = \mu/p,$$

for the smooth viscous solution inside each cell, and the other is the numerical one

$$\tau_n = \mu/p + \alpha_n \frac{|\rho_l/\lambda_l - \rho_r/\lambda_r|}{|\rho_l/\lambda_l + \rho_r/\lambda_r|}\Delta t,$$

for the discontinuity part evolution at a cell interface [Luo (2012); Luo *et al.* (2013)]. The physical one τ_p is used to describe the smooth NS solution inside each cell (hydrodynamic), and the numerical one τ_n takes into account the interface jump and interactions (kinetic). With the above understanding, the gas distribution function at a cell interface can be written as

$$\begin{aligned} f(x_{i+1/2}, y_j, t, u, v, \xi) = &(1 - e^{-t/\tau_n})g_0 \\ &+ \left(\tau_p(-1 + e^{-t/\tau_n}) + te^{-t/\tau_n}\right) \\ &\left(\bar{a}^l H[u] + \bar{a}^r(1 - H[u]) + \bar{b}v\right) ug_0 \\ &+\tau_p(t/\tau_p - 1 + e^{-t/\tau_n})\bar{A}g_0 \\ &+e^{-t/\tau_n}\left((1 - (t + \tau_p)(ua^l + vb^l))H[u]g^l\right. \\ &\left.+(1 - (t + \tau_p)(ua^r + vb^r))(1 - H[u])g^r\right) \\ &+e^{-t/\tau_n}\left(-\tau_p A^l H[u]g^l - \tau_p A^r(1 - H[u])g^r\right), \end{aligned}$$

where the numerical particle collision time only appears in the exponential parts. The use of the above definition of collision time takes into account the different physical mechanism for the time evolution of smooth and discontinuous parts, the performance of GKS can be enhanced in comparison with the method with a single collision time in Eq. (4.20) [Luo (2012)].

Smooth flow limit: Navier-Stokes solutions

In general, the underlying governing equation of GKS can be hardly obtained, especially in the discontinuity region. In a well-resolved smooth flow region, the property of the GKS can be still analyzed. Since the GKS is a direct modeling finite volume method, it is based on the conservation through the construction of the time-dependent interface flux. The conservation is a physical reality, which is valid in any scale, including any size of a control volume. Here, there is no any meaning to get the so-called modified governing for the discrete scheme if we don't have any intention to reduce the cell size to zero at all. What is important in determining the flow evolution is the interface flux. This flux is the key for the quality of the scheme. What is the physics or principle to construct such a flux function? The physics can be the kinetic or the Navier-Stokes evolution, which depend on the scale of the control volume. For the continuum flow computation, we implicitly assume that the cell size scale is much larger than the particle mean free path, and an equilibrium flow behavior is supposed to be simulated. Therefore, for the continuum flow study we need to figure out if the gas distribution function at a cell interface follows the time evolution of the NS solution.

In smooth flow region, even inside a well-resolved shock layer, once the discontinuities of flow variables at a cell interface disappear, the distribution function f_0 has $g^l = g^r, a^l = a^r$, and $b^l = b^r$. Consequently, Eq. (4.11) gives $g_0 = g^l = g^r$, and Eqs. (4.12) and (4.13) reduce to $\bar{a}^l = \bar{a}^r = a^l = a^r$ and $\bar{b}^l = \bar{b}^r = b^l = b^r$. As a result, $\bar{A}$ determined in Eq. (4.16) is exactly equal to A^l and A^r in Eq. (4.10). Therefore, without any further assumption, the gas distribution function f at a cell interface automatically reduces to

$$f = g_0 \left[1 - \tau(u\bar{a} + v\bar{b} + \bar{A}) + t\bar{A}\right], \tag{4.21}$$

where $-\tau(u\bar{a} + v\bar{b} + \bar{A})g_0$ is exactly the nonequilibrium state in the Chapman-Enskog expansion of the BGK model for the NS solution, and $g_0\bar{A}t$ is the time evolution part of the gas distribution function. Since the above time evolution part is obtained equivalently from the inviscid Euler equations, the above solution presents scheme with 2nd-order accuracy in space and time for the inviscid parts and second order in space and 1st-order in time for the viscous parts in the NS solution. The above equation (4.21) is the one used for the low Mach number viscous flow calculations [Su *et al.* (1999); Xu and He (2003); Guo *et al.* (2008)], where the accuracy of the above formulation is well established. The above formulation is similar to the Lax-Wendroff-type central difference scheme for the NS equations, but with a better performance in the coarse mesh case [Torrilhon and Xu (2006)]. The above limiting solution indicates that the GKS can make a transition from "upwind" scheme with the discontinuous initial data, to a "central difference" method once the initial discontinuity disappears. This is definitely a preferred property for any shock capturing scheme.

The boundary layer can be well-captured by GKS, once there are 5 to 10 grid points for a second-order scheme, and 3 points for a third-order one. Inside smooth boundary layer, the MUSCL-type reconstruction will introduce small jumps in the initial data at a cell interface. With such a weak discontinuity at a cell interface, Eq. (4.21) still has the main contribution in the gas distribution function f (4.15), because it can be obtained from the integration of the equilibrium state when $\Delta t \gg \tau$. This condition definitely holds for high Reynolds number boundary layer flow simulation. This is one of the main reason why the GKS performs extremely nice in the hypersonic viscous and heat conducting flow computations, since the hydrodynamic part from the integration of the equilibrium state plays a dominant role in the flux construction inside the dissipative layer [Ohwada and Kobayashi (2004); Luo *et al.* (2009)]. In other words, in the hydrodynamic limit with $\Delta t \gg \tau$, the GKS will become the same scheme as a central difference one, which has much less numerical dissipation than the upwind scheme for the viscous flow computation. For the Riemann solution based CFD method, once there is a discontinuous jump at a cell interface, the wave propagation will be generated even inside a dissipative boundary layer. As a result, these schemes are sensitive to the initial reconstruction in the viscous flow computation, especially in high Reynolds number cases. In order to reduce the jump, a high-order reconstruction and high quality mesh are usually required in the Godunov type schemes. There need at least 8 grid points inside the layer in order to avoid the cell interface discontinuities in the reconstruction. However, the GKS is not so sensitive to the limiter at all, and a boundary layer can be well captured with 4 to 5 grid points for a second order scheme with van Leer limiter. Instead of using the numerics to reduce the effect from the cell interface discontinuity, the dynamic evolution in GKS will quickly smooth out such a discontinuity. This is the key for the success of GKS [Ohwada and Kobayashi (2004)]. A comparison between GKS and Godunov-type method for the viscous flow computations is given in [Ilgaz and Tuncer (2009)], where it shows that for the inviscid flow computation the GKS has even less numerical dissipation than the Roe scheme with less entropy production around a low speed airfoil [Roe (1981)].

Collisionless limit: kinetic flux vector splitting (KFVS)

Once the solution is well-resolved (no cell interface discontinuity), the GKS presents a valid Navier-Stokes solution in both $\tau < \Delta t$ and $\tau > \Delta t$ limits, and the solution (4.21) is independent of the ratio between τ and Δt. This doesn't contradict with the direct modeling methodology, where flow physics in different regime should be captured properly, because even in the kinetic scale $\Delta t \simeq \tau$ the initial distribution function in GKS is constructed from the Chapman-Enskog expansion and the NS solution is obtained in the well-resolved region. But, the underlying dynamic model for the GKS is still a process from kinetic to hydrodynamic one. The particle collision effect has been taken into account explicitly. In the high Reynolds number flow computation with $\Delta t \gg \tau$, even with a cell interface discontinuity inside the boundary layer the GKS can still give an accurate NS solution since its gas evolution model converges to the hydrodynamic one exponentially. But, the kinetic flux vector splitting scheme (KFVS), which is another limiting solution of GKS, keeps the kinetic scale evolution model all the time even with the NS initial gas distribution function.

In comparison with KFVS, the collisionless Boltzmann solution is commonly used for the flux evaluation. [Pullin (1980); Reitz (1981); Deshpande (1986); Mandal and Deshpande (1994); Chou and Baganoff (1997)]. Based on the collision-less Boltzmann equation, i.e., $f_t + uf_x + vf_y = 0$, and the same initial reconstruction of the GKS, the GKS gas distribution function at a cell interface, i.e., (4.15), reduces to

$$\begin{aligned} f &= f_0(x-ut, y-vt) \\ &= \left[1-\tau(ua^l+vb^l+A^l)-t(ua^l+vb^l)\right] H[u]g^l \\ &\quad + \left[1-\tau(ua^r+vb^r+A^r)-t(ua^r+vb^r)\right](1-H[u])g^r \\ &= \left[1-(\tau+t)(ua^l+vb^l+A^l)+tA^l\right] H[u]g^l \\ &\quad + \left[1-(\tau+t)(ua^r+vb^r+A^r)+tA^r\right](1-H[u])g^r. \end{aligned} \tag{4.22}$$

This is the same scheme as the kinetic flux vector splitting scheme for the NS solution (KFVS-NS) [Chou and Baganoff (1997)]. In the smooth flow region, the above equation goes to

$$f = g_0\left[1-(\tau+t)(u\bar{a}+v\bar{b}+\bar{A})+t\bar{A}\right]. \tag{4.23}$$

In comparison with with Eq. (4.21), we can clearly observe that in the smooth region the above equation solves the NS equations with a dynamical viscosity coefficient $\mu_{kfvs-ns} = (\tau+t)p$ instead of $\mu_{ns} = \tau p$. As a result, additional numerical dissipation, which proportional to t or the time step Δt is introduced in KFVS-NS. From the GKS, we can clearly understand the limitation of the Chou-Baganoff's KFVS-NS method. Because Eq. (4.22) is the limiting case of Eq. (4.21) with $\tau \gg \Delta t$, and Eq. (4.22) is only valid for the Navier-Stokes solution under such a limiting condition. In other words, KFVS-NS scheme does give a Navier-Stokes solution accurately if the condition $\tau \gg \Delta t$ is satisfied, such as in a well-resolved shock layer, see the analysis below. In $\tau \leq \Delta t$ region, such as high Reynolds number boundary layer, the KFVS-NS solution will deviate from the Navier-Stokes solution significantly.

Due to the free transport mechanism in the KFVS scheme, the additional dissipation is proportional to the viscosity coefficient $\mu_{num} \simeq p\Delta t$, where p is the pressure and Δt is the time step. Since the time step is determined by the CFL condition, the artificial dissipation can also be written as $\mu_{num} \sim (\rho/(\gamma(1+M)))c\Delta x \sim \rho c\Delta x$, where Δx is the cell size, c is the sound speed, and M is the Mach number. Suppose that $N \simeq 10$ cells are needed to resolve a NS shock structure or a boundary layer. Since the shock thickness is proportional to the mean free path l_s, in the resolved shock region the mesh size must satisfy $\Delta x = l_s/N$. Then, in the stationary shock case, due to the conditions $M \geq 1$ and $Re \simeq 1$, we have

$$\frac{\mu_{num}}{\mu_{phys}} \simeq \frac{c\Delta x Re}{l_s U} \simeq \frac{Re}{M}\left(\frac{\Delta x}{l_s}\right) \simeq \frac{Re}{M}\left(\frac{1}{N}\right) \ll 1.$$

Therefore, the KFVS-NS scheme could give an accurate NS shock structure once the structure is well resolved. This has been numerically proved in [Chou and Baganoff (1997)]. However, if the boundary layer is resolved with the same number of grid points, such as $l_b = N\Delta x$, we have

$$\frac{\mu_{num}}{\mu_{phys}} \simeq \frac{c\Delta x Re}{LU} \simeq \frac{Re}{M}\left(\frac{\Delta x}{L}\right) \simeq \frac{\sqrt{Re}}{M}\left(\frac{1}{N}\right),$$

where L is the length of the flat plate. The boundary layer thickness $l_b \simeq \sqrt{\nu L/U}$ has been used in the above approximation. Therefore, for a subsonic boundary layer with $M < 1$ and $Re \gg 1$, i.e., $M = 0.3$ and $Re = 10^5$, the numerical dissipation could dominate the physical one $\mu_{num} >> \mu_{phys}$ when the grid point in the boundary layer is on the order of $N = 10$. So, the KFVS-NS scheme cannot be properly used in this case. Due to the lack of particle collisions in KFVS, the initial g^l and g^r take free transport for flux function, and there is no any tendency for them to form an equilibrium state. So, it is not surprising that for the viscous flow calculations many high-resolution flux vector splitting (FVS) schemes have difficulties to get accurate NS solutions [van Leer *et al.* (1987); Drikakis and Tsangaris (1993)]. The use of free transport mechanism for the flux evaluation, the so-called upwind treatment of the transport term in the Boltzmann equation, will make many asymptotic preserving (AP) schemes have difficulty to get NS solutions as well in the continuum flow limit [Chen and Xu (2013)].

If the above collisionless Boltzmann solution (4.22) is further simplified for the inviscid flow computation, similar to the Riemann solution, the KFVS distribution function goes to [Pullin (1980); Reitz (1981); Deshpande (1986); Mandal and Deshpande (1994); Perthame (1992)],

$$f = H[u]g^l + (1 - H[u])g^r, \tag{4.24}$$

which is equivalent to Steger-Warming flux vector splitting scheme [Steger and Warming (1981); van Leer (1982)]. The above KFVS scheme has been well studied and applied to many physical and engineering problems. The positivity of the above KFVS has been proved [Estivalezes and Villedieu (1996); Tang and Xu (1999)]. In order to prove the entropy condition for the 1st-order KFVS scheme, a new concept of distinguishable particles in different numerical cells has been introduced [Lui and Xu (2001)]. An earlier version of the above KFVS scheme is the beam scheme, where instead of using a full Maxwellian distribution function, g^l and g^r are replaced by three Delta functions or particles [Sanders and Prendergast (1974)]. As analyzed in [Tang and Xu (2001)], the Steger-Warming method can be represented as a "beam scheme" too [Steger and Warming (1981)]. However, due to their slight difference in the particle representation, such as the lack of internal energy in the second "particle" in the Steger-Warming method, it could be less robust than the beam scheme. The relation between the beam scheme and the lattice Boltzmann method (LBM) has been analyzed in [Xu and Luo (1998)]. With the above connection between the KFVS scheme and the Steger-Warming method, it is easy to understand the poor performance of many FVS schemes in the viscous boundary layer calculations [van Leer *et al.* (1987)]. For LBM [Chen and Doolen (1998)], due to the use of regular lattice and in the low speed limit, the numerical dissipative coefficient from the free transport has a fixed value, which can be absorbed into the physical one, such as $(\tau - \Delta t/2)$. However, with the inclusion of the particle collision in the transport, the LBM can be much improved in the low speed rarefied and continuum flow computation, such as the discrete unified gas-kinetic scheme (DUGKS) [Guo *et al.* (2013)], where the same number of discrete particle velocities as LBM has been used. But, the robustness of the DUGKS is much improved in comparison with LBM and MRT [Qian *et al.* (1992); Lallemand and Luo (2000); Guo and Shu (2013); Wang *et al.* (2014)].

Higher-order gas kinetic schemes

The above GKS has second order accuracy in space and time. There are two ways to extend the above GKS to higher-order. One is to use higher-order reconstruction, such as WENO, to get the cell interface values and their gradients, but still use the above GKS flux [Kumar *et al.* (2013)]. The second approach is to get higher-order reconstruction and follow its evolution directly, where the 2nd and 3rd-order derivatives will participate explicitly in the gas evolution and the flux evaluation. In the following, the higher-order flow evolution model will be introduced.

In order to extend the GKS to higher-order, the same integral solution (4.2) can be used for the flux evaluation, but the initial term f_0 and the equilibrium state g can include more terms for the higher-order spatial and time derivatives. Based on the high-order piecewise continuous initial reconstruction of the conservative variables inside each cell, the corresponding Chapman-Enskog expansion of a gas distribution function f_0 for the Navier-Stokes solution can be constructed. With the higher-order expansion of both f_0 and g, a time dependent gas evolution solution from a piecewise discontinuous parabolic macroscopic flow distributions can be obtained for the flux evaluation. For example, a directional splitting high-order gas kinetic scheme (HGKS) has been developed for the compressible viscous flow computations [Li *et al.* (2010)], followed by a multidimensional one [Luo and Xu (2013)]. Equipped with WENO reconstruction [Jiang and Shu (1996)], the WENO-GKS has been developed and compared with traditional WENO methods for the inviscid and viscous flow computations [Kumar *et al.* (2013); Luo (2012); Luo *et al.* (2013); Xuan and Xu (2013a)].

In comparison with the second-order GKS (4.15), in smooth flow region the third-order GKS has the distribution function [Luo and Xu (2013)]

$$\begin{aligned} f(0,y,t,\vec{u},\xi) = {} & g_0[1-\tau(a_1u+a_2v+A)] \\ & +g_0[a_2-\tau((a_1a_2+d_{12})u+(a_2^2+d_{22})v+Aa_2+b_2)]y \\ & +g_0[A-\tau((Aa_1+b_1)u+(Aa_2+b_2)v+A^2+B)]t \\ & +\frac{1}{2}g_0(a_2^2+d_{22})y^2 \\ & +\frac{1}{2}g_0(A^2+B)t^2+g_0(Aa_2+b_2)yt, \end{aligned} \tag{4.25}$$

where the coefficients are defined at the center of a cell interface

$$g_0=g(0,0,0), a_1=\left(\frac{\partial g}{\partial x}\Big/g\right)\Big|_{(0,0,0)}, a_2=\left(\frac{\partial g}{\partial y}\Big/g\right)\Big|_{(0,0,0)},$$

$$A=\left(\frac{\partial g}{\partial t}\Big/g\right)\Big|_{(0,0,0)},$$

$$d_{11}=\frac{\partial a_1}{\partial x}\Big|_{(0,0,0)}, d_{22}=\frac{\partial a_2}{\partial y}\Big|_{(0,0,0)}, d_{12}=\frac{\partial a_1}{\partial y}\Big|_{(0,0,0)}=\frac{\partial a_2}{\partial x}\Big|_{(0,0,0)},$$

$$b_1=\frac{\partial a_1}{\partial t}\Big|_{(0,0,0)}=\frac{\partial A}{\partial x}\Big|_{(0,0,0)}, b_2=\frac{\partial a_2}{\partial t}\Big|_{(0,0,0)}=\frac{\partial A}{\partial y}\Big|_{(0,0,0)}, B=\frac{\partial A}{\partial t}\Big|_{(0,0,0)},$$

which can be fully determined from the distribution of macroscopic flow variables. The gas distribution function depends on the location y of a cell interface, which needs to be integrated on y in order to evaluate the flux transport. In other words, besides time accuracy in (4.25) without using Runge-Kutta method, there is no Gaussian points needed for the spatial accuracy in the higher-order GKS as well. The above procedure can be continuously used to construct even higher order GKS flux function in [Liu and Tang (2014)]. In general case with discontinuous initial data, the multidimensional HGKS has been constructed in [Luo and Xu (2013)]. The comparison between HGKS and WENO-type scheme for viscous and inviscid flow computations in both compressible and incompressible limits are presented in [Luo *et al.* (2013)]. Recently, a direction splitting 4th-order GKS scheme has been obtained for flow computation [Liu and Tang (2014)].

Well-balanced gas-kinetic scheme

Since the GKS is a direct modeling method, sometimes it is more convenient to incorporate other physical effect into the scheme through the kinetic scale modeling, such as the design of a well-balanced scheme for the gravitational gas system. For the hyperbolic equation with source term, much attention has been paid to design well-balanced schemes [Greenberg and Leroux (1996); Bouchut (2004); Slyz and Prendergast (1999); Tian *et al.* (2007); Xing and Shu (2013); Vides *et al.* (2014)]. For example, many research work has been done for the shallow water equations with bottom topology [Zhou *et al.* (2001); Xu (2002b)]. However, for the gas dynamic system under gravitational field, to construct a well-balanced scheme encounters great difficulties.

For the gravitational gas dynamic system, the well balanced solution is defined in the following. For an isolated gas system under stationary gravitational field $\Phi(x)$, the equilibrium state is an isothermal one with constant temperature T, zero flow velocity, and density distribution $\rho \sim \exp[-\Phi(x)/kT]$. A well-balanced scheme will not only be able to keep the above state, certainly to capture the propagation of small perturbation around the above state as well, but also need to converge to the above state from any initial condition for an isolated gas system.

With a piecewise constant gravitational potential modeling inside each cell, a well-balanced GKS has been developed [Luo *et al.* (2011)]. In the GKS modeling, the particle transport across a cell interface under gravitational jump has to impose the following mechanism in the kinetic level precisely. (1) exact momentum (including force effect) and energy (kinetic plus potential) conservation in individual particle transport across a potential jump; (2) an exact Maxwellian distribution function, instead of other pseudo-approximations for the equilibrium state; (3) symplecticity preserving property (phase space volume preservation) for Hamiltonian particles system in order to connect the moments of the gas distribution function before and after passing through the gravitational potential barrier.

Starting from macroscopic governing equations of hydrodynamic modeling only, such as the gravitational Euler or NS equations, it will be hard, if not impossible, to develop such a well-balanced scheme. The use of kinetic formulation and the modeling in the kinetic level seem necessary.

For the shallow water equations, a well-balanced GKS can be developed as well, but with much straightforward and simple physical modeling [Xu (2002b)]. The construction of the well-balanced numerical schemes based on the macroscopic shallow water equations directly is very successful as well. A gigantic amount of numerical schemes have been developed and most of them perform equally well. Now it is an appropriate time to shift attention to the gas dynamic system.

Gas-kinetic scheme with turbulent modeling

The turbulent flow is characterized with chaotic change, which is the most difficult challenge problem in fluid mechanics. Most fluid flow is associated with turbulence, and the laminar one is exceptional, which appears at the leading edge of an airfoil or inner region of a boundary layer, followed by transition [Cardy *et al.* (2008); Lee and Wang (1995); Lee and Wu (2008); Zhang *et al.* (2013); Liu *et al.* (2012)]. The multiscale nature in turbulence makes the modeling difficult, except resolving all flow physics down to the smallest scale, like the direct numerical simulation (DNS) approach. To resolve all scales is basically impossible at current stage for high Reynolds flow. The validation of the NS equations in the smallest scale is questionable as well, because the local Knusden number there cannot be small anymore, especially for the high speed flow. As pointed out in [Cercignani (2000)], the ratio of the particle mean free path l to the dissipative scale of turbulence l_d can be

$$l/l_d = (l/L)(Re)^{3/4} = Kn(Re)^{3/4} = (Ma)^{3/4}(Kn)^{1/4},$$

where L is a macroscopic scale, Ma is the Mach number, Kn is the Knudsen number, and the relation $Re = Ma/Kn$ has been used. For a fully developed hypersonic turbulent flow with $Re = 10^4$, l and l_d will become the same order for $Kn = 10^{-3}$ and $Ma = 10$. So, for high speed flow, the rarefaction effect cannot be fully ignored and it does not hurt to consider the modeling in the kinetic level, because the NS modeling is not adequate for a correct description.

In most practical applications, it is necessary to model the unresolved scales of motion. Most of the turbulence models, such as RANS or LES, rely on the idea of eddy viscosity, which may have difficulties to reproduce many physical observations. The NS equations are the coarse-grained model for the flow description in comparison with the kinetic equation. From the modeling point of view, it may have advantages to do the modeling using the kinetic equation and extend its solution to the up level, especially for the cases without a clear scale separation between resolved and unresolved motion. From the kinetic equation, through the coarse graining process, or the Chapman-Enskog expansion, the macroscopic NS equations for the near equilibrium flows can be obtained. However, in the macroscopic equations-based approach, the direct turbulent modeling is more or less to do the opposite, where a subscale modeling is attempted for the eddy effect, such as splitting the turbulent velocity into the coarse grained and fluctuating ones, and the effect from the small scale fluctuation to large scale field is modeled through eddy viscosity. The fundamental basis for the above eddy modeling relies on the existence of scale separation between the motion of resolved large scale and those of unresolved fluctuating eddies. But, in reality, the scale for the turbulence is continuously changing and there is no any clear separation of scales.

The Boltzmann equation has fundamental advantages over the NS approaches [Chen *et al.* (2003)] in the turbulent modeling. The kinetic equation includes more information about flow motion and the hydrodynamic limit is only its projection. Both turbulent and thermodynamic fluctuations are treated in the same way, and time evolution instead of purely spatial averaging is used in the coarse-graining procedure, such as the physics represented in the integral solution of the BGK model. Also, the relaxation time τ_{turb} with the inclusion of turbulent effect naturally includes both the regular viscous effects and its higher order modification, which is similar to using a generalized particle collision time on the NS solution to include all high-order expansion in the rarefied flow study [Xu (2002a)]. We may imagine that $\tau_{turb} = \tau_{laminar}/[1 + \tau_{laminar}\langle D^2 g\rangle/\langle Dg\rangle]$, and $\langle ...\rangle$ are the moments on the spatial and temporal variations of a gas distribution function representation for the turbulence.

For the GKS, the most research for the turbulent modeling is to get a turbulent relaxation time derived from standard two-equation turbulence model or use the scheme as a DNS method. Through the collision time modification, the coarse graining effects that affect the transport properties of the flow can be included in the scheme. More precisely, the control of particle collision time is equivalent to the adaptation of different scale flow modeling in its evolution. The turbulent stress tensor may become non-linearly related to strain rate tensor, as the non-linear correction terms are activated by the scheme. Many researchers have conducted study with the implementation of turbulent modeling into the gas-kinetic scheme [Li and Fu (2003); Jiang and Qian (2012); Righi (2014); Kerimo and Girimaji (2007); Kumar *et al.* (2013); Suman and Girimaji (2013); Liao *et al.* (2009, 2010); Yang *et al.* (2014)]. The use of GKS and UGKS in turbulent flow study may discover new insight which cannot be described by the NS equations. It will become an attractive research topic. For example, we may consider to construct a turbulence relaxation model for the interface flux evaluation, such as a solution from the initial NS distribution function to the turbulent PDF integration, and the relaxation rate is controlled by the turbulent relaxation time. As a result, a distribution function with a mixed laminar and turbulent flow representation can be obtained at a cell interface.

4.4 Numerical Examples

Significant amount of test cases have been conducted using the GKS for the continuum flow computation. The GKS is an accurate and robust numerical scheme for the NS solutions when the flow structure is numerically resolved, and provides delicate numerical dissipation in the unresolved regions. The GKS is especially suitable for the hypersonic flow computations. In the following, a few test cases from the low speed cavity flow to the hypersonic viscous and heat conducting flow will be presented.

Case(1) Axis-symmetric double cone geometry with flow separation

The double-cone configuration has a first cone half-angle 25^0 and the second cone angle 55^0. Under the experimental condition (Run 28) [Holden and Wadhams (2003)], the incident flow has

$$\rho_\infty = 0.6545\times10^{-3} kg/m^3, U_\infty = 2664.0 m/s, T_\infty = 185.56K, T_{wall} = 293.33K.$$

The corresponding Mach and Reynolds number are

$$M_\infty = 9.59, R_e = 13090.$$

In all series of experiments, the RUN 28 with the above flow condition is the most difficult one to be calculated due to the large flow separation. Under this flow condition, the first cone produces an attached shock wave, and the second cone with large angle produces a detached bow shock. These two shocks interact to form a transmitted shock that strikes the second cone surface near the cone-cone juncture. The adverse pressure gradient due to the cone juncture and the transmitted shock generates a large region of separated flow that produces its own separation shock. This shock interacts with the attached shock from the first cone, altering the interaction with the detached shock from the second cone. This in turn effects the size of the separation region. The shock interaction produces very high surface pressure and heat transfer rates where the transmitted shock impinges on the second cone. As presented in [Nompelis *et al.* (2003)], the schematic flow structure is shown in Fig. 4.2. Many authors have conducted the simulation for this case [Gaitonde and Canupp (2002); Candler *et al.* (2001); Gnoffo (2001); Wright *et al.* (2000)].

The coupling between the shock waves and the separation zone makes this case very sensitive to the physical modeling of the flow and to the numerical method. In order to get grid refinement results, we have run this case with the following mesh

$$(250 \times 100, 500 \times 100, 1000 \times 100, 500 \times 200, 1000 \times 200, 1000 \times 400).$$

Basically there is no differences in the flow distributions when using 500×200 and 1000×400 grid points. The computed pressure and temperature contours with 500×200 points are shown in Fig. 4.3, where the cell Reynolds number on the order 1 is used in the computation. The Mach number and pressure distributions around the second cone surface are shown in Fig. 4.4, where the contact surfaces, transmitted shock, as well as supersonic shock can be observed clearly. Along the cone surfaces, the measured and computed pressure and heat flux are presented in Fig. 4.5, where the symbols are the experimental results [Holden and Wadhams (2003)]. As shown in these figures, the size of the separation region and the heat fluxes on the cone surfaces match the experimental data well. Especially along the surface of the 2nd-cone, the complicated flow structures are captured. As analyzed in [Nompelis *et al.* (2003)], the deviation in the heat flux before the separation shock along the first cone surface is due to the non-equilibrium nature of the incoming expansion gas in the experiment device. In our simulation, the ideal equilibrium incoming gas is assumed.

Case(2) Hypersonic flow passing through a double ellipsoid

The next test case is the hypersonic flow passing through a double ellipsoid at Mach number 8.02, total temperature $720K$, and the Reynolds number 1.98×10^7 with a characteristic length $L = 1m$. The wall has a temperature $300K$ and angle of attack 0 degree. The mesh size used in the computation is $150 \times 70 \times 60$, and the smallest cell's Reynolds number is 20. Figure 4.6 shows the flow patterns on the surface of the ellipsoid and the surface pressure on the symmetric line, which have good agreements with the experimental measurements [Li (2012)].

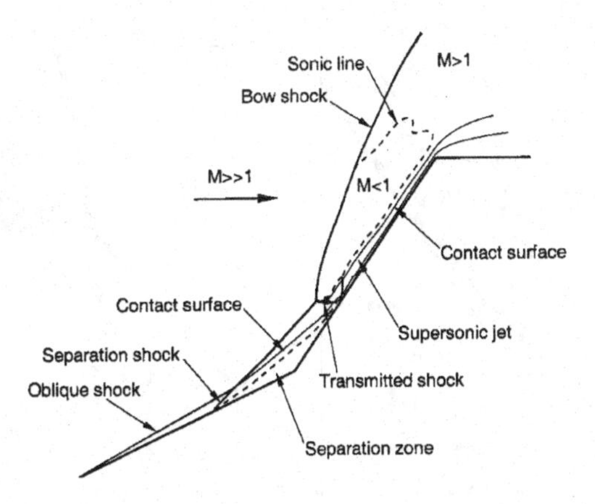

Fig. 4.2 Schematic flow structure for double-cone geometry [Nompelis *et al.* (2003)].

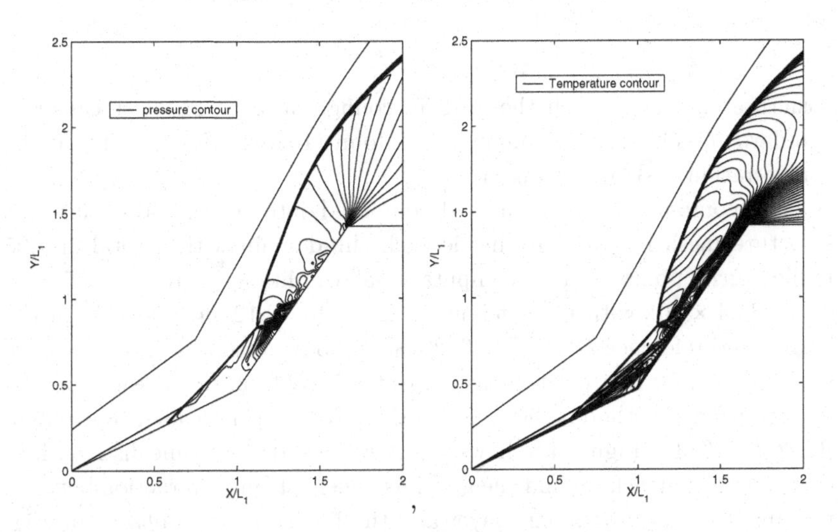

Fig. 4.3 Pressure and temperature distribution around double cone geometry ([Xu *et al.* (2005)]).

Case(3) Aerodynamics forces around a missile with grid fins

The study of aerodynamic forces and torques associated with missile configurations with grid fins in supersonic flight attracted much attention in the past decades. The CFD algorithm can be validated by comparing its computational results with the experimental measurement. In the supersonic regime, the nonlinearities are attributed to the complex shock

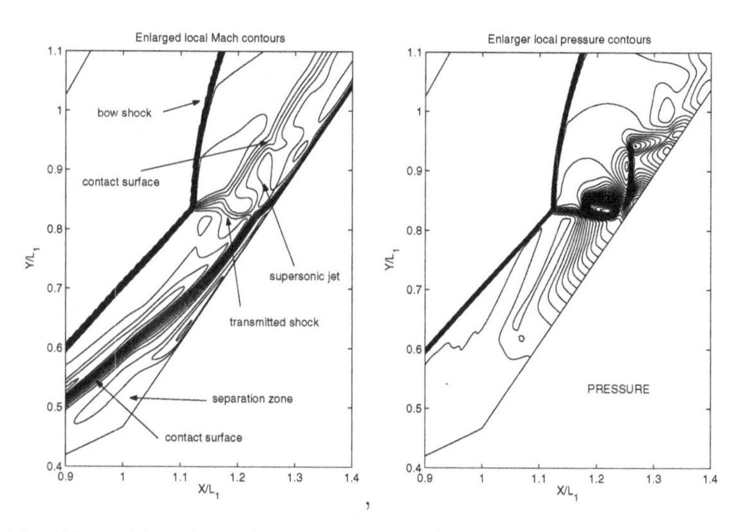

Fig. 4.4 Enlarged local Mach (left) and pressure (right) contours around the surface of second cone ([Xu *et al.* (2005)]).

structure associated with the grid fin configurations. The test case is a standard missile grid fin combination [John (1996)]. A schematic of the configuration is shown in Fig. 4.7.

The computation domain is decomposed into 619 sub-domains and structure mesh is generated inside each sub-domain with a total of 5.25 million grid points. The computational condition is $Ma = 2.5$ and $Re = 33.4 \times 10^7$ with the reference length $D = 0.127m$ (diameter of the body), and reference area $S = 0.012668m^2$ (body cross section). The reference location for the torque measurement is $5.2D$ from the missile tip. The computational method is the second order GKS, which is called BGK-NS in [Li *et al.* (2013)]. Figure 4.8 shows the pressure and streamline distributions at 0° angle of attack around the grid fins, where strong interaction between the missile surface boundary layer and the fin structure is observed. Figure 4.9 shows the normal force and pitching moments on the grid fins vs. angle of attack. Good agreements have been obtained between the GKS computation and experimental data.

The above two test cases are based on the second-order GKS schemes [Xu (2001); Xu *et al.* (2005)]. In the following, we will present the results from the third-order GKS flux function. With parabolic distributions for the conservative flow variables inside each control volume, and with the inclusion of cell interface discontinuity, a third-order GKS flux has been

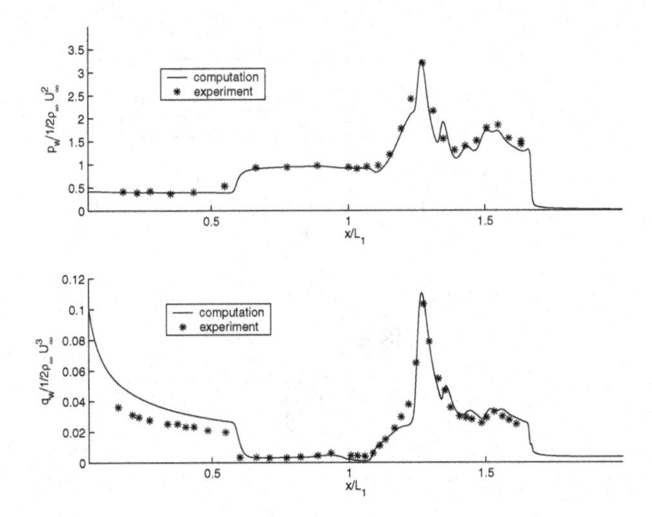

Fig. 4.5 Pressure (upper) and heat flux (lower) along cone surface, where the symbols are the experimental data [Holden and Wadhams (2003)] and the solid lines are the computational results ([Xu *et al.* (2005)]).

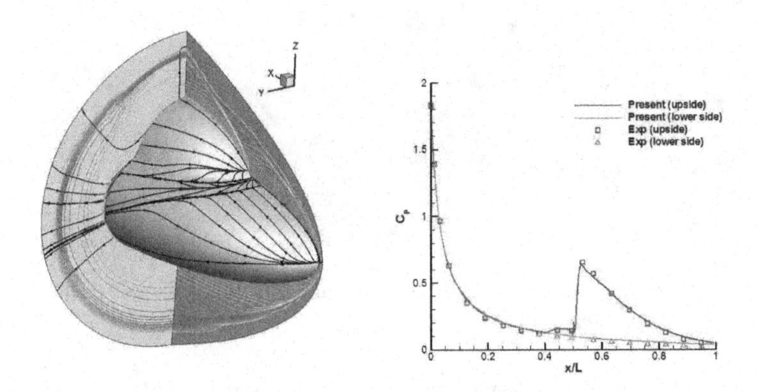

Fig. 4.6 Flow passing a double ellipsoid. left: Mach number contours and surface streamlines; right: surface pressure distribution along the symmetric lines [Li (2012)] (courtesy of Jin Li).

obtained [Li *et al.* (2010); Luo and Xu (2013)].

Case(4) Laminar boundary layer

This is a low-speed boundary layer test over a flat plate with length

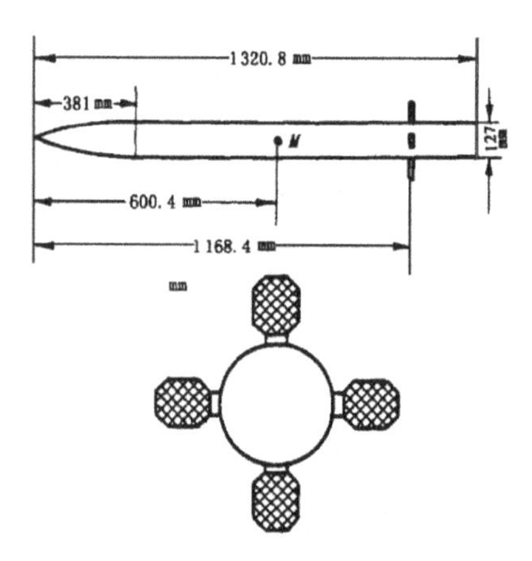

Fig. 4.7 A schematic configuration [Li *et al.* (2013)] (courtesy of Dingwu Jiang).

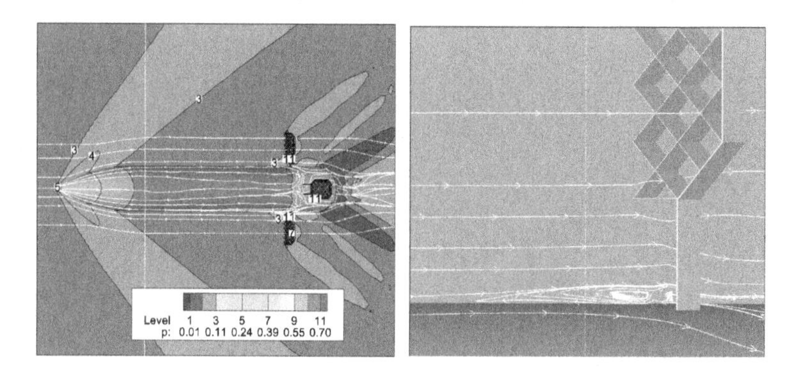

Fig. 4.8 (left) pressure distribution, (right) streamlines around the grid fin [Li *et al.* (2013)] (courtesy of Dingwu Jiang).

$L = 100$. The Mach number of the free-stream is 0.15 and the Reynolds number is $Re = U_\infty L/\nu = 10^5$. A total non-uniform 120×30 mesh points are adopted, where 40×30 cells are located ahead of the plate ($x < 0$). The minimal cell sizes are $\Delta x_m = 0.1$ and $\Delta y_m = 0.07$, respectively. The non-slip adiabatic boundary condition at the plate is used and a symmetry condition is imposed at the bottom boundary in front of the flat plate. The non-reflecting boundary condition based on the Riemann invariant is adopted for the other boundaries. The CFL number is set to be 0.6.

Figures 4.10 shows the calculated velocities vs. the non-dimensional

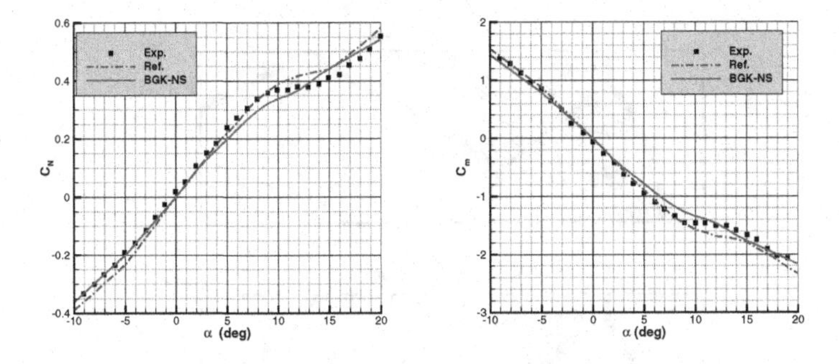

Fig. 4.9 (left) comparisons of normal force vs. angle of attack; (right) comparisons of pitching moment vs. angle of attack [Li *et al.* (2013)] (courtesy of Dingwu Jiang).

variables defined by $U^* = U/U_\infty$, $V^* = V/\sqrt{\nu U_\infty/x}$ and $y^* = y/\sqrt{\nu x/U_\infty}$. These results clearly show that both streamwise and transverse velocity components can be accurately predicted by the GKS scheme, even with as less as four mesh points in the viscous layer.

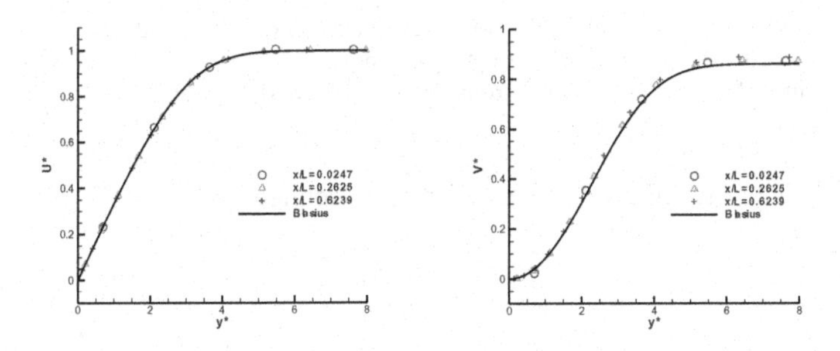

Fig. 4.10 Streamwise and transverse velocity profiles at different locations of a laminar boundary layer [Li *et al.* (2010)].

For the high-order GKS, the WENO-type reconstruction can be used to get the point-wise values at the left and right hand sides of a cell interface [Shu (1998)], then a parabola initial distribution can be obtained within each cell using the three conditions: two cell interface values and a cell averaged value. Assume that Q is the variable that needs to be reconstructed. $\bar{Q}_i$ is the cell averaged value in the ith cell. Q_i^l and Q_i^r are the two values obtained from the reconstruction at the left and right interfaces

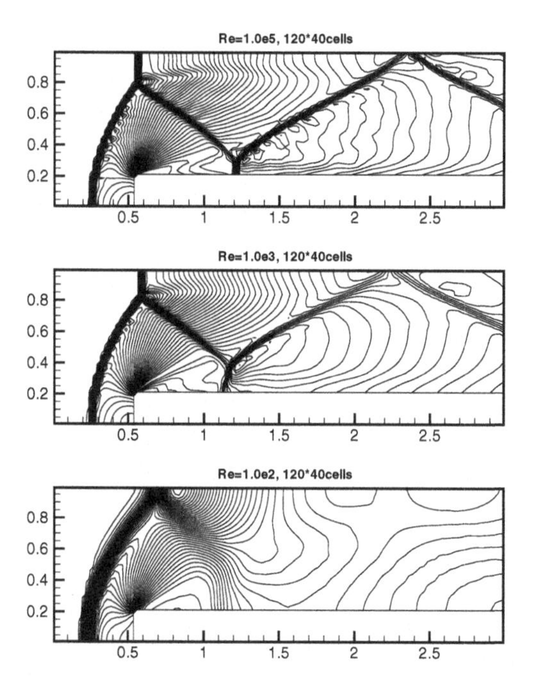

Fig. 4.11 Mach 3 step problem: density distribution for different Reynolds numbers $Re = 10^2, 10^3$, and 10^5 at $t = 4.0$ with 120×40 mesh points by the 3rd-order WENO-GKS. In each figure, there are 50 contours from 0.5 to 5 [Luo and Xu (2013)].

of the *ith* cell. The 5th WENO reconstruction is defined as,

$$Q_i^r = \sum_{s=0}^{2} w_s q_i^{(s)},\ Q_i^l = \sum_{s=0}^{2} \tilde{w}_s \tilde{q}_i^{(s)},$$

where

$$q_i^{(0)} = \frac{1}{3}\bar{Q}_i + \frac{5}{6}\bar{Q}_{i+1} - \frac{1}{6}\bar{Q}_{i+2}, q_i^{(1)} = -\frac{1}{6}\bar{Q}_{i-1} + \frac{5}{6}\bar{Q}_i + \frac{1}{3}\bar{Q}_{i+1},$$

$$q_i^{(2)} = \frac{1}{3}\bar{Q}_{i-2} - \frac{7}{6}\bar{Q}_{i-1} + \frac{11}{6}\bar{Q}_i,$$

$$\tilde{q}_i^{(0)} = \frac{11}{6}\bar{Q}_i - \frac{7}{6}\bar{Q}_{i+1} + \frac{1}{3}\bar{Q}_{i+2}, \tilde{q}_i^{(1)} = \frac{1}{3}\bar{Q}_{i-1} + \frac{5}{6}\bar{Q}_i - \frac{1}{6}\bar{Q}_{i+1},$$

$$\tilde{q}_i^{(2)} = -\frac{1}{6}\bar{Q}_{i-2} + \frac{5}{6}\bar{Q}_{i-1} + \frac{1}{3}\bar{Q}_i,$$

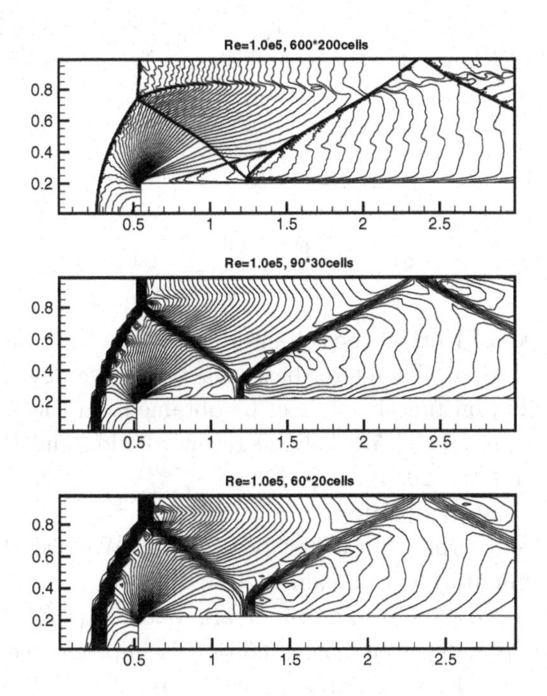

Fig. 4.12 Mach 3 step problem: density distribution with different number of mesh points at $t = 4.0$ with $Re = 10^5$ by the 3rd-order WENO-GKS [Luo and Xu (2013)].

$$w_s = \frac{\alpha_s}{\sum_{p=0}^{2} \alpha_p}, \ \alpha_s = \frac{d_s}{(\epsilon + \beta_s)^2}, \ \tilde{w}_s = \frac{\tilde{\alpha}_s}{\sum_{p=0}^{2} \tilde{\alpha}_p}, \ \tilde{\alpha}_s = \frac{\tilde{d}_s}{(\epsilon + \beta_s)^2}, \ s = 0, 1, 2,$$

$$\beta_0 = \tfrac{13}{12}(\bar{Q}_i - 2\bar{Q}_{i+1} + \bar{Q}_{i+2})^2 + \tfrac{1}{4}(3\bar{Q}_i - 4\bar{Q}_{i+1} + \bar{Q}_{i+2})^2,$$

$$\beta_1 = \tfrac{13}{12}(\bar{Q}_{i-1} - 2\bar{Q}_i + \bar{Q}_{i+1})^2 + \tfrac{1}{4}(\bar{Q}_{i-1} - \bar{Q}_{i+1})^2,$$

$$\beta_2 = \tfrac{13}{12}(\bar{Q}_{i-2} - 2\bar{Q}_{i-1} + \bar{Q}_i)^2 + \tfrac{1}{4}(\bar{Q}_{i-2} - 4\bar{Q}_{i-1} + 3\bar{Q}_i)^2,$$

$$d_0 = \tilde{d}_2 = \frac{3}{10}, \ d_1 = \tilde{d}_1 = \frac{3}{5}, \ d_2 = \tilde{d}_0 = \frac{1}{10}, \ \epsilon = 10^{-6}.$$

When the pointwise variables at both ends of a cell are provided, a third-order polynomial inside each cell can be written as

$$V(x) = V_i + S_i^1(x - x_i) + \frac{1}{2}S_i^2(x - x_i)^2, \quad x \in [x_{i-1/2}, x_{i+1/2}],$$

where x_i is the cell center, $x_{i-1/2}$ and $x_{i+1/2}$ are the left and right interfaces of the ith cell, V_i, S_i^1 and S_i^2 are three unknowns. From the three conditions

$$\frac{1}{x_{i+1/2}-x_{i-1/2}}\int_{x_{i-1/2}}^{x_{i+1/2}} V(x)dx = \bar{V}_i,\ V(x_{i-1/2}) = V_i^l,\ V(x_{i+1/2}) = V_i^r,$$

we can get

$$V_i = \frac{3}{2}\bar{V}_i - \frac{1}{4}(V_i^l + V_i^r),\ S_i^1 = \frac{V_i^r - V_i^l}{x_{i+1/2}-x_{i-1/2}},\ S_i^2 = \frac{6[(V_i^l + V_i^r) - 2\bar{V}_i]}{(x_{i+1/2}-x_{i-1/2})^2}.$$

So, the derivatives of $V(x)$ at the cell interface can be easily obtained from the above distribution. Therefore, based on the above flow variables, an initial gas distribution function f_0 can be obtained. In the following, the test cases are based on the 5th-WENO reconstruction and the 3rd-order GKS flux [Luo and Xu (2013)].

Case(5) Mach 3 step problem

The Mach 3 step problem was first proposed in [Woodward and Colella (1984)]. The computational domain is $[0, 3] \times [0, 1]$. A step with height 0.2 is located at $x = 0.55$. The upstream velocity is $(U, V) = (3, 0)$. The adiabatic slip Euler boundary condition is implemented at all solid boundaries. The results for different Reynolds number (Re) and different number of cells are given in figure 4.11 and 4.12.

Figure 4.11 shows the NS solutions at different Reynolds numbers. The combination of the 5th-order WENO reconstruction and the 3rd-order GKS-NS flux presents accurate solutions. Because of the use of both numerical and the physical collision times, an oscillation free transition can be obtained around a shock with different mesh sizes, see figure 4.12.

Case(6) Low speed cavity flow

The cavity flow at low Mach number is a standard test case for validating incompressible or low speed NS flow solvers. Since the benchmark solution is from incompressible NS equations, in order to avoid kinematic dissipation, most simulations in the past are based on either the methods for the incompressible equations or the artificial compressibility methods, where a continuous initial reconstruction across a cell interface is assumed. The second-order GKS with a continuous flow distribution at a cell interface, i.e., Eq. (4.21), has been used in cavity flow study previously [Su *et al.* (1999); Xu and He (2003); Guo *et al.* (2008)].

Here we are going to use the same 3rd-order shock capturing GKS for the Mach 3 step problem to the cavity simulation. This is a challenge for any shock capturing NS flow solver, because the cell interface discontinuity may generate a large amount numerical dissipation, especially in the coarse mesh case.

The flow simulated has a Mach number 0.3 at the Reynolds number 1000 and $\gamma = 2$. The fluid is bounded by a unit square and is driven by a uniform translation of the top boundary. All the boundaries are isothermal and nonslip. Figure 4.13 shows the calculated stream traces with 64×64 and 32×32 mesh points. The results of U-velocity and pressure along the vertical symmetric line at $x = 0.5$, and V-velocity and pressure along the horizontal symmetric line with the mesh sizes $33 \times 33, 65 \times 65$, and 101×101 at $y = 0.5$ are shown in figure 4.14. The benchmark solution is from [Ghia *et al.* (1982)]. And the reference solution for pressure is from [Botella and Peyret (1998)]. As shown in figure 4.14, even with 64×64 cells, the simulation results from WENO-GKS match with the exact solutions very well. This can be hardly achieved for a shock capturing scheme, especially for the schemes based on the directional splitting method.

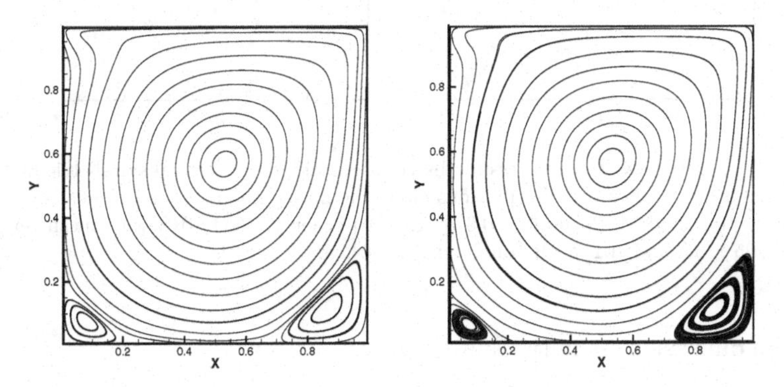

Fig. 4.13 Cavity flow: the streamlines at $Re = 1000$ calculated by the 3rd-order WENO-GKS. Uniform mesh is used. Left: 65×65 cells. Right: 33×33 cells [Luo and Xu (2013)].

Case(7) Reflecting shock-boundary layer interaction in a shock tube

This is a 2-D viscous shock tube problem studied in [Daru and Tenaud (2000, 2009)] and [Sjogreen and Yee (2003)]. An ideal gas is at rest inside a 2-D box $0 \leq x, y \leq 1$. Initially, a membrane is located at $x = 0.5$ which separates two different states of the gas. At time zero the membrane with a shock Mach number of 2.37 is removed and wave interaction occurs. Here, we solve the NS equations for this problem with non-slip boundary conditions at the adiabatic walls.

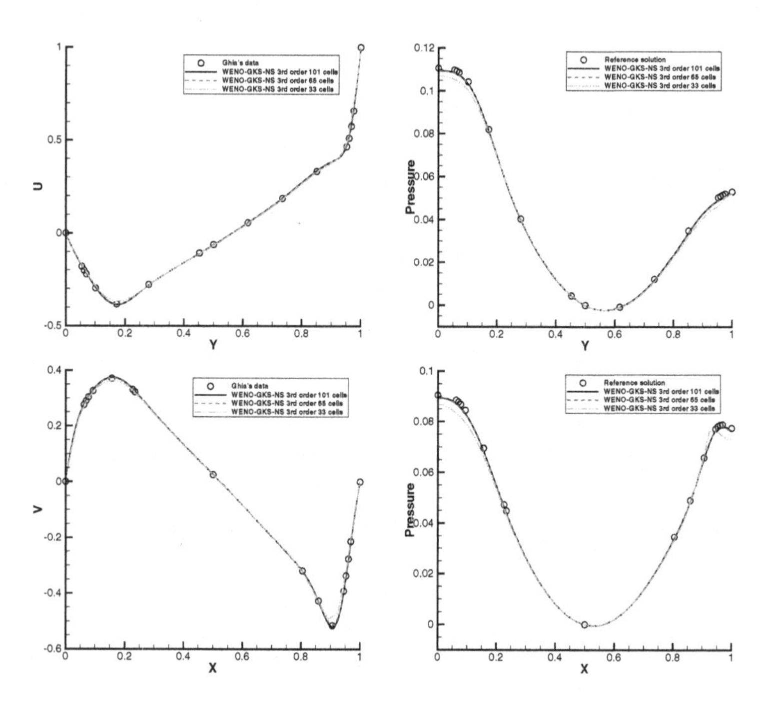

Fig. 4.14 Cavity flow: U and pressure distributions along the vertical symmetric line at $x = 0.5$ and V and pressure distributions along the horizontal symmetric line at $y = 0.5$ with $Re = 10^3$ and Mach number 0.3 by the 3rd-order WENO-GKS with mesh points $33 \times 33, 65 \times 65$, and 101×101 [Luo and Xu (2013)].

The dimensionless initial states are

$$(\rho, p) = \begin{cases} (120,\ 120/\gamma),\ 0 \leq x < 0.5, \\ (1.2,\ 1.2/\gamma),\ \ 0.5 < x \leq 1, \end{cases}$$

where $\gamma = 1.4$. The Prandtl number is 0.73 and Reynolds number is 1000. The sound speed is also 1. Due to the symmetry of the flow in the vertical direction, only half of the shock tube is calculated. Figure 4.15 shows the numerical results by 500×250 and 1000×500 uniform meshes.

4.5 Comparison of GKS and Godunov Method

Starting from a discontinuity, the Riemann solver and GKS take different physical process to model the gas evolution at a cell interface. In the following, the differences and similarities between Riemann solution and

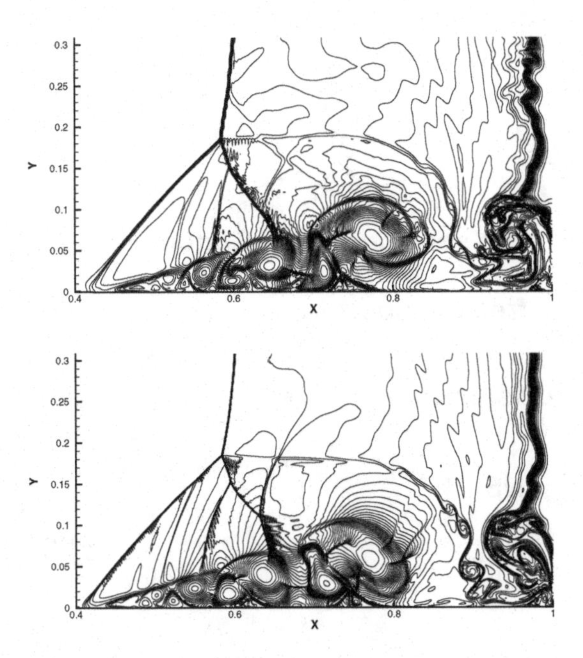

Fig. 4.15 Reflecting shock-boundary layer interaction in a shock tube: the density distributions at time $t = 1$. In each figure, there are 30 contours from 12 to 120. Upper: 500×250 cells. Lower: 1000×500 cells (courtesy of Jun Luo).

GKS will be pointed out. At the same time, the consequence from different dynamics on the numerical solutions will be analyzed.

(1) Different physical process from a discontinuity

Due to the mis-matching between the cell resolution and the physical layer thickness, in unresolved cases the use of nonlinear limiter is needed to avoid oscillation. For the compressible flow computation, a finite volume scheme has to incorporate a discontinuous initial condition at a cell interface. Based on the same discontinuity, as shown in figure 4.16, the GKS and Godunov-type method have different dynamics on the flow evolution from a discontinuity. Since the GKS covers a gas evolution process from the kinetic to the hydrodynamic scales, it first presents particle free transport, then due to particle collision the non-equilibrium dissipative layer will appear. With the increasing of particle collisions, the NS wave structure will be formed. The Euler solution becomes a limiting solution with the inclusion of infinite number of particle collisions and the equilibrium states are maintained due

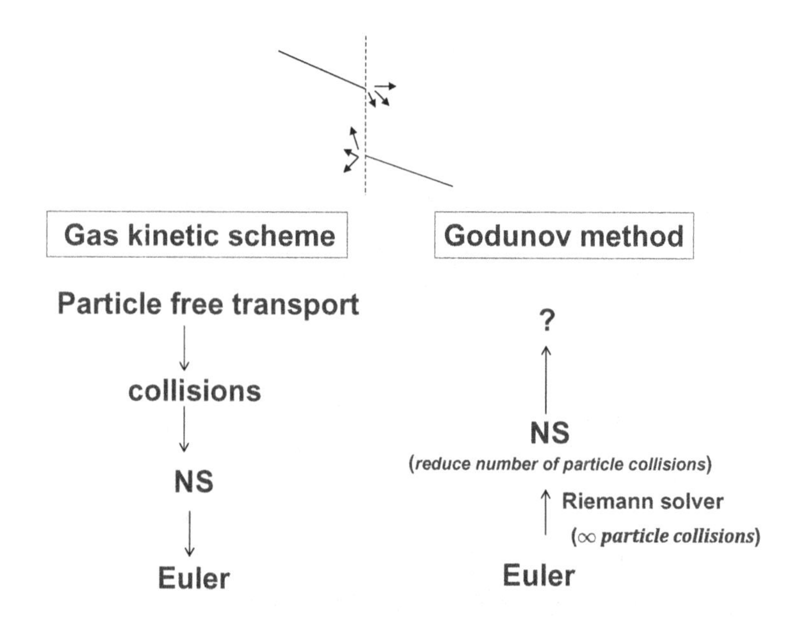

Fig. 4.16 GKS and Godunov-type scheme's physical process from a discontinuity.

to intensive particle collisions in the scale with $t \gg \tau$. As a result, distinctive wave structures, such as shock, contact, and rarefaction ones, emerge in this limiting solution. For the Godunov method, the Riemann solver is an exact solution of the Euler equations. Due to the Euler modeling, from the starting point, the use of the Riemann solution assumes the formation of distinctive wave structures with instantaneously infinite number of particle collisions. The discontinuous shock and contact wave in the Riemann solution do require intensive particle collisions, such as the non-penetration of particles across a contact discontinuity. In reality, this idealized solution cannot be achieved from a discontinuity if the particle transport and collision are taken into account, at least at the beginning of the time step. If CFD computation is targeting to simulate a physical flow, it needs to mimic the reality instead of sticking on a well-defined governing equation, where the condition for the validity of the equation may not be applicable. Along the methodology of the Godunov method, if the NS solution is required, the physical dissipation has to be added after the above Riemann solution. This is equivalent to reducing the number of particle collisions in order to keep a slightly non-equilibrium state. Dynamically, a process to reduce the particle collision is impossible once it has been implicitly taken into account in the formation of wave structures. Practically, for the Godunov method an operator splitting method, which separates the inviscid and viscous terms, is used in the construction of NS algorithm. But, the physical consistency and consequence of such an approach needs carefully analyzed. The transition from an equation based CFD to a modeling based CFD algorithm development may be needed because the numerical fluid in a discretized space may encounter situation, such as the interface discontinuity and a specific mesh size scale, which may not be included in the derivation of the theoretical fluid dynamic equations.

(2) Multi-dimensionality and space and time evolution coupling

In the GKS, on both sides of a cell interface the initial flow distributions keep derivatives in both normal and tangential directions. Both derivatives will participate in the gas evolution across a cell interface, such as the contribution of aug term in the normal direction and bvg term in the tangential direction in (4.15) for the flux evaluation. Therefore, even with a special orientation of the cell interface, the particle transport (from f_0) and wave propagation (from the integration of the equilibrium state g) can take place in any direction. In terms of the gas evolution the GKS is intrinsically a multidimensional scheme, and the mesh orientation effect on the numerical solution can be reduced to a minimal level if an appropriate initial reconstruction is obtained. However, the Riemann solution is an exact solution of the Euler equations with two constant states. The waves generated from the initial discontinuity always propagate in the normal direction of a cell interface. Therefore, the local wave propagation direction in the Godunov method depends solely on the mesh orientation. This is also true for the generalized Riemann solvers with second or third-order accuracy [Ben-Artzi and Falcovitz (2003); Qian *et al.* (2013); Wu *et al.* (2013)], where the slopes in the tangential direction can hardly be included the gas evolution. So, in the multidimensional case, the Godunov method is basically based on the numerical wave propagation instead of physical one and the mesh orientation largely affects its dynamics. Certainly, the numerical error due to the direction by direction splitting can be effectively removed with careful tuning of the numerical procedure and multiple sub-steps, and perfect numerical solutions can be obtained when the mesh is properly generated.

(3) Smooth transition from upwind to central difference

In GKS, associated with the multiscale modeling from the kinetic to the hydrodynamics, the corresponding flux makes a smooth transition from the upwind flux vector splitting (from f_0) to the central difference discretization (from the integration of the equilibrium state g). As a result, the kinetic scale upwind mechanism is used to capture the numerical discontinuity, such as the shock wave, and the hydrodynamical central difference is adopted to recover the NS solutions. Therefore, the GKS will not be so sensitive to the initial reconstruction in the viscous flow computation, especially in high Reynolds case, and it has intrinsic mechanism to quickly recover a smooth near equilibrium solution from an initial discontinuous data when $\Delta t \gg \tau$. In a barely resolved boundary layer, even with initial interface discontinuity in the reconstruction, the NS solution can be still obtained due to the dominant contribution from the central difference (equilibrium) part. As pointed out in [Ohwada and Kobayashi (2004)], this is the key for the success of the GKS method in the capturing of both discontinuous and continuous solutions, where the kinetic free transport mechanism is powerful to remove the shock instabilities which are intrinsically rooted in the Godunov-type shock capturing schemes [Ohwada *et al.* (2013)]. In the Godunov method, the upwind is intrinsically rooted in its wave propagation modeling from a discontinuity. The advantage is that the numerical dissipation associated with the upwind can be properly used to capture a crisp shock transition. The weakness is that the numerical dissipation may poison the NS solution in the smooth region, because even in the smooth region the dynamics from an artificially reconstructed initial discontinuity is still truthfully followed. The numerical dissipation is closely related to the initial jump at the cell interface. Therefore, the Godunov method is much sensitive to the initial reconstruction, the mesh orientation, and the limiters used in comparison with GKS. The current development of hybrid scheme, which combines the upwind in the discontinuous region and the central difference in the smooth region, is a meaningful approach. Fortunately, the GKS provides a dynamical transition between these two limits.

(4) High-order extension

The development of high-order CFD method has received considerable attention in the CFD community in the past two decades because of their potential in delivering accurate solution with lower cost than that from the 2nd-order scheme. There are many ways to extend the schemes from a 2nd-order to a higher one. Different researchers may have different understanding about the essentials for the development of high-order schemes. In order to construct a high-order compact scheme, an accurate capturing of both spatial and temporal evolution from an initial piecewise discontinuous polynomial is needed. Based on such an initial condition, a reasonable question is what the solution should be, instead of which solution is available to be used numerically. Even with highly complicated initial condition, the Riemann solution is still implemented because it is the only exact solution we know, even though it gives a 1st-order dynamics. A generalized Riemann solver based on the initial piecewise discontinuous linear flow distribution has been developed [Ben-Artzi and Falcovitz (1984, 2003); Ben-Artzi *et al.* (2006); Ben-Artzi and Li (2007)], which shows better numerical performance than the 1st-order Riemann-solution based methods. This result is physically funded. Even though it is difficult, it is fully legitimate to continue this track and to construct schemes based on the time evolution solution from a more complicated initial condition. The ADER method [Toro (2009)] is based on such a understanding, even though the linearized equations for the evolution of higher-order derivatives are adopted. The GKS has been extended to higher orders [Li *et al.* (2010); Luo and Xu (2013); Liu and Tang (2014)] along the same line with piecewise discontinuous initial polynomials. Since the high-order GKS couples the spatial and time evolution, the time dependent flux transport across a cell interface in multidimensional case depends on the location along the cell interface. In GKS, there are no Gaussian points used along the cell interface for spatial accuracy, no Runge-Kutta time stepping is necessary for time accuracy, and no separate numerical discretization for the inviscid and viscous terms. As a result, it is possible that a higher order (≥ 3) GKS can become efficient in comparison with Riemann solution-based high-order methods, where multi-stage Runge-Kutta time stepping with multiple Gaussian point integrations has to be used.

Instead of directly following the gas evolution from a high-order initial reconstruction, the Discontinuous Galerkin (DG) method uses mathematical transformation, mainly integration by parts, to get a weak solution of the governing equation. In this case, the use of the Riemann solution seems sufficient to get a high-order compact scheme. In the DG framework, besides the update of conservative flow variables, the gradients of the flow variables or the equivalent degree of freedoms, are updated as well. The equivalence between the DG and many other high-order methods has been established in recent investigations [Huynh (2007, 2009); Wang (2011); Wang and Gao (2009)]. The conservation for the conservative variables as a physical law is fully preserved in the DG method due to the update of cell averaged values, which is the same as the finite volume scheme. But, the update of other degrees of freedom in DG method needs to be carefully assessed.

All higher order methods need more degrees of freedom to record and evolve the solution. However, the DG is a compact scheme. Different from WENO approach, the additional degree of freedom in DG is not from reconstruction through the use of large stencils and the information from neighboring cells. Then, what kind of mechanism the DG methods can use to keep and evolve more degrees of freedom compactly. Even with the weak formulation and the integral form (moment) of the DG method, the only ingredient we can imagine is to make additional derivatives to the original PDEs for the derivation of discrete evolution equations of additional degree of freedom, such as creating governing equations for the high order derivatives over the cell size scale, even though it has been done obscurely in the DG approach through mathematical integrations. Theoretically, it is perfectly valid to take additional derivatives to the original PDE in the smooth flow region, but this approach has problem in discontinuous region for the creation of evolution laws for the non-existing physical laws. In the discontinuous region, physically there is no and will not have any evolution equation for the slope or gradient of flow variable in differential or discretized forms. Intrinsically, the use of the weak solution and the evolution equation (averaged over the control volume) for the gradient in the DG methods, or other equivalences, must have fundamental problems in the discontinuous region. This is why the updated gradients (or other equivalent variables) in DG have been continuously limited in order to remove the inappropriate solutions. Certainly, for all finite volume schemes, the limiter has to be used in the reconstruction. But for the finite volume scheme the use of the limiter has not been reached to such a sensitive level as that in the DG method. The most unfortunate observation is that it is hard to get the physical and mathematical reasons once the DG method fails. It is always associated with mysterious circumstances.

The GKS is based on the strong solution in its algorithm development. In order to get a "compact" higher-order scheme, besides updating the conservative variables through the cell interface fluxes, we believe that the subcell solutions at many points inside each control volume have to be updated as well if they will not be reconstructed from information in neighboring cells. This update will be different from the common practice of re-interpolating the cell interface flux differences into inner cell points [Huynh (2007, 2009)], but it is a dynamic solution with piecewise discontinuous high-order initial condition. For GKS, besides getting the integral solution at the cell interface, a similar integral solution has to be obtained as well at a few points inside each cell to obtain the time evolution of conservative flow variables there. As a result, besides the cell averaged value, the subcell point values, including the values at the cell interfaces at the next time level, can be updated and used locally to reconstruct the initial condition at the next time level directly. The condition for a valid strong solution at different subcell points at the next time step is important to construct high-order compact scheme, and this feasibility critically depends on the high-order dynamics of the gas evolution model. For example, the cell interface value has been updated and used in the 2nd-order compact GKS method [Xu (2002c)]. The exact Riemann solution doesn't provide the necessary dynamics to construct this kind of solution, because it has only first-order dynamics.

4.6 Principle of CFD

The advancement of CFD has been driven by industry requirements for robust and accurate prediction of flow motion from incompressible to hypersonic one, such as the design of air vehicles. A major achievement in the past decades was the development of high resolution shock capturing schemes, which have second order accuracy in the smooth flow region, and become first or even zero order accuracy in the shock regions [LeVeque (2002); van Leer (2006)]. The main concept for the success of modern shock capturing scheme is the introduction of nonlinear limiter in the flux or the initial data reconstruction [Boris and Book (1973); Kuzmin *et al.* (2012)], where the numerical dissipation is implicitly added. This can be quantitatively evaluated by comparing total kinetic energy within a cell before and after the use of nonlinear limiter. As a result, the redistribution between kinetic and thermal energy in the initial condition provides the major source of numerical dissipation in most shock capturing schemes if the Euler solutions are precisely implemented in the flux evaluation, such as the exact Riemann solver.

At low and modest flow speed, the robustness and accuracy of shock capturing schemes for the Euler and Navier-Stokes solutions have been well established. However, the hypersonic flow computations are still challenging due to emerging of shock instabilities, i.e., the so-called carbuncle phenomena [Quirk (1994)], when strong shocks are nearly aligned with the grid. Even though it is still unclear about the exact reason for the instability, the inconsistency of numerical dissipation in the Godunov method in the multi-dimensional case may cause the problem. Based on the extensive testing and analysis, the instability seems depend on the geometry, mesh size, Mach number, and many other factors. It is still an open question whether the carbuncle phenomenon is a numerical artifact, which can be simply cured. Instead of figuring out the fundamental reason for the shock instability, many researchers take great effort to construct and test different flux functions in hope of getting reliable schemes, such as the continuous efforts on the search of AUSM-type methods [Liou (2006); Kim *et al.* (2001); Kitamura and Shima (2013)]. All these attempts are based on a belief that a perfect Riemann solver can be obtained with more careful analysis of physical and numerical shock wave. Certainly, some authors may think that the shock instability is intrinsically rooted in the Godunov method and is incurable if the CFD is targeting on the Euler solutions [Xu (1999); Elling (2009)].

Through the comparison between the GKS and the Riemann solution, a different point of view on the fundamental basis for the shock instability will be proposed. The shock instability comes from the inconsistency between the Euler equations modeling and limited resolution in a discretized space, i.e. the modeling flaws of using the Euler solutions in the shock capturing schemes.

A numerical shock structure should mimic the physical process of a real shock layer, but with a much enlarged thickness due to the limited resolution in a discretized space. The numerical shock layer should be formed through a non-equilibrium process on the scale of mesh size. The use of the Euler equations cannot provide such a dissipative mechanism for the numerical shock formation.

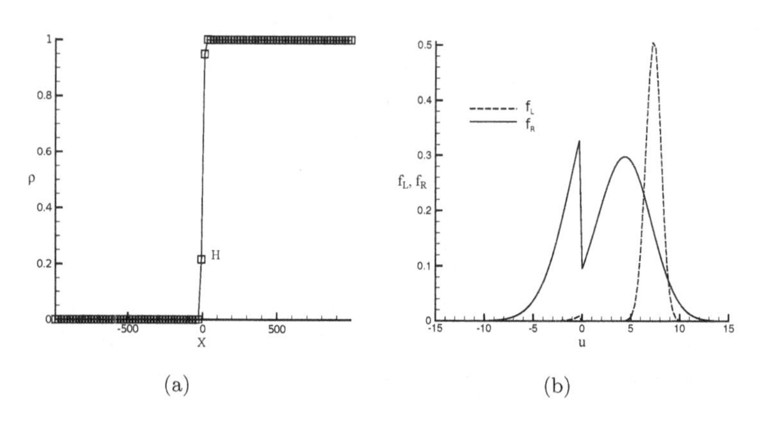

Fig. 4.17 (a) Mach 8 stationary shock wave calculated by GKS; (b) The gas distribution functions at the left and right cell interfaces of point (H) for the flux evaluations.

In the Euler modeling, an equilibrium state is always assumed. Even inside the shock layer, the flux evaluation for the update of flow variables inside such a layer is based on the equilibrium assumption at the cell interface. In 1980s much effort had been paid to construct a numerical shock transition as sharp as possible. The successful story is that if the shock front is exactly located at the cell interface, both exact Riemann solver and the Roe's approximate Riemann solver can precisely preserve such a shock transition without any intermediate point [Godunov (1959); Roe (1981)]. However, even in this case when solving the NS equations, the added viscous and heat conduction terms through the central difference discretization will make the above perfect shock front oscillate, because these additional terms will generate perturbation on the Rankine-Hugonit shock condition. As a result, there must be post-shock oscillation for the NS solution even with the exact Riemann solver. Even for the Euler equations, if the shock front is not precisely located at the cell interface, the averaged intermediate states will appear and this state will not stay on the Hugoniot curve of the shock wave [Arora and Roe (1997)]. Therefore, the post-shock oscillation is unavoidable.

Figure 4.17 shows a Mach 8 stationary shock wave calculated using GKS. As presented in the figure, there is no post-shock oscillation. In order to understand the reason why a stationary numerical shock structure can be maintained from the GKS, the gas distribution function for the flux evaluation around the intermediate transition point H inside the shock layer is plotted in the figure. As shown in the figure, at the left cell interface of the intermediate point H, the state is close to the upstream, and an equilibrium state with low temperature and high speed appears. At the right cell interface of point H, the state is inside the shock layer and a bi-Maxwellian gas distribution, which is similar to Mott-Smith model for the real physical shock construction [Mott-Smith (1951)], emerges automatically. The dissipation provided through this bi-Maxwellian gas distribution function mimics the real non-equilibrium gas distribution function inside a real physical shock layer [Pham-Van-Diep *et al.* (1989)], which will be studied in later chapters using UGKS scheme. This kind of dissipation or non-equilibrium mechanism is critically important to totally remove the shock instability [Ohwada *et al.* (2013)]. And this dissipation can be only provided through the kinetic scale modeling, such as the limited number of particle collisions. For the Godunov method, corresponding Maxwellian distribution functions always appear at both cell interfaces of the point H due to the assumption of equilibrium states in the Riemann solution, or the direct implementation of the Euler solutions. Therefore, there is no dissipation in its numerical flux function for the update of state H inside the shock layer.

The numerical dissipation provided in the Godunov method is mainly coming from the initial data reconstruction, which continuously transfers the kinetic energy into thermal one with the application of limiter. This kind of dissipation depends closely on the mesh orientation [Xu (2001); Xu and Li (2001)]. At a high Mach number, the mesh-dependent numerical dissipation in the Godunov method will be inconsistent with the physical one, which triggers the carbuncle phenomena. Actually, an attempt to fix the shock instability is to change the underlying dynamics of the Riemann solution [Loh and Jorgenson (2009)], which is equivalent to implicitly using different governing equations in the flux construction. But, the fact of using different governing equations underlying these approximate Riemann solvers has never been explicitly pointed out, because most CFD participants only have access to the Euler and NS equations from the starting point. The training on the shock capturing scheme makes the Godunov method on a superstitious status. Without setting up a fundamental principle on the algorithm development, any attempt to cure the weakness in the Godunov method can be only successful case by case. The direct modeling in GKS somehow enriches the content of CFD research.

The Riemann solver for CFD plays the role of an engine in a CFD vehicle. The current study shows that this engine may get malfunction when the speed of vehicle gets to a high level. The numerical evidence indicates that the direct adaptation of the Riemann solution in an algorithm development is not fully valid in a discretized space for the recovering of the non-equilibrium shock transition. A numerical fluid motion should mimic the physical one even in a discretized space, instead of targeting on any specific governing equation with different modeling scale. The mechanism to form a physical shock structure should be implemented numerically in the algorithm development. Unfortunately, the Euler equations don't provide such a mechanism to construct an enlarged physically sounded shock layer. Since a physical shock thickness is on the order of a few particle mean free path, an enlarged numerical shock thickness should also be on the order of numerical particle mean free path, which is the cell size. Therefore, inside the shock layer the mesh size should be on the order of numerical particle mean free path, and the non-equilibrium flow physics needs to be presented in the shock region. When a particle passes through a physical shock layer, it only takes a few particle collisions. A non-equilibrium shock structure is formed through the competition between particle free transport and collisions. This requires a flow modeling in the "kinetic" scale. Once the numerical shock layer is extended to a few grid points, inside the shock layer the kinetic scale mechanism requires that the cell size and time step are on the same order as the particle mean free path and collision time. This mechanism is intrinsically rooted in the kinetic flux vector splitting (KFVS) scheme [Ohwada *et al.* (2013)], which is exactly the kinetic scale part f_0 of the GKS method. So, for a valid shock capturing scheme, in the shock region (the thickness of the physical shock layer is on the order of the numerical cell size) a kinetic scale mechanism should be used in the flux modeling. In the continuous region (the flow structure is well-resolved), a hydrodynamic NS flux mechanism is needed. But, between these two limits, it is hard to say which governing equations should be solved, or what kind of flow mechanism should be adopted. The flow needs to stay in different regimes when the cell size (Δx) is used to represent a physical reality with structure thickness δ, under the conditions of $\Delta x \gg \delta$ or $\Delta x \ll \delta$. Therefore, even for the continuum flow computation, in a discretized space a multiple scale modeling is required in the construction of shock capturing scheme, which recovers different dynamics in different regions. The gas-kinetic scheme provides such a way to construct this kind of multiscale and multiphysics method.

The CFD algorithm development follows a path from simple to complicated PDEs. Since the NS equations seem to be the most complete governing equations for the continuum flow, their numerical algorithms become an assembling of numerical discretization of many simple parts of the original PDEs. For the compressible inviscid flow, due to its hyperbolicity, the linear advection equation $U_t + CU_x = 0$ becomes a stereo-type equation for compressible flow algorithm development. As pointed out in [Roe (1986)], it seems that the development of algorithm for the linear advection equation is the central task of CFD research. Historically, this linear advection equation was also called ICASE equation, because many scientists at the Institute of Computers Applications in Science and Engineering (ICASE) in NASA Langley took great efforts in 70s and 80s to develop algorithms for this equation. Many concepts, such as upwind, nonlinear limiters, artificial dissipation, TVD, ENO, and WENO [Harten (1983); Harten *et al.* (1987); Shu (1998); Wang (2011); Shyy (2006); LeVeque (2002)], had been developed and validated through the discretizing of this equation. Along the same line, the Riemann solver of the inviscid Euler equations comes to play a dominant role in CFD algorithm construction for gas dynamics. Many approximate Riemann solvers have been continuously developed, even at the current stage [Toro (2009)]. The shock instability at high speed flow computation appears to become a "black cloud" hanging over the clear

CFD sky. It appears that the foundation of the current CFD methodology may not be as solid as people think of.

Instead of going from simple to complete governing equations, the CFD principle should be based on the modeling of the physical process in a discretized space directly. This modeling needs to include different scale physics to represent a physical solution in the mesh size scale. In other words, the numerical flow evolution in $\Delta x \gg \delta$ and $\Delta x \ll \delta$, where δ is the physical structure thickness, should be controlled by different governing equations with different dynamics. The direct multiscale modeling in a discretized space is necessary because the final goal of CFD is to construct the "governing equations" in a numerical mesh size scale, and capture the corresponding physical evolution solution in such a scale. The coarse grained averaging due to mesh size limitation must have dynamic influence on the numerical flow evolution. This methodology is different from the current CFD approach. The typical CFD tries to separate the mesh size from the dynamics and tries to reduce the so-called truncation error as small as possible in the limiting case of $\Delta x \to 0$ and $\Delta t \to 0$. But, as analyzed before, the discontinuous and the smooth regions relative to the mesh size do need different governing equations. With the free choices of numerical resolution, the cell size needs actively participate in the dynamics of CFD and in the construction of grid-dependent fluid dynamic equations.

We believe that the CFD principle for the algorithm development is to directly model the flow evolution in the mesh size scale. Due to the variation of the mesh size relative to the physical flow structure, a continuum spectrum governing equations, or modeling mechanism, need to be constructed. The existing fluid dynamics equations, such as the Euler and NS with their own distinguishable modeling scales, are not enough to get a complete description of the "numerical" fluid evolution. The GKS flux covers a physical process from kinetic to the hydrodynamic scale flow evolution with distinguishable dissipative mechanism. At the current stage, the CFD research on the algorithm development seems saturated. Without establishing a fundamental principle, it will become more and more difficult to get any remarkable break-through. But, it also provides opportunities at current stage for CFD researchers to think and laid the foundation of CFD.

Chapter 5

Unified Gas Kinetic Scheme

In the last chapter, the gas-kinetic scheme (GKS) for the continuum flow computation is introduced. Due to the well-defined gas distribution function in the continuum flow regime, such as the Chapman-Enskog expansion for the NS solution, only macroscopic flow variables inside each control volume are updated in GKS, and the flux across a cell interface is based on a gas evolution model which covers flow physics from the kinetic to the hydrodynamic scales. Even in the continuum flow limit, this multiscale modeling is still needed in order to mimic the physical process in both continuous and discontinuous flow regions in terms of the mesh resolution. The different mechanism between GKS and Riemann solver may enhance the debate on the fundamental principle of CFD, such as the direct modeling or numerical PDEs. This understanding may assist the further development of shock capturing scheme.

In this chapter, we are going to present the unified gas-kinetic scheme (UGKS), which targets to simulate the flow all the way from rarefied to continuum one. Except at two end limits, such as the NS in the hydrodynamic regime and the Boltzmann in the kinetic regime, there is no well-defined governing equation for the description of flow physics in the scale between them. But, this doesn't prevent us from simulating the flow dynamics in the transition regime through the direct modeling. Due to the complexity of the non-equilibrium gas distribution function in the rarefied flow regime, besides the update of macroscopic flow variables, the time evolution of a gas distribution function has to be followed as well. But, the physical mechanism for the particle transport across a cell interface is the same in both GKS and UGKS, except different initial condition for the distribution function at the beginning of each time step. The goal of the UGKS is to capture a continuous flow physics from the kinetic to the hydrodynamic scale, which has not been fully explored before.

5.1 Introduction

With discretized particle velocity space, a multi-scale unified gas-kinetic scheme (UGKS) for entire Knudsen number flow is presented in this chapter. As an extension of the GKS in the previous chapter, where only conservative flow variables are updated for the continuum flow simulations, the UGKS updates both macroscopic conservative flow variables and the microscopic gas distribution function within a time step, see Eqs. (3.3) and (3.4). Instead of reconstructing the gas distribution function in GKS using the Chapman-Enskog expansion, due to the peculiarity of gas distribution in non-equilibrium regime, the update of the distribution function becomes necessary. In comparison with many existing kinetic solvers for the Boltzmann equation, the distinguishable feature of the UGKS is the coupling of the particle transport and collision in the evaluation of a gas distribution function at a cell interface for the flux calculation. Due to the wide variation of the ratio of the particle collision time over the numerical time step, or the particle mean free path over the cell size, the flow evolution mechanism in different regimes can be recovered in the flux modeling of UGKS. Therefore, the UGKS has no difficulty to get accurate Navier-Stokes (NS) solutions in the continuum regime, where intensive particle collisions take place within a time step, and the rarefied flow solution, even the free molecular one, where the non-equilibrium states can be preserved due to inadequate particle collisions. The un-splitting treatment of the particle transport and collision in the evolution model provides a smooth transition from the kinetic to hydrodynamic dynamics, and the proper choice at each interface depends on the ratio of the time step to the local particle collision time. For the methods without coupling the particle transport and collision in the flux function, the modeling mechanism remains in the kinetic scale. Therefore, the time step has to be less than the particle collision time. This constraint is used in many Boltzmann solvers and the Direct Simulation Monte Carlo (DSMC) method. In the continuum flow regime, especially in the high Reynolds number flow, the above constraint makes these operator splitting methods be extremely expensive, if not impossible.

The general formulation for the gas evolution at a cell interface in the unified scheme is still modeled by the integral solution of kinetic models, i.e. Eq. (4.2), which is the same equation used in GKS. In order to update both gas distribution function and macroscopic flow variables, the keys in UGKS are to obtain a time-dependent gas distribution function at a cell interface to evaluate the fluxes, and to model particle collisions inside each cell. As emphasized before, the Boltzmann equation itself has its intrinsic modeling scale, which is the particle mean free path and mean collision time. However, the cell size and time step used in any numerical scheme are basically free parameters, which could be much larger than the modeling scale of the kinetic equation. For example, for a flight simulation from the ground to $100km$ altitude, with a limited total number of mesh points, such as $10^3 \times 10^3 \times 10^3$ in the physical space, the ratio of the mesh size to the local particle mean free path varies significantly at different altitude and different location around the space vehicle. As a result, different flow physics needs to be identified during the whole flight passage. The UGKS is targeting to do it.

In comparison with GKS for the continuum flow computation, the construction of UGKS is straightforward and requires less knowledge of gas kinetic theory. In GKS, only macroscopic flow variables are updated, where the kinetic theory, such as the Chapman-Enskog expansion, is needed for the construction of the gas distribution function f_0 at the beginning of each time step. However, for the UGKS, both the gas distribution function and macroscopic flow variables will be followed. As a result, the updated gas distribution function can be used directly in UGKS as the initial condition f_0 in the next time step. Therefore, the kinetic theory, such as the Chapman-Enskog expansion, is not needed at all in the construction of UGKS. The Chapman-Enskog NS gas distribution function in the continuum flow regime will be recovered automatically from the competition between particle transport and collision in this regime, where the time step can be much larger than the local particle collision time. In the rarefied flow region, a highly non-equilibrium gas distribution function will be obtained due to inadequate particle collisions.

The UGKS presents a general framework for the study of transport phenomena. It depends heavily on the modeling of the physical process within a time step. Even though in the following presentation the Shakhov model is used to construct the gas evolution, the full Boltzmann collision term can also be adopted to do the modeling, but only in a subset of a time step when $t \leq \tau$. When $\Delta t \geq \tau$, the evolution solution from the full Boltzmann collision term and model equations is indistinguishable. Not only to the gas dynamics, the UGKS framework can be applied to radiative and neutron transport as well due to its multiscale modeling mechanism.

5.2 Unified Gas Kinetic Scheme in One-dimensional Space

In this section, the unified scheme in one-dimension will be presented. The key of the UGKS is to construct a cell interface gas evolution, where different kinetic equations, such as BGK, ES-BGK, and Shakhov, can be used. The UGKS formulation for all these model equations are similar. Here we present the scheme from the Shakhov model. The UGKS for the full Boltzmann collision term will be presented before the summary section of this chapter.

In one dimensional case, the BGK-Shakhov model [Shakhov (1968)] is

$$\frac{\partial f}{\partial t} + u\frac{\partial f}{\partial x} = \frac{f^+ - f}{\tau}, \tag{5.1}$$

where f is the distribution function, u is particle velocity, $\tau = \mu/p$ is particle collision time, μ is the dynamic viscosity coefficient, p is the pressure, and f^+ is the modified equilibrium distribution function, see Chapter 2 for the model construction.

The modified equilibrium distribution is given by

$$f^+ = g\left[1 + (1 - \mathrm{Pr})\mathbf{c}\cdot\mathbf{q}\left(\frac{c^2}{RT} - 5\right)\Big/(5pRT)\right] = g + g^+, \tag{5.2}$$

where g is the Maxwellian distribution, Pr is the Prandtl number, $\mathbf{c}$ is the peculiar velocity, $\mathbf{q}$ is heat flux, R is gas constant, and T is the temperature.

The Maxwellian distribution in 1D case is

$$g = \rho\left(\frac{\lambda}{\pi}\right)^{\frac{K+1}{2}} e^{-\lambda((u-U)^2+\xi^2)},$$

where ρ is density, $\lambda = m/2kT$, m is molecule mass, k is Boltzmann constant, U is the macroscopic velocity, K is the number of internal degree of freedom, and $\xi^2 = \xi_1^2 + \xi_2^2 ... + \xi_K^2$. A monatomic gas in 1D has $K = 2$ to account for the particle random motion in the y and z directions, such as $\xi^2 = v^2 + w^2$. The relation between K and the ratio of specific heats is $\gamma = (K + 3)/(K + 1)$.

The dynamic viscosity coefficient can be calculated from the Sutherland's law. In most time, in order to have the same viscosity coefficient as that in the DSMC method, the viscosity for the hard-sphere (HS) or variable hard-sphere model (VHS) models are used,

$$\mu = \mu_{ref}\left(\frac{T}{T_{ref}}\right)^{\omega}, \tag{5.3}$$

where μ_{ref} is the reference viscosity coefficient, T_{ref} is the reference temperature, and ω is the index related to HS or VHS models. The collision term meets the requirement of conservative constraint or compatibility condition

$$\int (f^+ - f)\psi d\Xi = 0, \tag{5.4}$$

where $\psi = (1, u, 1/2(u^2 + \xi^2))^T$ is the collision invariants and $d\Xi = dud\xi$ with $d\xi = d\xi_1 d\xi_2$ for a monatomic gas.

The macroscopic conservative variables can be calculated via

$$\mathbf{W} = \begin{pmatrix} \rho \\ \rho U \\ \rho E \end{pmatrix} = \int \psi f d\Xi,$$

$$p = \frac{1}{3}\int [(u-U)^2 + \xi^2] f d\Xi,$$

$$q = \frac{1}{2}\int (u-U)[(u-U)^2 + \xi^2] f d\Xi,$$

where ρE is total energy, p is the pressure, and q is the heat flux.

The flux evaluation of the unified scheme is based on the the time dependent solution f at a cell interface. Same as the GKS, the integral solution of the BGK-Shakhov model can be constructed by the method of characteristics [Kogan (1969); Xu (1993); Prendergast and Xu (1993)],

$$\begin{aligned} f(x, t, u, \xi) = & \frac{1}{\tau}\int_{t^n}^{t} f^{+}(x', t', u, \xi) e^{-(t-t')/\tau} dt' \\ & + e^{-(t-t^n)/\tau} f_0^n(x - u(t - t^n), t^n, u, \xi), \end{aligned} \tag{5.5}$$

where $x' = x - u(t - t')$ is the particle trajectory and f_0^n is the initial gas distribution function at t^n. The unified scheme will reconstruct the initial condition f_0^n in space and model the "equilibrium" distribution f^+ in space and time in order to fully determine the above solution. In comparison with the full Boltzmann equation, due to the inclusion of explicit equilibrium state the BGK-type models provide a clear path about the transition from the kinetic to hydrodynamic solution. The unified scheme is targeting to simulate both continuum and rarefied flow. In the continuum limit, the BGK-type model equations present an explicit equilibrium state, which can be directly used to recover the NS solutions. In the transition regime, the difference between the kinetic model solution and the full Boltzmann solution may not be as large as people think of, especially for the low speed miscroflows, where the solutions are not sensitive to the inter-molecular potential and the specific molecular scattering rules [Wu (2013)], but to the particle collision rate and the collision with the solid wall. Also, when the mesh size and time step are a few times of the particle mean free path and collision time, the dynamics in the mesh size scale will not be sensitive to the full Boltzmann collision term, but to the overall evolution mechanism which is beyond the kinetic scale modeling.

Numerical algorithm in 1-D

Different from the GKS in the previous chapter for the continuum flow computation, besides the discretization in physical space, the particle velocity space is also discretized in UGKS. The distribution function is represented by the value at discrete particle velocity, which is supposed to recover the whole velocity space. The moments of the non-equilibrium distribution function are calculated through numerical integration, and the moments of an equilibrium distribution can be calculated using analytical integration, see Appendix C for moments evaluation of a Maxwellian. The discretization of the velocity space is determined by the choice of numerical integration method, see Chapter 3.

In the update of the gas distribution function in UGKS, when trapezoidal rule is used for the approximation of collision term, Eq. (3.3) becomes,

$$\begin{aligned} f_{i,k}^{n+1} &= f_{i,k}^{n} + \frac{1}{\Delta x}\int_{t^n}^{t^{n+1}} u_k(f_{i-1/2,k} - f_{i+1/2,k})dt \\ &+\frac{\Delta t}{2}\left(\frac{f_{i,k}^{+(n+1)} - f_{i,k}^{n+1}}{\tau^{n+1}} + \frac{f_{i,k}^{+(n)} - f_{i,k}^{n}}{\tau^{n}}\right), \end{aligned} \tag{5.6}$$

where $f_{i,k}^n$ and $f_{i,k}^{n+1}$ are cell averaged distribution function in the i-th cell and k-th discrete particle velocity u_k at time $t = t^n$ and $t = t^{n+1}$ respectively, Δx is the cell size and Δt is the time step. Here $f_{i-1/2,k}$ and $f_{i+1/2,k}$ are the gas distribution functions at the cell interfaces which are based on the solution in Eq. (5.5). In the collision part, $f_{i,k}^{+(n)}$ and $f_{i,k}^{+(n+1)}$ are modified equilibrium states, and τ^n and τ^{n+1} are particle collision times.

Multiplying the conservative moments to Eq. (5.6) and making integration over the velocity space, the evolution equation for the conservative variables becomes

$$\mathbf{W}_i^{n+1} = \mathbf{W}_i^n + \frac{1}{\Delta x}(\mathbf{F}_{i-1/2} - \mathbf{F}_{i+1/2}), \tag{5.7}$$

where $\mathbf{F}_{i+1/2} = \displaystyle\int_{t^n}^{t^{n+1}} u\psi f_{i+1/2} dt d\Xi$ is the flux transport within a time step across a cell interface.

In order to update the distribution function in Eq. (5.6), three unknowns need to be addressed: the interface gas distribution function $f_{i+1/2}$, the modified equilibrium distribution $f^{+(n+1)}$ and collision time τ^{n+1} at the next time level. The function $f_{i+1/2}$ is calculated by using the integral solution Eq. (5.5) at the cell interface. Since $f^{+(n+1)}$ and τ^{n+1} have one-to-one correspondence with the macroscopic variables, which can be obtained after the updating of the conservative flow variables through Eq. (5.7).

In order to save computational cost, for the relaxation kinetic models, the dependence of distribution functions on the internal degree of freedom ξ can be removed. The reduced distribution functions [Chu (1965)] are used in computation, which are defined by

$$h = \int_{-\infty}^{\infty} f d\xi, \quad b = \int_{-\infty}^{\infty} \xi^2 f d\xi,$$

and the reduced equilibrium states become

$$h^+ = H + H^+, \quad b^+ = B + B^+,$$

where the corresponding reduced Maxwellian distributions go to

$$H = \int_{-\infty}^{\infty} g d\xi = \rho \left(\frac{\lambda}{\pi}\right)^{1/2} e^{-\lambda(u-U)^2}, \quad B = \int_{-\infty}^{\infty} \xi^2 g d\xi = \frac{K}{2\lambda} H, \tag{5.8}$$

and the reduced additional terms from g^+ are

$$\begin{aligned} H^+ &= \int_{-\infty}^{\infty} g^+ d\xi = \frac{4(1-\mathrm{Pr})\lambda^2}{5\rho}(u-U)q(2\lambda(u-U)^2 + K - 5)H, \\ B^+ &= \int_{-\infty}^{\infty} \xi^2 g^+ d\xi = \frac{4(1-\mathrm{Pr})\lambda^2}{5\rho}(u-U)q(2\lambda(u-U)^2 + K - 3)B. \end{aligned} \tag{5.9}$$

As a result, both the integral solution (5.5) and the numerical scheme (5.6) need to be integrated in terms of $d\xi$ and $\xi^2 d\xi$ as well for the updates of h and b. The interface integral solutions for h and b can be directly obtained from

$$\begin{aligned} \frac{\partial h}{\partial t} + u\frac{\partial h}{\partial x} &= \frac{h^+ - h}{\tau}, \\ \frac{\partial b}{\partial t} + u\frac{\partial b}{\partial x} &= \frac{b^+ - b}{\tau}, \end{aligned} \tag{5.10}$$

which have the same integral solutions as that in Eq. (5.5). The detailed interface flux and source term evaluation for the update of h and b inside each control volume will be presented next.

With the definition of h and b, the expressions for the macroscopic variables become

$$\begin{aligned} \rho &= \int h \mathrm{du} = \sum \alpha_k h_k, \\ \rho U &= \int h u \mathrm{du} = \sum \alpha_k h_k u_k, \\ \rho E &= \frac{1}{2}\left(\int h u^2 \mathrm{du} + \int b \mathrm{du}\right) = \frac{1}{2}\left(\sum \alpha_k h_k u_k^2 + \sum \alpha_k b_k\right), \end{aligned} \tag{5.11}$$

and

$$(K+1)p = \int (u-U)^2 h\mathrm{du} + \int b\mathrm{du} = \sum \alpha_k (u_k - U)^2 h_k + \sum \alpha_k b_k,$$

$$q = \frac{1}{2}\left[\int (u-U)(u-U)^2 h\mathrm{du} + \int (u-U) b\mathrm{du}\right]$$
$$= \frac{1}{2}\left[\sum \alpha_k (u_k - U)(u_k - U)^2 h_k + \sum \alpha_k (u_k - U) b_k\right], \quad (5.12)$$

where α_k is the weight of the numerical integration at the k-th particle velocity. The summation is over all discrete particle speed.

In UGKS, the time step is determined by the CFL condition

$$\Delta t = \mathrm{CFL}\frac{\Delta x}{u_{max}}$$

where CFL is the CFL number, $u_{max} = \max(|u_k|)$ is the largest discretized particle velocity.

The general framework for interface flux evaluation

With the notation of a cell interface $x_{i+1/2} = 0$ at $t^n = 0$, the time-dependent interface distribution function in 1D becomes,

$$f(0,t,u_k,\xi) = \frac{1}{\tau}\int_0^t f^+(x',t',u_k,\xi)e^{-(t-t')/\tau}dt' + e^{-t/\tau}f_0(-u_k t, 0, u_k, \xi). \quad (5.13)$$

When taking moments of $d\xi$ and $\xi^2 d\xi$ to the above equation, the corresponding reduced distribution functions can be obtained. Therefore, in the following we will present first the unified scheme for the update of the distribution function. The corresponding reduced distribution functions can be easily obtained after the determination of the full solution $f(0,t,u_k,\xi)$.

For the unified scheme, the initial cell averaged gas distribution function is known, i.e., $f_{i,k}^n = f(x_i, u_k, t^n)$. To the second-order accuracy, the initial gas distribution function f_0 can be reconstructed as

$$f_0(x,0,u_k,\xi) = \begin{cases} f_{i+1/2,k}^L + \sigma_{i,k} x, & x \le 0, \\ f_{i+1/2,k}^R + \sigma_{i+1,k} x, & x > 0, \end{cases} \quad (5.14)$$

where $f_{i+1/2,k}^L, f_{i+1/2,k}^R$ are the reconstructed initial distribution functions at the left and right hand sides of a cell interface. In the reconstruction, the van Leer limiter is used. For example, the slope of h at the i-th cell and k-th particle velocity is

$$\sigma_{i,k}^h = (\mathrm{sign}(s_1) + \mathrm{sign}(s_2))\frac{|s_1||s_2|}{|s_1| + |s_2|},$$

where $s_1 = (h_{i,k} - h_{i-1,k})/(x_i - x_{i-1})$, $s_2 = (h_{i+1,k} - h_{i,k})/(x_{i+1} - x_i)$. The slope of b is calculated in a similar way. So, different from the GKS in the last chapter, the Chapman-Enskog theory is not needed in the construction of f_0 in UGKS.

The equilibrium distribution function around a cell interface, which is used in the integral part of the solution (5.13), is constructed in the same way as that in the GKS. Through a Taylor expansion, the equilibrium state can be approximated locally as

$$g(x,t,u,\xi) = g_0[1 + (1 - H[x])a^L x + H[x]a^R x + At], \tag{5.15}$$

where g_0 is the Maxwellian distribution at $x = 0, t = 0$ and $H[x]$ is the Heaviside function. Here a^L, a^R and A are from the Taylor expansion of a Maxwellian, which depend on the particle velocity through the following,

$$a^{L,R} = a_1^{L,R} + a_2^{L,R}u + a_3^{L,R}\frac{1}{2}(u^2 + \xi^2),$$

and

$$A = A_1 + A_2 u + A_3\frac{1}{2}(u^2 + \xi^2),$$

where $a_1, a_2, ..., A_3$ are local constants. Since there is one to one correspondence between the equilibrium state and the macroscopic flow variables, based on macroscopic flow variables and their slopes all above coefficients can be fully determined. The procedures for their determination are identical to those in the GKS, which are repeated here again.

From the initial gas distribution function f_0 at the cell interface, the left and right moving particles across the interface can be explicitly evaluated. These opposite moving particles will presumably take particle collision, and the corresponding equilibrium state at a cell interface can be obtained based on the conservation. Note that this equilibrium is a state, where the real gas will evolve to, but not necessarily get to it. Therefore, based on the compatibility condition, the equilibrium state g_0 and the corresponding macroscopic conservative variables $\mathbf{W}_0$ at the cell interface ($x = 0, t = 0$) can be determined as

$$\int (f^+ - f)|_{x=0,t=0}\psi d\Xi = 0,$$

which gives

$$\mathbf{W}_0 = \int g_0\psi d\Xi = \int f_0\psi d\Xi = \int_{u_k>0} f_{i+1/2,k}^L \psi d\Xi + \int_{u_k<0} f_{i+1/2,k}^R \psi d\Xi. \tag{5.16}$$

After the determination of the equilibrium state at the cell interface, its spatial slopes a^L, a^R of g_0 can be obtained from the slope of conservative variables on both sides of a cell interface,

$$\left(\frac{\partial \mathbf{W}}{\partial x}\right)^L = \int a^L g_0 \psi d\Xi, \quad \left(\frac{\partial \mathbf{W}}{\partial x}\right)^R = \int a^R g_0 \psi d\Xi. \tag{5.17}$$

Both a^L and a^R in the above equations can be obtained from

$$\frac{1}{\rho_0}\left(\frac{\partial \mathbf{W}}{\partial x}\right) = M\mathbf{a}, \tag{5.18}$$

where $M_{\alpha,\beta} = (1/\rho_0)\int g_0\psi_\alpha\psi_\beta d\Xi$ and $\mathbf{a} = (a_1, a_2, a_3)^T$. The solution $\mathbf{a}$ can be obtained by using M^{-1}, see appendix C, or directly be given later.

The time derivative A of g_0 is related to the temporal variation of conservative variables,

$$\frac{\partial \mathbf{W}}{\partial t} = \int A g_0 \psi d\Xi. \tag{5.19}$$

The time derivative of $\mathbf{W}$ can be calculated via the derivatives of the compatibility condition,

$$\frac{\mathrm{d}}{\mathrm{d}t}\int (f^+ - f)\psi d\Xi\bigg|_{x=0,t=0} = 0,$$

which gives

$$-\int \left(a^L H[u] + a^R(1 - H[u])\right) u g_0 \psi d\Xi = \frac{\partial \mathbf{W}}{\partial t} = \int A g_0 \psi d\Xi, \tag{5.20}$$

from which the coefficients $\mathbf{A} = (A_1, A_2, A_3)^T$ can be obtained from the following equations,

$$\frac{1}{\rho_0}\frac{\partial \mathbf{W}}{\partial t} = M\mathbf{A}.$$

The evaluation of $\mathbf{A}$ in the second order GKS is through the averaging of the time derivative over a whole time step. Here we apply the compatibility condition initially. In the well resolved region, both methods for $\mathbf{A}$ are equivalent. In the un-resolved region, f_0 will play a dominant role, and the difference in the evaluation of $\mathbf{A}$ in the equilibrium part will have marginal effect.

Inserting Eq. (5.14) and Eq. (5.15) into Eq. (5.13), one obtains

$$\begin{aligned} f(0,t,u_k,\xi) &= (1 - e^{-t/\tau})(g_0 + g^+) \\ &\quad + (\tau(-1 + e^{-t/\tau}) + te^{-t/\tau})(a^L H[u_k] + a^R(1 - H[u_k]))u_k g_0 \\ &\quad + \tau(t/\tau - 1 + e^{-t/\tau})A g_0 \\ &\quad + e^{-t/\tau}\Big[(f^L_{i+1/2,k} - u_k t\sigma_{i,k})H[u_k] \\ &\quad + (f^R_{i+1/2,k} - u_k t\sigma_{i+1,k})(1 - H[u_k])\Big] \\ &= \tilde{g}_{i+1/2,k} + \tilde{f}_{i+1/2,k}, \end{aligned} \tag{5.21}$$

where $\tilde{g}_{i+1/2,k}$ is the first three terms related to equilibrium distribution, $\tilde{f}_{i+1/2,k}$ is the last two terms from the initial non-equilibrium distribution. In the following, the detailed formulation for the evaluation of each term in UGKS will be presented.

The detailed formulation

Reconstruction of initial distribution f_0

Since the unified scheme uses reduced distribution function h and b, the reconstruction of h and b inside each cell can be done similarly. The initial condition f_0 is fully determined from the reconstructed h and b. Dynamically, only the particles passing through the cell interface will be counted, such as $h^L_{i+1/2,k}$ for $u_k \geq 0$ and $h^R_{i+1/2,k}$ for $u_k < 0$ (see Eq. (5.21)). There is no need to store other moving particles. After reconstruction of h inside each cell, at the cell interface the distribution has the form

$$h_{i+1/2,k} = \begin{cases} h_{i,k} + (x_{i+1/2} - x_i)\sigma^h_{i,k}, & u_k \geq 0, \\ h_{i+1,k} - (x_{i+1} - x_{i+1/2})\sigma^h_{i+1,k}, & u_k < 0, \end{cases}$$

or

$$\begin{aligned} h_{i+1/2,k} = & \left(h_{i,k} + (x_{i+1/2} - x_i)\sigma^h_{i,k}\right) H[u_k] \\ & + \left(h_{i+1,k} - (x_{i+1} - x_{i+1/2})\sigma^h_{i+1,k}\right)(1 - H[u_k]). \end{aligned}$$

Similar construction can be done for b term.

Calculation of macroscopic variables $\mathbf{W}_0$ on the cell interface

After determining $h_k = h_{i+1/2,k}$ and $b_k = b_{i+1/2,k}$ at a cell interface, the macroscopic variables $\mathbf{W}_0$ from the colliding particles is calculated from Eq. (5.11). Then, the exact Maxwellian distribution corresponding to $\mathbf{W}_0$ can be obtained. The primary variables in the equilibrium state g_0 are determined from the following relations

$$\rho_0 = \rho_0, \quad U_0 = \frac{\rho_0 U_0}{\rho_0}, \quad \lambda_0 = \frac{(K+1)\rho_0}{4\left(\rho_0 E_0 - \frac{1}{2}\rho U_0^2\right)}.$$

The heat flux is calculated by Eq. (5.12) with the input of $h_k = h_{i+1/2,k}, b_k = b_{i+1/2,k}$ and $U = U_0$.

Calculation of the slopes a^L, a^R in the expansion of an equilibrium state

After determining $\mathbf{W}_0$ at the cell interface, the slopes of macroscopic flow variables on the left and right hand sides of a cell interface are given by

$$\left(\frac{\partial \mathbf{W}}{\partial x}\right)^L \approx \frac{\mathbf{W}_0 - \mathbf{W}_i}{x_{i+1/2} - x_i}, \quad \left(\frac{\partial \mathbf{W}}{\partial x}\right)^R \approx \frac{\mathbf{W}_{i+1} - \mathbf{W}_0}{x_{i+1} - x_{i+1/2}},$$

and the three components in a^L, a^R can be calculated from the direct Maxwellian expansion,

$$\begin{aligned}
a_3 &= \frac{4\lambda_0^2}{(K+1)\rho_0}\left[2\frac{\partial(\rho E)}{\partial x} + \left(U_0^2 - \frac{K+1}{2\lambda_0}\right)\frac{\partial \rho}{\partial x} - 2U_0\frac{\partial(\rho U)}{\partial x}\right], \\
a_2 &= \frac{2\lambda_0}{\rho_0}\left(\frac{\partial(\rho U)}{\partial x} - U_0\frac{\partial \rho}{\partial x}\right) - U_0 a_3, \\
a_1 &= \frac{1}{\rho_0}\frac{\partial \rho}{\partial x} - U_0 a_2 - \frac{1}{2}\left(U_0^2 + \frac{K+1}{2\lambda_0}\right) a_3,
\end{aligned} \tag{5.22}$$

or the method in the Appendix C by obtaining the inverse of matrix M of Eq. (5.18).

Calculation of $\partial \mathbf{W}/\partial t$ and time derivative $\mathbf{A}$ of the equilibrium state

From Eq. (5.20), the time derivative of $\mathbf{W}$ is calculated from the spatial derivatives of a Maxwellian,

$$\frac{\partial \mathbf{W}}{\partial t} = -\rho_0\left(< a^L u\psi >_{>0} + < a^R u\psi >_{<0}\right) = \rho_0 M \mathbf{A}, \tag{5.23}$$

where $< ... >$ is the moments of equilibrium state g_0. The moments evaluation and the inverse of M are given in Appendix C. Or, $\mathbf{A}$ can be calculated by using the same way as a^L, a^R in Eq. (5.22), where the spatial derivative is replaced by time derivative in Eq. (5.23).

Some time integrals used in Eq. (5.21) for the flux transport evaluation are listed below

$$
\begin{aligned}
T_4 &= \int_0^{\Delta t} e^{-t/\tau} dt = \tau(1 - e^{-\Delta t/\tau}), \\
T_5 &= \int_0^{\Delta t} t e^{-t/\tau} dt = -\tau \Delta t e^{-\Delta t/\tau} + \tau T_4, \\
T_1 &= \int_0^{\Delta t} (1 - e^{-t/\tau}) dt = \Delta t - T_4, \\
T_2 &= \int_0^{\Delta t} (\tau(-1 + e^{-t/\tau}) + t e^{-t/\tau}) dt = -\tau T_1 + T_5, \\
T_3 &= \int_0^{\Delta t} \tau(t/\tau - 1 + e^{-t/\tau}) dt = \frac{1}{2}\Delta t^2 - \tau T_1.
\end{aligned}
\tag{5.24}
$$

Flux calculation for macroscopic flow variable update from integration of g_0

There is an analytic solution for the integral evaluation of

$$\int_0^{\Delta t} \int \tilde{g}_{i+1/2} u \psi d\Xi dt.$$

However, the integration of g^+ for the additional heat flux correction term is complicated. The terms related to g_0 can be integrated analytically, such as

$$\mathbf{F}_{g_0} = T_1 \rho_0 < u\psi > + T_2 \rho_0 \left(< a^L u^2 \psi >_{>0} + < a^R u^2 \psi >_{<0}\right) + T_3 \rho_0 < Au\psi >,$$

and the terms related to g^+ can be evaluated numerically.

Flux calculation for macroscopic flow variable update from g^+ and f_0

First, at discrete particle velocity point u_k, evaluate H_k, B_k corresponding to g_0 by Eq. (5.8),

$$H_k = \rho_0 \left(\frac{\lambda_0}{\pi}\right)^{1/2} e^{-\lambda_0 (u_k - U_0)^2}, \quad B_k = \frac{K}{2\lambda_0} H_k,$$

and then evaluate H_k^+, B_k^+ corresponding to g^+ by Eq. (5.9)

$$
\begin{aligned}
H_k^+ &= \frac{4(1 - \mathrm{Pr})\lambda_0^2}{5\rho_0} (u_k - U_0) q \left(2\lambda_0 (u_k - U_0)^2 + K - 5\right) H_k, \\
B_k^+ &= \frac{4(1 - \mathrm{Pr})\lambda_0^2}{5\rho_0} (u_k - U_0) q \left(2\lambda_0 (u_k - U_0)^2 + K - 3\right) B_k.
\end{aligned}
\tag{5.25}
$$

The fluxes of conservative variables related to g^+ become

$$\mathbf{F}_{g^+} = T_1 \begin{pmatrix} \sum \alpha_k u_k H_k^+ \\ \sum \alpha_k u_k^2 H_k^+ \\ \frac{1}{2}\left(\sum \alpha_k u_k^3 H_k^+ + \sum \alpha_k u_k B_k^+\right) \end{pmatrix}.$$

The fluxes of conservative variables related to f_0 are

$$\begin{aligned} \mathbf{F}_{f_0} = T_4 & \begin{pmatrix} \sum \alpha_k u_k h_{i+1/2,k} \\ \sum \alpha_k u_k^2 h_{i+1/2,k} \\ \frac{1}{2}\left(\sum \alpha_k u_k^3 h_{i+1/2,k} + \sum \alpha_k u_k b_{i+1/2,k}\right) \end{pmatrix} \\ -T_5 & \begin{pmatrix} \sum \alpha_k u_k^2 \sigma_{i+1/2,k}^h \\ \sum \alpha_k u_k^3 \sigma_{i+1/2,k}^h \\ \frac{1}{2}\left(\sum \alpha_k u_k^4 \sigma_{i+1/2,k}^h + \sum \alpha_k u_k^2 \sigma_{i+1/2,k}^b\right) \end{pmatrix}. \end{aligned}$$

The total flux for the conservative variable update is

$$\mathbf{F}_{i+1/2} = \int_0^{\Delta t} \int f_{i+1/2} u \psi d\Xi dt = \mathbf{F}_{g_0} + \mathbf{F}_{g^+} + \mathbf{F}_{f_0},$$

from which the conservative variables inside each cell can be obtained,

$$\mathbf{W}_i^{n+1} = \mathbf{W}_i^n + \frac{1}{\Delta x}(\mathbf{F}_{i-1/2} - \mathbf{F}_{i+1/2}).$$

Fluxes for the update of reduced gas distribution functions h and b

The flux transport across a cell interface for the update of the cell averaged reduced distribution function h at particle velocity u_k is

$$\begin{aligned} f_{i+1/2,k}^h &= \int_0^{\Delta t} u_k \left(\int f_{i+1/2,k} d\xi \right) dt \\ &= T_1 u_k (H_k + H_k^+) \\ &\quad + T_2 u_k^2 \left(a_1^L H_k + a_2^L u_k H_k + \frac{1}{2} a_3^L (u_k^2 H_k + B_k) \right) H[u_k] \\ &\quad + T_2 u_k^2 \left(a_1^R H_k + a_2^R u_k H_k + \frac{1}{2} a_3^R (u_k^2 H_k + B_k) \right) (1 - H[u_k]) \\ &\quad + T_3 u_k \left(A_1 H_k + A_2 u_k H_k + \frac{1}{2} A_3 (u_k^2 H_k + B_k) \right) \\ &\quad + T_4 u_k h_{i+1/2,k} - T_5 u_k^2 \sigma_{i+1/2,k}^h. \end{aligned} \tag{5.26}$$

Similarly, the flux for the update of the cell averaged reduced distribution function b is

$$
\begin{aligned}
f^b_{i+1/2,k} &= \int_0^{\Delta t} u_k \left(\int f_{i+1/2,k} \xi^2 d\xi \right) dt \\
&= T_1 u_k (B_k + B_k^+) \\
&\quad + T_2 u_k^2 \left(a_1^L B_k + a_2^L u_k B_k + \frac{1}{2} a_3^L (u_k^2 B_k + < \xi^4 > H_k) \right) H[u_k] \\
&\quad + T_2 u_k^2 \left(a_1^R B_k + a_2^R u_k B_k + \frac{1}{2} a_3^R (u_k^2 B_k + < \xi^4 > H_k) \right) (1 - H[u_k]) \\
&\quad + T_3 u_k \left(A_1 B_k + A_2 u_k B_k + \frac{1}{2} A_3 (u_k^2 B_k + < \xi^4 > H_k) \right) \\
&\quad + T_4 u_k b_{i+1/2,k} - T_5 u_k^2 \sigma^b_{i+1/2,k}.
\end{aligned} \tag{5.27}
$$

Update of cell averaged h and b

The detailed formulations for updating h^{n+1} and b^{n+1} from Eq. (5.6) are the following,

$$
\begin{aligned}
h^{n+1}_{i,k} &= \left(1 + \frac{\Delta t}{2\tau^{n+1}}\right)^{-1} \left[h^n_{i,k} + \frac{1}{\Delta x}(f^h_{i-1/2,k} - f^h_{i+1/2,k}) \right. \\
&\quad \left. + \frac{\Delta t}{2} \left(\frac{h^{+(n+1)}_{i,k}}{\tau^{n+1}} + \frac{h^{+(n)}_{i,k} - h^n_{i,k}}{\tau^n} \right) \right] \\
b^{n+1}_{i,k} &= \left(1 + \frac{\Delta t}{2\tau^{n+1}}\right)^{-1} \left[b^n_{i,k} + \frac{1}{\Delta x}(f^b_{i-1/2,k} - f^b_{i+1/2,k}) \right. \\
&\quad \left. + \frac{\Delta t}{2} \left(\frac{b^{+(n+1)}_{i,k}}{\tau^{n+1}} + \frac{b^{+(n)}_{i,k} - b^n_{i,k}}{\tau^n} \right) \right].
\end{aligned}
$$

The above description presents the numerical procedure of UGKS in one dimensional case.

5.3 Unified Gas Kinetic Scheme in Two-dimensional Space

The basic procedures of UGKS in 2D are similar to the above formulation in 1D case. The main difference is the inclusion of discrete particle velocity v in the y-direction explicitly in the velocity space. With the account of the normal direction of a cell interface in a 2D domain, the mechanism of the flux transport across the cell interface and particle collision term discretization inside each control volume are identical. For an accurate computation of a 2D flow, the tangential derivatives of the flow variables have to be included in the scheme as well, the so-called multidimensional method.

In a 2D case, even with unstructured mesh in a physical space the flux has to be evaluated across each cell interface. The general kinetic formulation in a 2D domain with unstructured mesh is given in Appendix D. In UGKS, both macroscopic variables and the gas distribution functions will be updated. The particle velocity is decomposed into interface normal and tangential direction. Here, in order to present the unified scheme in a general coordinate framework, two coordinates systems, such as a local one with unit normal $\vec{e}_{x'}$ and tangential direction $\vec{e}_{y'}$ of a cell interface, and a global one with $\vec{e}_x$ and $\vec{e}_y$ in the $x - y$ coordinate system, will be introduced. The transformation of coordinate axes between the local and global ones are

$$\vec{e}_{x'} = \vec{e}_x(\vec{e}_{x'} \cdot \vec{e}_x) + \vec{e}_y(\vec{e}_{x'} \cdot \vec{e}_y) = \vec{e}_x \cos\alpha + \vec{e}_y \sin\alpha,$$

$$\vec{e}_{y'} = \vec{e}_x(\vec{e}_{y'} \cdot \vec{e}_x) + \vec{e}_y(\vec{e}_{y'} \cdot \vec{e}_y) = \vec{e}_x(-\sin\alpha) + \vec{e}_y \cos\alpha,$$

where α is the intersection angle between two coordinate axes. With the above coordinate transformation, the particle velocity (u, v) in the global x and y directions can be transformed into the local coordinate system with velocity (u', v') and

$$u' = u \cos\alpha + v \sin\alpha,$$

$$v' = u(-\sin\alpha) + v \cos\alpha.$$

For the discrete particle velocity (u_k, v_l), it becomes (u'_k, v'_l). Across a cell interface, the flux for particle transport, i.e., $u' f_{interface} du'dv'dw'$, is a scalar, which is the same in both local and global coordinate system. Therefore, once all flow variables around a cell interface are transformed into a local coordinate system $(\vec{e}_{x'}, \vec{e}_{y'})$, such as macroscopic velocity $(U, V) \to (U', V')$ and particle velocity $(u, v) \to (u', v')$, we can evaluate the gas evolution in the local coordinate system $(\vec{e}_{x'}, \vec{e}_{y'})$. Then, after obtaining the particle transport in the normal direction, the corresponding microscopic and macroscopic flow variable transport in a global coordinate system can be obtained as well, which are used to the update f and $\mathbf{W}$ in UGKS in a global coordinate system.

In the following, we are going to present a multidimensional unified scheme in a two-dimensional space. For simple presentation, we will consider the x and y directions as the normal and tangential directions of a local cell interface. The way to update the flow variables in a global one inside each control volume will be given later.

The two-dimensional gas-kinetic BGK-Shakhov equation in the normal-tangential coordinate system can be written as

$$f_t + uf_x + vf_y = \frac{f^+ - f}{\tau}, \tag{5.28}$$

where f is the gas distribution function and f^+ is the heat flux modified equilibrium state which is approached by f,

$$f^+ = g\left[1 + (1 - \mathrm{Pr})\mathbf{c}\cdot\mathbf{q}\left(\frac{c^2}{RT} - 5\right)\Big/(5pRT)\right] = g + g^+,$$

with random velocity $\mathbf{c} = \mathbf{u} - \mathbf{U}$ and the heat flux $\mathbf{q}$. Both f and f^+ are functions of space (x, y), time t, particle velocity (u, v) in x and y directions, and the random velocity w in z-direction. For a monatomic gas in 2D case, the equilibrium Maxwellian distribution is,

$$g = \rho\left(\frac{\lambda}{\pi}\right)^{\frac{3}{2}} e^{-\lambda((u-U)^2+(v-V)^2+w^2)},$$

where ρ is the density, (U, V) is the macroscopic velocity in the x and y directions. The relation between mass ρ, momentum $(\rho U, \rho V)$, and energy ρE densities with the distribution function f is

$$\begin{pmatrix} \rho \\ \rho U \\ \rho V \\ \rho E \end{pmatrix} = \int \psi_\alpha f d\Xi, \qquad \alpha = 1, 2, 3, 4, \tag{5.29}$$

where ψ_α is the component of the conservative moments

$$\boldsymbol{\psi} = (\psi_1, \psi_2, \psi_3, \psi_4)^T = \left(1, u, v, \frac{1}{2}(u^2 + v^2 + w^2)\right)^T,$$

and $d\Xi = dudvdw$ is the volume element in the phase space. Similar to the 1D case, reduced distribution functions with the particle velocity integration in the z-direction can be used for the development of the unified scheme. With the definition

$$h = \int_{-\infty}^{\infty} f dw, \quad b = \int_{-\infty}^{\infty} w^2 f dw,$$

and the reduced equilibrium state become

$$h^+ = H + H^+, \quad b^+ = B + B^+,$$

where the corresponding Maxwellians are

$$H = \rho\left(\frac{\lambda}{\pi}\right) e^{-\lambda[(u-U)^2+(v-V)^2]}, \quad B = \frac{1}{2\lambda}H,$$

and the corresponding terms in g^+ are

$$H^+ = H(1-\text{Pr})[(u-U)q_x+(v-V)q_y]\left(\frac{(u-U)^2+(v-V)^2}{RT}-4\right)\Big/(5pRT),$$

$$B^+ = B(1-\text{Pr})[(u-U)q_x+(v-V)q_y]\left(\frac{(u-U)^2+(v-V)^2}{RT}-2\right)\Big/(5pRT).$$

Based on the distribution function f, all macroscopic flow variables can be evaluated, such as the stress P_{ij} and heat fluxes q_i,

$$P_{ij} = \int (u_i - U_i)(u_j - U_j) f d\Xi = \int (u_i - U_i)(u_j - U_j) h du dv,$$

$$\begin{aligned} q_i &= \int \frac{1}{2}(u_i - U_i)\left((u-U)^2 + (v-V)^2 + w^2\right) f d\Xi \\ &= \int \frac{1}{2}(u_i - U_i)\left(((u-U)^2 + (v-V)^2)h + b\right) du dv. \end{aligned}$$

Since mass, momentum, and energy are conserved during particle collisions, f and g satisfy the conservation constraint,

$$\int (f^+ - f)\psi_\alpha d\Xi = 0, \qquad \alpha = 1,2,3,4, \tag{5.30}$$

at any point in space and time.

In the unified scheme, at the center of a cell interface $(x_{i+1/2}, y_j)$ the solution $f_{i+1/2,j,k,l}$ is constructed from an integral solution of the kinetic model (5.28) using the method of characteristics,

$$\begin{aligned} &f_{i+1/2,j,k,l}(x_{i+1/2}, y_j, t, u_k, v_l, w) \\ &\quad = \frac{1}{\tau}\int_{t^n}^{t^{n+1}} f^+(x', y', t', u_k, v_l, w) e^{-(t-t')/\tau} dt' \\ &\qquad + e^{-(t-t^n)/\tau} f^n_{0,k,l}(x_{i+1/2} - u_k(t-t^n), y_j - v_l(t-t^n), t^n, u_k, v_l, w), \end{aligned} \tag{5.31}$$

with $f^+ = g + g^+$. Here $(x' = x_{i+1/2} - u_k(t-t'), y' = y_j - v_l(t-t'))$ is the particle trajectory and $f^n_{0,k,l}$ is the initial gas distribution function f at time $t = t^n$ around the cell interface center $(x_{i+1/2}, y_j)$ at the particle velocity (u_k, v_l), i.e., $f^n_{0,k,l} = f^n_0(x, y, t^n, u_k, v_l, w)$. In order to fully determine the integral solution, the terms related to the initial distribution and equilibrium states have to be modeled, especially in the case with discontinuous initial data.

In the above equation, inside each control volume, $f^n_{0,k,l}$ is known at the beginning of each time step t^n. In the following, $t^n = 0$ and $(x_{i+1/2}, y_j) = (0, 0)$ will be used. A high-order reconstruction can be used to reconstruct its subcell resolution using TVD and ENO methods. If the solutions are well resolved, the discontinuous reconstruction will become a continuous one automatically. For example, around each cell interface $x_{i+1/2}$, at time step t^n the initial distribution function becomes,

$$f_0(x, y, 0, u_k, v_l, w) = f_{0,k,l}(x, y, 0)$$
$$= \begin{cases} f^L_{i+1/2,k,l} + \sigma_{i,k,l}x + \theta_{i,k,l}y, & x \le 0, \\ f^R_{i+1/2,k,l} + \sigma_{i+1,k,l}x + \theta_{i+1,k,l}y, & x > 0, \end{cases} \tag{5.32}$$

where nonlinear limiter is used to reconstruct $f^L_{i+1/2,k,l}$, $f^R_{i+1/2,k,l}$ and the corresponding slopes $\sigma_{i,k,l}, \sigma_{i+1,k,l}, \theta_{i,k,l}$, and $\theta_{i+1,k,l}$.

There is one-to-one correspondence between an equilibrium state and macroscopic flow variables. The equilibrium state in the integral term of Eq. (5.31) around a cell interface ($x_{i+1/2} = 0, y_j = 0, t = 0$) can be expanded as,

$$g = g_0 \left[1 + (1 - \mathrm{H}[x])\bar{a}^L x + \mathrm{H}[x]\bar{a}^R x + \bar{b}y + \bar{A}t\right]. \tag{5.33}$$

The dependence of $\bar{a}^L$, $\bar{a}^R$ and $\bar{A}$ on the particle velocity can be obtained from a Taylor expansion of a Maxwellian and have the following form,

$$\bar{a}^L = \bar{a}^L_1 + \bar{a}^L_2 u + \bar{a}^L_3 v + \bar{a}^L_4 \frac{1}{2}(u^2 + v^2 + w^2) = \bar{a}^L_\alpha \psi_\alpha,$$

$$\bar{a}^R = \bar{a}^R_1 + \bar{a}^R_2 u + \bar{a}^R_3 v + \bar{a}^R_4 \frac{1}{2}(u^2 + v^2 + w^2) = \bar{a}^R_\alpha \psi_\alpha,$$

$$\bar{b} = \bar{b}_1 + \bar{b}_2 u + \bar{b}_3 v + \bar{b}_4 \frac{1}{2}(u^2 + v^2 + w^2) = \bar{b}_\alpha \psi_\alpha,$$

$$\bar{A} = \bar{A}_1 + \bar{A}_2 u + \bar{A}_3 v + \bar{A}_4 \frac{1}{2}(u^2 + v^2 + w^2) = \bar{A}_\alpha \psi_\alpha,$$

where $\alpha = 1, 2, 3, 4$ and all coefficients $\bar{a}^L_1, \bar{a}^L_2, ..., \bar{A}_4$ are local constants.

The determination of g_0 depends on the local macroscopic values of ρ_0, U_0, V_0 and λ_0 in g_0, i.e.,

$$g_0 = \rho_0 \left(\frac{\lambda_0}{\pi}\right)^{\frac{3}{2}} e^{-\lambda_0((u-U_0)^2 + (v-V_0)^2 + w^2)},$$

which is determined from $\mathbf{W}_0$

$$\mathbf{W}_0 = \int g_0 \psi d\Xi = \sum \left(f^L_{i+1/2,k,l} H[u_k] + f^R_{i+1/2,k,l}(1 - H[u_k])\right)\psi, \tag{5.34}$$

where $\mathbf{W}_0 = (\rho_0, \rho_0 U_0, \rho_0 V_0, \rho_0 E_0)^T$ are the corresponding conservative macroscopic flow variables. Since $f^L_{i+1/2,k,l}$ and $f^R_{i+1/2,k,l}$ have been reconstructed as the initial distribution function f_0 around a cell interface, the above moments can be evaluated explicitly. Therefore, the conservative variables $\rho_0, \rho_0 U_0, \rho_0 V_0$, and $\rho_0 E_0$ at the cell interface can be obtained, from which g_0 is uniquely determined. In the equilibrium state, λ_0 in g_0 is given by

$$\lambda_0 = 3\rho_0 \Big/ \left(4(\rho_0 E_0 - \frac{1}{2}\rho_0 (U_0^2 + V_0^2)) \right).$$

Then, $\bar{a}^L$ and $\bar{a}^R$ of g in Eq. (5.33) can be obtained through the relation of

$$\frac{\bar{\mathbf{W}}_{j+1}(x_{j+1}) - \mathbf{W}_0}{\rho_0 \Delta x^+} = \frac{1}{\rho_0} \int \bar{a}^R g_0 \psi d\Xi = \bar{M}^0_{\alpha\beta} \begin{pmatrix} \bar{a}_1^R \\ \bar{a}_2^R \\ \bar{a}_3^R \\ \bar{a}_4^R \end{pmatrix} = \bar{M}^0_{\alpha\beta} \bar{a}^R_\beta, \quad (5.35)$$

and

$$\frac{\mathbf{W}_0 - \bar{\mathbf{W}}_j(x_j)}{\rho_0 \Delta x^-} = \frac{1}{\rho_0} \int \bar{a}^L g_0 \psi d\Xi = \bar{M}^0_{\alpha\beta} \begin{pmatrix} \bar{a}_1^L \\ \bar{a}_2^L \\ \bar{a}_3^L \\ \bar{a}_4^L \end{pmatrix} = \bar{M}^0_{\alpha\beta} \bar{a}^L_\beta, \quad (5.36)$$

where the matrix $\bar{M}^0_{\alpha\beta} = \int g_0 \psi_\alpha \psi_\beta d\Xi / \rho_0$ is known, and $\Delta x^+ = x_{i+1} - x_{i+1/2}$ and $\Delta x^- = x_{i+1/2} - x_i$ are the distances from the cell interface to cell centers. For the tangential derivative $\bar{b}$ in the equilibrium state g, it can be obtained from

$$\int \bar{b} \psi g_0 d\Xi = \sum \left(\theta_{i,k,l} H[u_k] + \theta_{i+1,k,l}(1 - H[u_k]) \right) \psi.$$

Then, $(\bar{a}_1^R, \bar{a}_2^R, \bar{a}_3^R, \bar{a}_4^R)^T$ $(\bar{a}_1^L, \bar{a}_2^L, \bar{a}_3^L, \bar{a}_4^L)^T$, and $(\bar{b}_1, \bar{b}_2, \bar{b}_3, \bar{b}_4)^T$ can be found by calculating M^{-1}, which is shown in the Appendix C. Or, these coefficients can be obtained directly from the Taylor expansion of a Maxwellian, such as

$$\frac{1}{g}\frac{\partial g}{\partial x} = a_1 + a_2 u + a_3 v + a_4 \frac{1}{2}(u^2 + v^2 + w^2),$$

which gives

$$\begin{aligned}
a_4 &= -2\lambda_x \\
a_3 &= 2\lambda V_x + 2V\lambda_x \\
a_2 &= 2\lambda U_x + 2U\lambda_x \\
a_1 &= \frac{\rho_x}{\rho} - 2\lambda(UU_x + VV_x) + \left(\frac{K+2}{2\lambda} - (U^2 + V^2) \right) \lambda_x,
\end{aligned}$$

where $K = 1$ for monatomic gas in 2D case, and the derivatives of ρ_x, U_x, V_x and λ_x can be obtained from the derivatives of macroscopic variables $\partial \mathbf{W}/\partial x$.

In order to evaluate the time evolution part $\bar{A}$ in the equilibrium state, we can apply the following condition

$$\frac{d}{dt}\int (g-f)\psi\Xi = 0,$$

at $(x=0, y=0, t=0)$ and get

$$\begin{aligned}\bar{M}^0_{\alpha\beta}\bar{A}_\beta &= \frac{1}{\rho_0}(\partial\rho/\partial t, \partial(\rho U)/\partial t, \partial(\rho V)/\partial t, \partial(\rho E)/\partial t)^T \\ &= -\frac{1}{\rho_0}\int \left[\left(\bar{a}^L \mathrm{H}[u] + \bar{a}^R(1-\mathrm{H}[u])\right)u + \bar{b}v\right] g_0\psi d\Xi. \quad (5.37)\end{aligned}$$

With the determination of equilibrium state and the heat flux at the cell interface, the additional term g^+ in the Shakhov model can be well-defined.

Up to this point, we have determined all parameters in the initial gas distribution function f_0 and the state f^+ in space and time locally. After substituting Eq. (5.32) and Eq. (5.33) into Eq. (5.31) and taking $(u = u_k, v = v_l)$ in $g_0, \bar{a}^L, \bar{a}^R, \bar{b}$, and $\bar{A}$, the gas distribution function $f(x_{i+1/2}, y_j, t, u_k, v_l, w)$ at the discretized particle velocity (u_k, v_l) can be expressed as

$$\begin{aligned}&f_{i+1/2,j,k,l}(x_{i+1/2}, y_j, t, u_k, v_l, w) \\ &\quad= (1-e^{-t/\tau})(g_0 + g^+) \\ &\qquad+\left(\tau(-1+e^{-t/\tau}) + te^{-t/\tau}\right)\left((\bar{a}^L \mathrm{H}[u_k] + \bar{a}^R(1-\mathrm{H}[u_k]))u_k + \bar{b}v_l\right) g_0 \\ &\qquad+\tau(t/\tau - 1 + e^{-t/\tau})\bar{A}g_0 \\ &\qquad+e^{-t/\tau}\Big((f^L_{i+1/2,k,l} - u_k t\sigma_{i,k,l} - v_l t\theta_{i,k,l})\mathrm{H}[u_k] \\ &\qquad+(f^R_{i+1/2,k,l} - u_k t\sigma_{i+1,k,l} - v_l t\theta_{i+1,k,l})(1-\mathrm{H}[u_k])\Big) \\ &\quad= \tilde{g}_{i+1/2,j,k,l} + \tilde{f}_{i+1/2,j,k,l}, \qquad (5.38)\end{aligned}$$

where $\tilde{g}_{i+1/2,j,k,l}$ is all terms related to the integration of the equilibrium state g and g^+, and $\tilde{f}_{i+1/2,j,k,l}$ is the terms from initial condition f_0. The collision time τ in the above distribution function is determined by $\tau = \mu(T_0)/p_0$, where T_0 is the temperature and p_0 is the pressure, and both can be evaluated from $\mathbf{W}_0$ at the cell interface.

With the evaluation of the gas distribution function in a local coordinate system $(\vec{e}_{x'}, \vec{e}_{y'})$, we can transfer it to the global coordinate system $(\vec{e}_x, \vec{e}_y)$. Since (u'_k, v'_l) and (u_k, v_l) refer to the same particle, but with different projected velocity components, such as $u'_k = u_k \cos\alpha + v_l \sin\alpha$, and $v'_l =$

$u_k(-\sin\alpha)+v_l\cos\alpha$, the number of particle transport across a cell interface will be the same for the particle velocity (u'_k, v'_l) and (u_k, v_l), which is used to update the gas distribution function $f^{n+1}(u_k, v_l)$ at a global velocity (u_k, v_l). Therefore, for the update of gas distribution function at particle velocity (u_k, v_l), the flux for the particle number transport is

$$\mathbf{f}_{k,l} \cdot \vec{e}_{x'} = \int_0^{\Delta t} \int u'_k f(u'_k, v'_l, w) dw dt, \tag{5.39}$$

which can be explicitly evaluated using the distribution function of a local system in Eq. (5.38).

For the unified scheme, besides updating the gas distribution function at discretized particle velocity, we also need update the conservative variables $\mathbf{W}^{n+1}$ inside each control volume in order to evaluate equilibrium state at the next level for the collision term. For the update of conservative variables inside each cell, the global coordinate system has to be used for the common definition of macroscopic momentum and energy. Therefore, with the particle transport across the cell interface $u'_k f(u'_k, v'_l)$, these particles will carry the mass, momentum, and energy in a global reference of frame, i.e., $(1, u_k, v_l, \frac{1}{2}(u_k^2 + v_l^2 + w^2))^T$. The fluxes across the cell interface for the update of conservative flow variables become

$$\begin{aligned} \mathbf{F} \cdot \vec{e}_{x'} &= \int_0^{\Delta t} \int\int u'_k f(u'_k, v'_l, w) \begin{pmatrix} 1 \\ u_k \\ v_l \\ \frac{1}{2}(u_k^2 + v_l^2 + w^2) \end{pmatrix} du' dv' dw dt \\ &= \int_0^{\Delta t} \int\int u'_k f(u'_k, v'_l, w) \begin{pmatrix} 1 \\ u'_k \cos\alpha - v'_l \sin\alpha \\ u'_k \sin\alpha + v'_l \cos\alpha \\ \frac{1}{2}(u'^2_k + v'^2_l + w^2) \end{pmatrix} du' dv' dw dt, \end{aligned} \tag{5.40}$$

which can be fully determined using numerical summation for the non-equilibrium part $\tilde{f}_{i+1/2,j,k,l}$ and direct integration for the equilibrium part $\tilde{g}_{i+1/2,j,k,l}$ in the local coordinate system.

In order to discretize the collision term efficiently, the unified formulation will update the macroscopic variables first. The update of conservative flow variables inside the control volume Ω_{ij} is

$$\mathbf{W}_{i,j}^{n+1} = \mathbf{W}_{i,j}^{n} + \frac{1}{\Omega_{ij}} \sum_{\sigma} S_\sigma \mathbf{F} \cdot \vec{e}_\sigma, \tag{5.41}$$

where the transport $\mathbf{F} \cdot \vec{e}_\sigma$ is the flux in Eq. (5.40), S_σ is the length of the cell interface with unit out-normal direction $\vec{e}_\sigma$ for the cell interface σ. The current scheme has second order accuracy. Based on the above updated conservative variables, we can immediately obtain the equilibrium gas distribution function $g^{n+1}_{i,j,k,l}$ inside each cell and the additional term $f^{+(n+1)}$ in the Shakhov model, therefore based on Eq. (3.3) the unified kinetic scheme for the update of gas distribution function in 2D becomes

$$f^{n+1}_{i,j,k,l} = f^n_{i,j,k,l} + \frac{1}{\Omega_{i,j}} \sum_\sigma S_\sigma \mathbf{f}_{k,l} \cdot \vec{e}_\sigma + \frac{\Delta t}{2} \left(\frac{f^{+(n+1)}_{i,j,k,l} - f^{(n+1)}_{i,j,k,l}}{\tau^{n+1}_{i,j}} + \frac{f^{+(n)}_{i,j,k,l} - f^{(n)}_{i,j,k,l}}{\tau^n_{i,j}} \right), \tag{5.42}$$

where the flux for the particle at velocity (u_k, v_l) is given in Eq. (5.39), and the trapezoidal rule has been used for the time integration of collision term. So, from the above equation, the UGKS for the update of gas distribution function is

$$f^{n+1}_{i,j,k,l} = \left(1 + \frac{\Delta t}{2\tau^{n+1}}\right)^{-1} \left[f^n_{i,j,k,l} + \frac{1}{\Omega_{i,j}} \sum_\sigma S_\sigma \mathbf{f}_{k,l} \cdot \vec{e}_\sigma + \frac{\Delta t}{2} \left(\frac{f^{+(n+1)}{}_{i,j,k,l}}{\tau^{n+1}_{i,j}} + \frac{f^{+(n)}{}_{i,j,k,l} - f^n_{i,j,k,l}}{\tau^n_{i,j}} \right) \right], \tag{5.43}$$

where no iteration is needed for the update of the above solution. The particle collision times $\tau^n_{i,j}$ and $\tau^{n+1}_{i,j}$ are defined based on the temperature and pressure in the cell, i.e., $\tau^n_{i,j} = \mu(T^n_{i,j})/p^n_{i,j}$ and $\tau^{n+1}_{i,j} = \mu(T^{n+1}_{i,j})/p^{n+1}_{i,j}$, which are known due to the updated macroscopic flow variables in Eq. (5.41).

5.4 Unified Gas Kinetic Scheme in Three-dimensional Space

In three dimensional space, for monatomic gas there is no need to introduce reduced gas distribution function. In the 3D case, under a global coordinate system $(\vec{e}_x, \vec{e}_y, \vec{e}_z)$, a local coordinate system at a cell interface is needed, where $\vec{e}_{x'}$ is the normal direction, and $\vec{e}_{y'}$ and $\vec{e}_{z'}$ are two orthogonal unit vectors in the tangential plane. The transformation between the local and global coordinates are

$$\vec{e}_{x'} = \vec{e}_x(\vec{e}_{x'} \cdot \vec{e}_x) + \vec{e}_y(\vec{e}_{x'} \cdot \vec{e}_y) + \vec{e}_z(\vec{e}_{x'} \cdot \vec{e}_z),$$

$$\vec{e}_{y'} = \vec{e}_x(\vec{e}_{y'} \cdot \vec{e}_x) + \vec{e}_y(\vec{e}_{y'} \cdot \vec{e}_y) + \vec{e}_z(\vec{e}_{y'} \cdot \vec{e}_z),$$

$$\vec{e}_{z'} = \vec{e}_x(\vec{e}_{z'} \cdot \vec{e}_x) + \vec{e}_y(\vec{e}_{z'} \cdot \vec{e}_y) + \vec{e}_z(\vec{e}_{z'} \cdot \vec{e}_z),$$

from which all flow variables, such as the gas distribution function

$$f_{i+1/2,j,k,l,m,n} = f(x_{i+1/2}, y_j, z_k, u_l, v_m, w_n)$$

and the macroscopic variables

$$\mathbf{W} = (\rho, \rho U, \rho V, \rho W, \rho E)^T,$$

can be transformed to a local coordinate system.

In the following, we consider a local coordinate system and evaluate the fluxes in the normal x-direction. The three-dimensional gas-kinetic BGK-Shakhov equation can be written as

$$f_t + uf_x + vf_y + wf_z = \frac{f^+ - f}{\tau}, \tag{5.44}$$

where f is the gas distribution function and f^+ is the heat flux modified equilibrium state which is approached by f,

$$f^+ = g\left[1 + (1 - \mathrm{Pr})\mathbf{c} \cdot \mathbf{q}\left(\frac{c^2}{RT} - 5\right)\Big/(5pRT)\right] = g + g^+,$$

with random velocity $\mathbf{c} = \mathbf{u} - \mathbf{U}$ and the heat flux $\mathbf{q}$. Both f and f^+ are functions of space (x, y, z), time t, particle velocity (u, v, w) in $x-$, $y-$, and $z-$direction. For a monatomic gas in 3D case, the equilibrium Maxwellian distribution is,

$$g = \rho\left(\frac{\lambda}{\pi}\right)^{\frac{3}{2}} e^{-\lambda((u-U)^2+(v-V)^2+(w-W)^2)},$$

where ρ is the density, (U, V, W) is the macroscopic velocity in the x, y and z directions. The relation between mass ρ, momentum $(\rho U, \rho V, \rho W)$, and energy ρE densities with the distribution function f is

$$\begin{pmatrix} \rho \\ \rho U \\ \rho V \\ \rho W \\ \rho E \end{pmatrix} = \int \psi_\alpha f d\Xi, \qquad \alpha = 1, 2, 3, 4, 5, \tag{5.45}$$

where ψ_α is the component of the conservative moments

$$\boldsymbol{\psi} = (\psi_1, \psi_2, \psi_3, \psi_4, \psi_5)^T = \left(1, u, v, w, \frac{1}{2}(u^2 + v^2 + w^2)\right)^T,$$

and $d\Xi = dudvdw$ is the volume element in the phase space.

In the unified scheme, at the center of a cell interface ($x_{i+1/2} = 0, y_j = 0, z_k = 0$) the solution $f_{i+1/2,j,k,l,m,n}$ is constructed from an integral solution of the BGK-Shakhov model (5.44),

$$\begin{aligned}
&f_{i+1/2,j,k,l,m,n}(0,0,0,t,u_l,v_m,w_n)\\
&= \frac{1}{\tau}\int_{t^n}^{t^{n+1}} f^+(x',y',z',t',u_l,v_m,w_n)e^{-(t-t')/\tau}dt'\\
&+e^{-(t-t^n)/\tau}f_{0,l,m,n}(-u_l(t-t^n),-v_m(t-t^n),-w_n(t-t^n),t^n,u_l,v_m,w_n),
\end{aligned} \tag{5.46}$$

where $f^+ = g + g^+$ will be approximated separately. Here $x' = -u_l(t-t')$, $y' = -v_m(t-t')$, and $z' = -w_n(t-t')$ are particle trajectory and $f^n_{0,l,m,n}$ is the initial gas distribution function of f at time $t = t^n$ around the center of the cell interface ($x_{i+1/2} = 0, y_j = 0, z_k = 0$) at the particle velocity (u_l, v_m, w_n), i.e., $f^n_{0,l,m,n} = f^n_0(x,y,z,t^n,u_l,v_m,w_n)$. In the following, $t^n = 0$ will be used.

In the above equation, inside each control volume $f^n_{0,l,m,n}$ is known at the beginning of each time step t^n. After reconstruction, around each cell interface $x_{i+1/2}$, at time step t^n the initial distribution function becomes,

$$\begin{aligned}
&f_0(x,y,z,t^n,u_l,v_m,w_n)\\
&\quad= f_{0,l,m,n}(x,y,z,0)\\
&\quad= \begin{cases} f^L_{i+1/2,l,m,n} + \sigma_{i,l,m,n}x + \theta_{i,l,m,n}y + \epsilon_{i,l,m,n}z, & x \le 0,\\ f^R_{i+1/2,l,m,n} + \sigma_{i+1,l,m,n}x + \theta_{i+1,l,m,n}y + \epsilon_{i+1,l,m,n}z, & x > 0, \end{cases}
\end{aligned} \tag{5.47}$$

where nonlinear limiter is used in the above reconstruction.

For an equilibrium state g around a cell interface ($x_{i+1/2} = 0, y_j = 0, z_k = 0, t^n = 0$), it can be expanded with two slopes,

$$g = g_0\left[1 + (1 - \mathrm{H}[x])\bar{a}^L x + \mathrm{H}[x]\bar{a}^R x + \bar{b}y + \bar{c}z + \bar{A}t\right]. \tag{5.48}$$

The dependence of $\bar{a}^L$, $\bar{a}^R$, $\bar{b}$, $\bar{c}$, and $\bar{A}$ on the particle velocity can be obtained from a Taylor expansion of a Maxwellian and have the following forms,

$$\bar{a}^L = \bar{a}^L_1 + \bar{a}^L_2 u + \bar{a}^L_3 v + \bar{a}^L_4 w + \bar{a}^L_5 \frac{1}{2}(u^2+v^2+w^2) = \bar{a}^L_\alpha \psi_\alpha,$$

$$\bar{a}^R = \bar{a}^R_1 + \bar{a}^R_2 u + \bar{a}^R_3 v + \bar{a}^R_4 w + \bar{a}^R_5 \frac{1}{2}(u^2+v^2+w^2) = \bar{a}^R_\alpha \psi_\alpha,$$

$$\cdots$$

$$\bar{A} = \bar{A}_1 + \bar{A}_2 u + \bar{A}_3 v + \bar{A}_4 w + \bar{A}_5 \frac{1}{2}(u^2 + v^2 + w^2) = \bar{A}_\alpha \psi_\alpha,$$

where $\alpha = 1, 2, 3, 4, 5$ and all coefficients $\bar{a}_1^L, \bar{a}_2^L, ..., \bar{A}_5$ are local constants.

The determination of g_0 depends on the local macroscopic variables ρ_0, U_0, V_0, W_0 and λ_0 in g_0, i.e.,

$$g_0 = \rho_0 \left(\frac{\lambda_0}{\pi}\right)^{\frac{3}{2}} e^{-\lambda_0((u-U_0)^2+(v-V_0)^2+(w-W_0)^2)},$$

which is determined from colliding particles at $(x = 0, y = 0, z = 0, t = 0)$,

$$\mathbf{W}_0 = \int g_0 \psi d\Xi = \sum \left(f_{i+1/2,l,m,n}^L H[u_k] + f_{i+1/2,l,m,n}^R (1 - H[u_k]) \right) \psi, \tag{5.49}$$

where $\mathbf{W}_0 = (\rho_0, \rho_0 U_0, \rho_0 V_0, \rho_0 W_0, \rho_0 E_0)^T$ are the conservative macroscopic flow variables located at the cell interface at time $t = 0$. Since $f_{i+1/2,l,m,n}^L$ and $f_{i+1/2,l,m,n}^R$ have been obtained earlier in the initial distribution function f_0, the above moments can be evaluated explicitly. Therefore, the conservative variables $\rho_0, \rho_0 U_0, \rho_0 V_0, \rho_0 W_0$, and $\rho_0 E_0$ at the cell interface can be obtained, from which g_0 is uniquely determined. For the equilibrium state, λ_0 in g_0 can be found from

$$\lambda_0 = 3\rho_0 \Big/ \left(4 \left(\rho_0 E_0 - \frac{1}{2}\rho_0 (U_0^2 + V_0^2 + W_0^2)\right)\right).$$

Then, $\bar{a}^L$ and $\bar{a}^R$ of g in Eq. (5.33) can be obtained through the relation of

$$\frac{\bar{\mathbf{W}}_{i+1}(x_{i+1}) - \mathbf{W}_0}{\rho_0 \Delta x^+} = \frac{1}{\rho_0} \int \bar{a}^R g_0 \psi d\Xi = \bar{M}_{\alpha\beta}^0 \begin{pmatrix} \bar{a}_1^R \\ \bar{a}_2^R \\ \bar{a}_3^R \\ \bar{a}_4^R \\ \bar{a}_5^R \end{pmatrix} = \bar{M}_{\alpha\beta}^0 \bar{a}_\beta^R, \tag{5.50}$$

and

$$\frac{\mathbf{W}_0 - \bar{\mathbf{W}}_i(x_i)}{\rho_0 \Delta x^-} = \frac{1}{\rho_0} \int \bar{a}^L g_0 \psi d\Xi = \bar{M}_{\alpha\beta}^0 \begin{pmatrix} \bar{a}_1^L \\ \bar{a}_2^L \\ \bar{a}_3^L \\ \bar{a}_4^L \\ \bar{a}_5^L \end{pmatrix} = \bar{M}_{\alpha\beta}^0 \bar{a}_\beta^L, \tag{5.51}$$

where the matrix $\bar{M}_{\alpha\beta}^0 = \int g_0 \psi_\alpha \psi_\beta d\Xi / \rho_0$ is known, and $\Delta x^+ = x_{i+1} - x_{i+1/2}$ and $\Delta x^- = x_{i+1/2} - x_i$ are the distances from the cell interface to cell centers. Therefore, $(\bar{a}_1^R, \bar{a}_2^R, \bar{a}_3^R, \bar{a}_4^R, \bar{a}_5^R)^T$ and $(\bar{a}_1^L, \bar{a}_2^L, \bar{a}_3^L, \bar{a}_4^L, \bar{a}_5^L)^T$ can be found from the above equations with the inverse of the matrix M, which is presented in Appendix C. Another direct way to get these coefficients is to use the Taylor expansion of a Maxwellian,

$$\frac{1}{g}\frac{\partial g}{\partial x} = a_\alpha \psi_\alpha,$$

which gives

$$\begin{aligned}
a_5 &= -2\lambda_x \\
a_4 &= 2\lambda W_x + 2W\lambda_x \\
a_3 &= 2\lambda V_x + 2V\lambda_x \\
a_2 &= 2\lambda U_x + 2U\lambda_x \\
a_1 &= \frac{\rho_x}{\rho} - 2\lambda(UU_x + VV_x + WW_x) + \left(\frac{3}{2\lambda} - (U^2 + V^2 + W^2)\right)\lambda_x,
\end{aligned}$$

and the derivatives of ρ_x, U_x, V_x, W_x and λ_x can be obtained from the derivatives of conservatives variables $\partial \mathbf{W}/\partial x$. For the tangential derivatives of $\bar{b}$ and $\bar{c}$ in the expansion of equilibrium state g, it can be obtained from

$$\int \bar{b}\psi g_0 d\Xi = \sum \left(\theta_{i,l,m,n} H[u_l] + \theta_{i+1,l,m,n}(1 - H[u_l])\right)\psi,$$

and

$$\int \bar{c}\psi g_0 d\Xi = \sum \left(\epsilon_{i,l,m,n} H[u_l] + \epsilon_{i+1,l,m,n}(1 - H[u_l])\right)\psi.$$

In order to evaluate the time evolution part $\bar{A}$ in the equilibrium state, we can apply the following condition

$$\frac{d}{dt}\int (g - f)\psi \Xi = 0,$$

at $(x = 0, y = 0, z = 0, t = 0)$ and get

$$\begin{aligned}
\bar{M}^0_{\alpha\beta}\bar{A}_\beta &= \frac{1}{\rho_0}(\partial\rho/\partial t, \partial(\rho U)/\partial t, \partial(\rho V)/\partial t, \partial(\rho W)/\partial t, \partial(\rho E)/\partial t)^T \\
&= -\frac{1}{\rho_0}\int \left[\left(\bar{a}^L \mathrm{H}[u] + \bar{a}^R(1 - \mathrm{H}[u])\right)u + \bar{b}v + \bar{c}w\right] g_0\psi d\Xi.
\end{aligned} \tag{5.52}$$

With the determination of equilibrium state and the heat flux at the cell interface, the additional term g^+ in the Shakhov model can be well-determined as well.

Now we have determined all parameters in the initial gas distribution function f_0 and the state f^+ in space and time locally. After substituting Eq. (5.47) and Eq. (5.48) into Eq. (5.46) and taking $(u = u_l, v =$

$v_m, w = w_n$) in $g_0, \bar{a}^L, \bar{a}^R$, $\bar{b}$, $\bar{c}$, and $\bar{A}$, the gas distribution function $f(x_{i+1/2}, y_j, z_k, t, u_l, v_m, w_n)$ at the discretized particle velocity (u_l, v_m, w_n) can be expressed as

$$\begin{aligned}
&f_{i+1/2,j,k,l,m,n}(x_{i+1/2}, y_j, z_k, t, u_l, v_m, w_n)\\
&= (1 - e^{-t/\tau})(g_0 + g^+)\\
&\quad + \left(\tau(-1 + e^{-t/\tau}) + te^{-t/\tau}\right)\left[\left(\bar{a}^L \mathrm{H}[u_l] + \bar{a}^R(1 - \mathrm{H}[u_l])\right)u_l + \bar{b}v_m + \bar{c}w_n\right]g_0\\
&\quad + \tau(t/\tau - 1 + e^{-t/\tau})\bar{A}g_0\\
&\quad + e^{-t/\tau}\Big((f^L_{i+1/2,l,m,n} - u_l t\sigma_{i,l,m,n} - v_m t\theta_{i,l,m,n} - w_n t\epsilon_{i,l,m,n})\mathrm{H}[u_l]\\
&\quad + (f^R_{i+1/2,l,m,n} - u_l t\sigma_{i+1,l,m,n} - v_m t\theta_{i+1,l,m,n} - w_n t\epsilon_{i+1,l,m,n})(1 - \mathrm{H}[u_l])\Big)\\
&= \tilde{g}_{i+1/2,j,k,l,m,n} + \tilde{f}_{i+1/2,j,k,l,m,n},
\end{aligned} \tag{5.53}$$

where $\tilde{g}_{i+1/2,j,k,l,m,n}$ is all terms related to the integration of the equilibrium state g and g^+, and $\tilde{f}_{i+1/2,j,k,l,m,n}$ is the term from initial condition f_0. The collision time τ in the above distribution function is determined by $\tau = \mu(T_0)/p_0$, where T_0 and p_0 are given in $\mathbf{W}_0$ at the cell interface.

After evaluating the gas distribution function in a local coordinate system $(\vec{e}_{x'}, \vec{e}_{y'}, \vec{e}_{z'})$, the flux for the particle transport at velocity $({u'}_l, {v'}_m, {w'}_n)$ in a global coordinate system is given by

$$\mathbf{f}_{l,m,n} \cdot \vec{e}_{x'} = {u'}_l f_{i+1/2,j,k,l,m,n}(0, 0, 0, {u'}_l, {v'}_m, {w'}_n), \tag{5.54}$$

which can be explicitly evaluated using the distribution function in Eq. (5.53). The above particle flux at velocity $({u'}_l, {v'}_m, {w'}_n)$ in a local coordinate system will be the same as the particle flux at velocity (u_l, v_m, w_n) in the global coordinate system.

For the update of conservative variables, a global coordinate system has to be used in order to have the same definition of momentum and energy in different control volumes. Therefore, with the particle transport across the cell interface surface $\mathbf{f}_{l,m,n} \cdot \vec{e}_{x'}$, these particles will carry the mass, momentum, and energy in a global reference of frame, i.e., $(1, u_l, v_m, w_n, \frac{1}{2}(u_l^2 + v_m^2 + w_n^2))^T$. The fluxes across the cell interface for the update of macroscopic flow variables in a global coordinate system become

$$\mathbf{F} \cdot \vec{e}_{x'} = \int_0^{\Delta t} \int {u'}_l f_{i+1/2,j,k,l,m,n}({u'}_l, {v'}_m, {w'}_n) \begin{pmatrix} 1 \\ u_l \\ v_m \\ w_n \\ \frac{1}{2}(u_l^2 + v_m^2 + w_n^2) \end{pmatrix} d\Xi' dt \tag{5.55}$$

where $d\Xi' = du'dv'dw'$ and

$$\begin{pmatrix} 1 \\ u_l \\ v_m \\ w_n \\ \frac{1}{2}(u_l^2 + v_m^2 + w_n^2) \end{pmatrix}$$

can be changed to

$$\begin{pmatrix} 1 \\ u'_l(\vec{e}_x \cdot \vec{e}_{x'}) + v'_m(\vec{e}_x \cdot \vec{e}_{y'}) + w'_n(\vec{e}_x \cdot \vec{e}_{z'}) \\ u'_l(\vec{e}_y \cdot \vec{e}_{x'}) + v'_m(\vec{e}_y \cdot \vec{e}_{y'}) + w'_n(\vec{e}_y \cdot \vec{e}_{z'}) \\ u'_l(\vec{e}_z \cdot \vec{e}_{x'}) + v'_m(\vec{e}_z \cdot \vec{e}_{y'}) + w'_n(\vec{e}_z \cdot \vec{e}_{z'}) \\ \frac{1}{2}(u'^2_l + v'^2_m + w'^2_n) \end{pmatrix}$$

in the above flux evaluations. Certainly, the above integrations need to be evaluated using the numerical summation for the non-equilibrium part $\tilde{f}_{i+1/2,j,k,l,n,m}$ and direct analytic integration for the equilibrium part $\tilde{g}_{i+1/2,j,k,l,m,n}$.

The update of the conservative flow variables inside each control volume Ω_{ijk} is

$$\mathbf{W}^{n+1}_{i,j,k} = \mathbf{W}^{n}_{i,j,k} + \frac{1}{\Omega_{ijk}} \sum_{\sigma} S_\sigma \mathbf{F} \cdot \vec{e}_\sigma, \tag{5.56}$$

where the transport $\mathbf{F} \cdot \vec{e}_\sigma$ is the flux given in Eq. (5.55), and S_σ is the surface area with unit normal direction $\vec{e}_\sigma$. Based on the above updated conservative variables, an equilibrium state $g^{n+1}_{i,j,k,l,m,n}$ inside each control volume and the additional term $f^{+(n+1)}$ in the Shakhov model can be constructed. Therefore, the UGKS updates the gas distribution function

$$f^{n+1}_{i,j,k,l,m,n} = \left(1 + \frac{\Delta t}{2\tau^{n+1}}\right)^{-1} \left[f^{n}_{i,j,k,l,m,n} + \frac{1}{\Omega_{i,j,k}} \sum_{\sigma} S_\sigma \mathbf{f}_{l,m,n} \cdot \vec{e}_\sigma \right.$$
$$\left. + \frac{\Delta t}{2} \left(\frac{f^{+(n+1)}_{i,j,k,l,m,n}}{\tau^{n+1}_{i,j}} + \frac{f^{+(n)}_{i,j,k,l,m,n} - f^{n}_{i,j,k,l,m,n}}{\tau^{n}_{i,j}} \right) \right]. \tag{5.57}$$

The particle collision times $\tau^n_{i,j}$ and $\tau^{n+1}_{i,j}$ are defined based on the temperature and pressure inside each control volume, i.e., $\tau^n_{i,j} = \mu(T^n_{i,j})/p^n_{i,j}$ and $\tau^{n+1}_{i,j} = \mu(T^{n+1}_{i,j})/p^{n+1}_{i,j}$, which are known due to the update of macroscopic flow variables in Eq. (5.56).

Same as GKS with moving mesh [Jin and Xu (2007); Hui and Xu (2012)], the above finite volume UGKS can be extended to the moving mesh cases as well [Chen *et al.* (2012b)], see Appendix D.

5.5 Boundary Conditions

The interaction between the gas flow and the solid boundary has been studied by many authors, see [Patterson (1956)] and [Cercignani (1988)]. This section is mainly about how to implement these conditions numerically in the gas-kinetic scheme.

In the near continuum flow regime, even for the Navier-Stokes equations the application of slip boundary condition becomes necessary. Since the unified gas-kinetic schemes are based on the time evolution of a gas distribution function, the slip boundary condition can be naturally obtained for both GKS and UGKS methods.

(i). Boundary condition at solid walls

Maxwell boundary condition in transition regime

Here we first discuss the isothermal wall boundary condition with complete accommodation for the flow in the transition regime, where the cell size next to the wall is less than the local particle mean free path. Assuming the left wall is located at $x = 0$ for the 2D or 3D simulations. The boundary condition described here is quiet simple, the incoming distribution function is directly obtained through interpolation.

For example, in the normal direction we can obtain h_k^{in}, b_k^{in} by one-sided interpolation from the interior region to the wall surface, i.e.,

$$h_k^{in} = h_{1,k} - \sigma_{1,k}^h \frac{\Delta x}{2},$$

$$b_k^{in} = b_{1,k} - \sigma_{1,k}^b \frac{\Delta x}{2}.$$

Then, evaluating the density at the wall with the condition that no particle penetrating through the wall

$$\int_{t^n}^{t^{n+1}} \int_{u>0} u g_w d\Xi dt + \int_{t^n}^{t^{n+1}} \int_{u<0} u f^{in} d\Xi dt = 0, \tag{5.58}$$

which gives

$$\rho_w = -\frac{\sum \alpha_k u_k h_k^{in}}{\left(\frac{\lambda_w}{\pi}\right)^{1/2} \sum \alpha_k u_k e^{-\lambda_w (u_k - U_w)^2}},$$

where $g_w, \rho_w, \lambda_w, U_w$ are the variables of the equilibrium state on the wall. Consequently, the corresponding reduced Maxwellian distribution with

H_k^w, B_k^w on the wall can be obtained. Then, the distribution functions at the boundary can be expressed as

$$h_k = H_k^w H[u_k] + h_k^{in}(1 - H[u_k]),$$

$$b_k = B_k^w H[u_k] + b_k^{in}(1 - H[u_k]).$$

Finally, the fluxes across the wall are calculated by

$$\mathbf{F}_{1/2} = \Delta t \begin{pmatrix} \sum \alpha_k u_k h_k \\ \sum \alpha_k u_k^2 h_k \\ \sum \alpha_k \frac{1}{2}\left(u_k^3 h_k + u_k b_k\right) \end{pmatrix}$$

for the macroscopic variables update, and

$$f_{1/2,k}^h = \Delta t u_k h_k$$
$$f_{1/2,k}^b = \Delta t u_k b_k$$

for the update of the gas distribution functions.

With the above boundary condition, the slip velocity at the wall surface will appear automatically. Since there is no explicit inclusion of particle collision among incoming and outgoing particles before and after hitting the wall, the above boundary condition is valid in the transition flow regime, where the cell size next to the wall is less than the local particle mean free path. In the continuum flow regimes, where the cell size is much larger than the local particle mean free path, the above boundary condition is not appropriate due to its free transport mechanism for the incoming molecules. As a consequence, if the above boundary condition is implemented in the continuum flow computation, the Knudsen layer will be much enlarged, since the wall-colliding particles from interior region take free transport without suffering inter-molecule collisions and the cell size becomes the pseudo-particle mean free path. On the other hand, in the continuum flow regime, the NS equations require the non-slip boundary condition. For example, in order to satisfy the non-slip boundary condition, the bounce back boundary treatment is widely used in the Lattice Boltzmann method. But this kind of bounce-back boundary condition cannot be extended to the rarefied flow regimes, where velocity slip appears. Therefore, a multiscale boundary condition (MBC) which is valid in all range of Knudsen number is required. For UGKS, the non-slip boundary condition for the NS solution can be easily set up using the ghost cells, which is the same approach as other CFD methods based on macroscopic flow equations [Xu (2001)].

Multiscale boundary condition (MBC) for all flow regimes

Here, we propose a multiscale boundary condition for UGKS [Xu and Li (2004); Chen and Xu (2013)]. The ideal is to use local solution again, namely, f^{in} in Eq. (5.58) is replaced by the full gas evolution distribution function in Eq. (5.53) for 3D case, where the corresponding 2D case is given by Eq. (5.38), where the equilibrium state in the integral solution is obtained from the macroscopic flow variables and their initial dervatives close to the boundary.

The initial gas distribution is reconstructed as usual and the incoming particles hitting the wall are computed from the gas evolution solution with the consideration of particle collisions. The local equilibrium state from the wall g_w is constructed using Eq. (5.58), where the wall temperature and velocity are supposed known. This multiscale boundary condition only changes the incoming distribution function. The particles emitted from the boundary also obey the fully diffusion assumption. The validation for this boundary condition doesn't require the cell size close to the boundary is less than the local particle mean free path.

In the following we will discuss the inlet and outlet boundary conditions. Since the low speed micro-channel flow and high speed rarefied flows need different boundary treatment, here we list two kinds of inlet and outlet boundary treatments.

(ii). Inlet and outlet boundary for micro-channel flows

Inlet and outlet boundary conditions for the pressure driven microchannel flows are the following. At both inlet and outlet faces, the pressure, temperature, and transverse flow velocity are specified there, leaving the streamwise velocity to be obtained using second-order interpolation from the interior region. The inflow boundary conditions are given as

$$\rho_{in} = \rho_+, p_{in} = p_+, U_{in} = 2U_1 - U_2, V_{in} = V_+,$$

where 1 refers to the first cell inside the computational domain, 2 is the second cell, and p_+ and ρ_+ are the given inlet pressure and density while the temperature for the incoming gas is determined from them. For the outflow boundary condition, only the pressure is specified and the remaining variables are obtained by the second-order extrapolation from interior region, the outflow boundary conditions are

$$\rho_{out} = 2\rho_{out-1} - \rho_{out-2}, p_{out} = p_-, U_{out} = 2U_{out-1} - U_{out-2},$$
$$V_{out} = 2V_{out-1} - V_{out-2},$$

where $(out - 1)$ is the last cell inside the computational domain, and p_- is the given outlet pressure.

(iii). Inlet and outlet boundary for hypersonic rarefied flows

For high speed external flows passing through a flying vehicle, the inlet and outlet boundary condition may need to cover a wide range of flow regimes. The flow can be mainly categorized into two kinds by the Mach number, i.e., supersonic flow and subsonic one. In the continuum flow computations, the Riemann invariants are usually used to construct the quantities on the boundary. However, when the Knudsen number becomes large, the Euler and Navier-stokes equations are invalid. As a result, the boundary condition based on Riemann invariant will not be accurate for rarefied flow. Therefore, we propose a semi-empirical boundary condition to include rarefied effect [Chen *et al.* (2012b)]. A global Knudsen number is used to construct a weight function $\alpha = e^{-1/\text{Kn}}$. The uniform incoming flow is presented by $\mathbf{W}_\infty$, and the macroscopic quantities constructed by Riemann invariants are denoted by $\mathbf{W}_R$. Then the macroscopic quantities on the boundary is constructed as,

$$\mathbf{W}_b = \alpha \mathbf{W}_\infty + (1 - \alpha)\mathbf{W}_R. \tag{5.59}$$

In the two limiting cases of free molecular and continuum flow, the above boundary condition is consistent with traditional approach. Meanwhile, the above formulation provides a smooth transition connecting the limiting cases.

5.6 Analysis of Unified Gas Kinetic Scheme

The traditional continuum and rarefied flow simulations are based on different numerical methods which are valid in their own modeling regime, such as the NS and direct Boltzmann solver. The UGKS provides a smooth transition to cover the whole flow regime. In the previous section, we present the UGKS in one, two, and three dimensions. The UGKS uses the time evolution solution of a gas distribution function to construct flux transport across a cell interface. The UGKS becomes a multiscale method, which connects the flow evolution from the kinetic to the hydrodynamic scale. The solutions obtained from UGKS depend on the ratio of the time step to the local particle collision time. In the following, we are going to analyze properties of the UGKS.

Dynamic coupling of different scales

The UGKS is a multiscale method to simulate both rarefied and continuum flows with the update of both macroscopic flow variables and microscopic gas distribution function. Many hybrid approaches separate different regions geometrically and treat them differently, where buffer zones are used to connect these regions. Instead of solving different governing equations, the UGKS captures the flow physics through a time evolution of a gas distribution function across different scales.

In the continuum flow regime, the intensive particle collision with $\Delta t \gg \tau$ will drive the gas system close to equilibrium state. Therefore, the part based on the integration of equilibrium state $\tilde{g}_{j+1/2,k}$ in Eq. (5.21) will automatically take a dominant contribution. It can be shown that in smooth flow region $\tilde{g}_{j+1/2,k}$ gives precisely the NS gas distribution function when $\Delta t \gg \tau$. Because there is one-to-one correspondence between macroscopic flow variables and the equilibrium state, the integration of the equilibrium part can be also fairly considered as a macroscopic component of the scheme to capture the flow physics in hydrodynamic scale. In the free molecule limit with inadequate particle collisions, the integral solution at the cell interface will automatically present a purely upwind scheme, where the particle free transport from $\tilde{f}_{j+1/2,k}$ in Eq. (5.21) will give the main contribution when $\Delta t \ll \tau$. Therefore, the UGKS captures the flow physics in the collisionless limit as well.

The UGKS targets to solve a whole spectrum of discrete governing equations from rarefied to continuum flow. Certainly, in order to save computational time we may also develop a hybrid method which uses the UGKS as a non-equilibrium flow solver and adopts the GKS presented in the last chapter as a continuum flow solver. The main differences between UGKS and GKS are the initial condition of the gas distribution function f_0. In UGKS, it is updated and used directly, while it is reconstructed in GKS from macroscopic flow variables through the Chapman-Enskog expansion. In the continuum flow regime, the update of the distribution function is not necessary because the corresponding NS distribution function is well defined. Due to the intensive particle collisions, the flow evolution in the continuum regime is coherent which can be captured through a few degree of freedom of macroscopic flow variables, or equivalently a few moments of the distribution function.

In UGKS simulation, the ratio between the time step Δt to the local particle collision time τ can cover a wide range of values. Here, the time step is determined by the maximum particle velocity, such as $\Delta x/\max(|\mathbf{u}|)$, which is equivalent to

$$\Delta t = \text{CFL}\frac{\Delta x}{|U| + c},$$

where CFL is the Courant-Friedrichs-Lewy (CFL) number and c is the sound speed. On the other hand, the particle collision time is defined by

$$\tau = \frac{\mu}{p} = \frac{\rho|U|\Delta x}{p\text{Re}},$$

where $\text{Re} = \rho\Delta x|U|/\mu$ is the cell's Reynolds number. With the approximation,

$$|U| + c \simeq c,$$

which is true for low speed flow and is approximately correct even for hypersonic flow because the flow velocity is on the same order of sound speed, we have

$$\frac{\Delta t}{\tau} = \frac{\text{Re}}{M},$$

where M is the Mach number. So, in the region close to equilibrium, even for a modest cell Reynolds number, such as 10^2, the local time step for the UGKS can be much larger than the local particle collision time. The above equation can be also written in the following form

$$\Delta t = \frac{\tau}{\text{Kn}}, \tag{5.60}$$

where $\text{Kn} = M/\text{Re}$ is the local cell Knudsen number. So, in a computation which covers both continuum and rarefied regimes, such as the nozzle flow, the ratio of time step over local particle collision time can be changed significantly. For example, the time step Δt used can be much larger than the particle collision time, i.e., $\Delta t \gg \tau$ inside the nozzle, and less than the collision time, i.e., $\Delta t \ll \tau$ outside. In comparison with kinetic solvers with the requirement of $\Delta t < \tau$ everywhere, the UGKS is efficient. Also, for steady state calculation, the use of local time step will enhance the efficiency of the scheme further.

A variation of governing equations from UGKS

As pointed out early, the dynamics provided by UGKS depends on the mesh resolution to identify the flow physics. For a physical reality, such as the fully resolved solution of the exact Boltzmann equation, this solution can be represented with different cell resolution through the coarse-graining process, such as the purely numerical averaging of the full Boltzmann solution within different cell size. In order to recover the averaged solution in different scale or resolution, a different governing equation from the Boltzmann equation needs to be solved, and the modeling scale of this equation should be the varying mesh size. The UGKS just provides such as a scheme from the fully resolved kinetic scale Boltzmann solver to a mesh size dependent solver, up to the hydrodynamic one for the Navier-Stokes solutions. In the following example, the shock structure at $M = 6$ of argon gas is fully resolved by solving the full Boltzmann equation with a mesh size on the order of 1/4 of the particle mean free path, see Fig. 5.1. Based on such a fully resolved resolution, we can get the cell averaged solution for both macroscopic flow variables and the particle distribution function through a purely averaging in each cell with an enlarged cell size. For the UGKS, besides using the same mesh size as that of the full Boltzmann solution, a different mesh size can be used as well to get the corresponding solutions, and these solutions can be compared with the averaged Boltzmann solutions. As shown in Fig. 5.1, even with the variations of the mesh size from 1 to 4 mean free path with the corresponding time step from $\Delta t < \tau$ to $\Delta t > \tau$, the UGKS does recover the solution of the full Boltzmann equation with different mesh size. This example illustrates the variation of the governing equations underlying the UGKS, which recover the corresponding solutions in different scale. This example also presents the insensitivity of the UGKS solution on the mesh size variations due to its adaptive numerical governing equations.

Asymptotic preserving property

The numerical scheme, which is capable of capturing the characteristic behavior in different scales with a fixed discretization both in space and time, is called asymptotic preserving (AP) scheme [Jin (2012)]. Specifically, for a gas system, it requires the scheme to recover the NS limit in a fixed time step and mesh size as the Knudsen number goes to zero. A standard explicit scheme for kinetic equation always requires the space and time discretizations to resolve the smallest scale in the system, such as the particle mean free path and collision time. It makes the scheme be extremely expensive when the system is close to the continuum limit. In recent years, many studies contribute to the development of AP schemes. It has been shown that the delicate time [Caflisch *et al.* (1997); Gabetta *et al.* (1997)] and space [Jin and Levermore (1996)] discretizations should be adopted in order

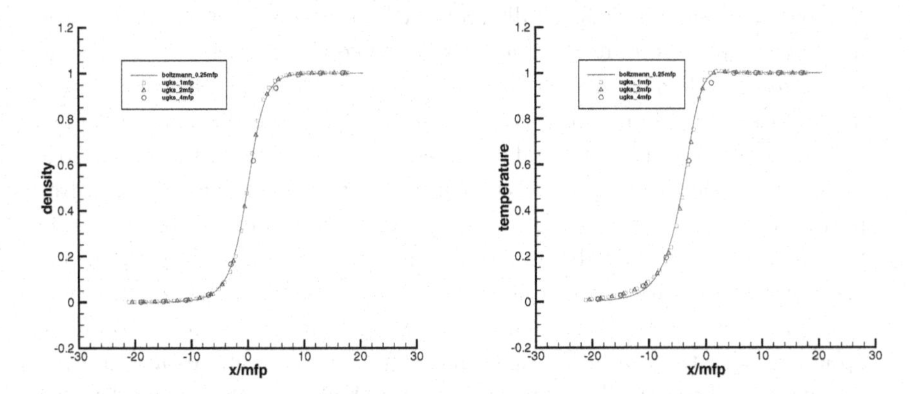

Fig. 5.1 Argon shock structure at Mach number 6 and $\mu \sim T^{0.81}$ calculated with the full Boltzmann equation (solid line and $\Delta x = 0.25$ mean free path (mfp)) and UGKS solutions with different cell sizes ($\Delta x = 1, 2, 4$mfp). (left) density, (right) temperature.

to achieve AP property. From a physical point of view, the continuum limit is achieved through intensive particle collisions. The local velocity distribution function evolves rapidly to an equilibrium state. Based on this fact, it is clear that any plausible approximation to the collision process must project the nonequilibrium data to the local equilibrium one in the continuum limit. Previous results [Jin (1995); Caflisch *et al.* (1997); Gabetta *et al.* (1997)] show that an effective condition for recovering the correct continuum limit is that the numerical scheme projects the distribution function to the local equilibrium, which has a discrete analogue of the asymptotic expansion for the continuous equations. In these studies, implicit time discretization is implemented to meet requirement of the numerical stability and AP property. On the other hand, the space discretization is not as crucial as the time discretization in the sense of AP property [Jin (2012)]. However, the sophisticated space discretization is still necessary for the correct long-time behavior for hyperbolic systems with stiff relaxation terms [Jin and Levermore (1996)]. The detailed analysis and numerical comparison between a standard AP method and the current UGKS has been conducted recently in [Chen and Xu (2013)].

Asymptotic preserving for the NS solution in the continuum limit is a preferred property for the kinetic schemes. Before designing such an AP method, we have to realize that the continuum and non-continuum flow behavior depends closely on our numerical cell resolution. Specifically, it depends on the ratio of numerical cell size to the particle mean free path.

It is basically meaningless to talk about AP method without sticking to the discretized space resolution. The Boltzmann equation itself is a dynamical model in particle mean free path and mean collision time scale. The direct upwind discretization of the transport part of the kinetic equation cannot go beyond such a modeling scale, where Δx needs to be on the same order as particle mean free path. If Δx is on the hundreds of particle mean free path, this discretization is problematic. Certainly, the flow physics from the kinetic to the hydrodynamic scale can be still captured by the direct discretization method of kinetic scale through the brutal force approach. Then, the concern is the efficiency, because in many cases the brutal force approach with the resolution up to mean free path is unnecessary and too expensive. Theoretically, in most cases we don't need to get so detailed information of the flow system. In a real application, the cell size used can be varied significantly in different regimes. As a result, with the variation of the ratio between the cell size and particle mean free path, the corresponding physical behavior needs to be captured. The purely upwind approach for the kinetic scale transport has to be avoided for a valid AP method. Unfortunately, most kinetic solvers are commonly using upwind discretization for the flux evaluation without doubt.

The distinguishable feature of the UGKS is that a time dependent solution of the kinetic model equation with the inclusion of collision effect is used for the flux evaluation at a cell interface. This solution itself covers different flow regimes from the initial free molecular transport to the Navier-Stokes formulation. The final solution depends on the ratio of time step to the particle mean collision time. When $\Delta t \gg \tau$, the intensive particle collision will converge the distribution function to

$$f_{j+1/2}(t) = g_0 + g^+ - (\tau(\bar{a}u + \bar{A}) + t\bar{A})g_0, \tag{5.61}$$

which is exactly the Chapman-Enskog expansion of the NS solution with the correct Prandtl number. Here due to the massive particle collision, the free transport part f_0 disappears. In the continuum flow limit, the UGKS will pick up the NS solution automatically. For the update of the conservative flow variables, the unified scheme recovers the GKS for NS solution in the hydrodynamic limit. On the other hand, in the free molecular limit, i.e., $\tau \gg \Delta t$, the UGKS presents a free transport,

$$f_{j+1/2,k}(x=0,t) = \begin{cases} f^l_{j+1/2,k} - u_k t\sigma_{j,k}, & u_k \geq 0, \\ f^r_{j+1/2,k} - u_k t\sigma_{j+1,k}, & u_k < 0. \end{cases} \tag{5.62}$$

In the region with comparable values of time step and local particle collision time, such as in the transitional flow regime, both the kinetic scale free transport and the hydrodynamic scale NS evolution will contribute to the dynamical evolution. The UGKS provides a continuum transition with the variation of the ratio $\Delta t/\tau$. The validity of the UGKS can be checked with benchmark solutions from the full Boltzmann equation or DSMC solution.

In order to develop an AP method, most current kinetic equation solvers explicitly or implicitly use Eq. (5.62) to evaluate the flux transport at a cell interface. This flow physics is valid in the rarefied flow regime only. Even though, based on Eq. (5.62) the fluxes for the macroscopic flow variables can be evaluated as well to update the conservative flow variables inside each control volume, such as

$$\mathbf{W}^{n+1} = \mathbf{W}^n + \frac{1}{\Delta x}\int \sum u_k \psi (f_{j-1/2,k} - f_{j+1/2,k})dt,$$

from which, same as the UGKS, the equilibrium state at next level g^{n+1} can be obtained. However, in the continuum flow regime, as analyzed in previous chapter the above flux function does not go back to the NS solution at all. It is equivalent to the Kinetic Flux Vector Splitting (KFVS) scheme, where the numerical dissipation is proportional to the time step Δt instead of the particle collision time τ. When $\Delta t \gg \tau$, the numerical dissipation in KFVS will overwhelm the physical one. So, if the above method is used for the update of macroscopic flow variables, we can only get an AP method for the Euler solutions in the continuum flow regime. Practically, there are few rigorous tests which have been used to validate the above AP methods in capturing of the NS solutions in 2D and 3D cases, especially for the high Reynolds number flow.

Many AP schemes have difficulties to capture the Navier-Stokes limit. In a recent paper [Chen and Xu (2013)], a simplified implicit-explicit (IMEX) AP scheme and UGKS are investigated for the Navier-Stokes limit. The IMEX AP scheme and the UGKS are identical except the flux evaluation at a cell interface. For the UGKS scheme, the flux is based on the time evolution of the gas distribution function (5.13), where both transport and collision effect are coupled in this solution. However, for the IMEX AP scheme, the flux at a cell interface is based on the solution of Eq. (5.62), where only free transport is taken into account in this process. The reason for the comparison is that most AP schemes in the literatures do use Eq. (5.62) for the interface flux transport, and its dynamic effect has never been recognized clearly. The lid-driven cavity flow at Reynolds number 1000 is simulated by these two schemes, and the results are presented in the Fig. 5.2 to Fig. 5.4. The numerical results show that UGKS captures the NS viscous solutions with a successive mesh refinement. The velocity profiles are very close to the classical NS benchmark solutions. However, significant error appears in the IMEX AP scheme. In order to develop a valid AP method, the collision effect must be included in the flux transport. Otherwise, any attempt to get AP scheme with NS limit in the continuum flow regime is in vain, except with a fully resolved solution down to the mean free path scale.

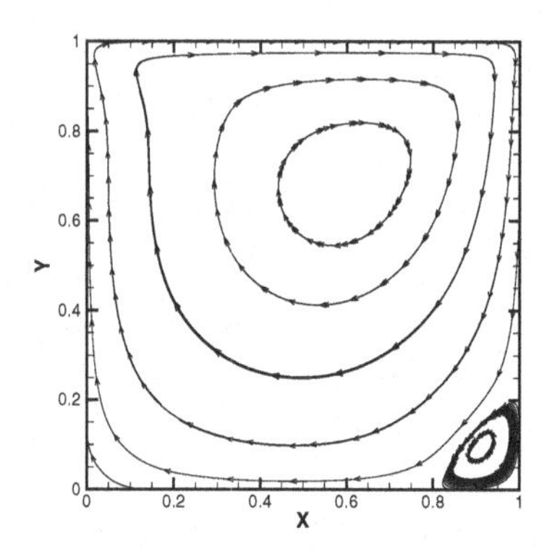

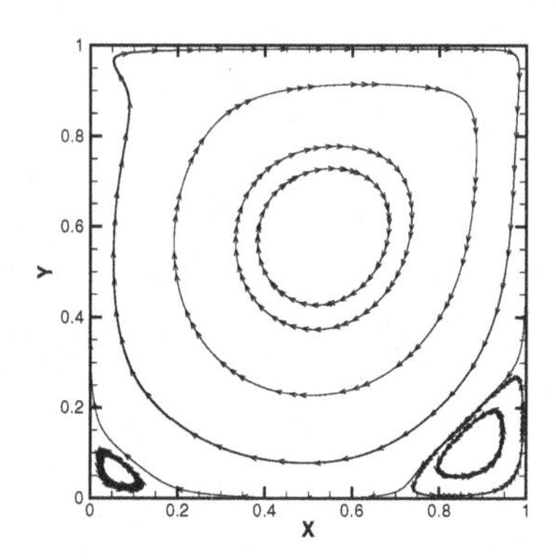

Fig. 5.2 The stream lines and flow field for lid-driven cavity flow at Re = 1000 with 61×61 mesh points in physical space and 20×20 mesh points in the velocity space for both IMEX AP (left) and UGKS scheme (right).

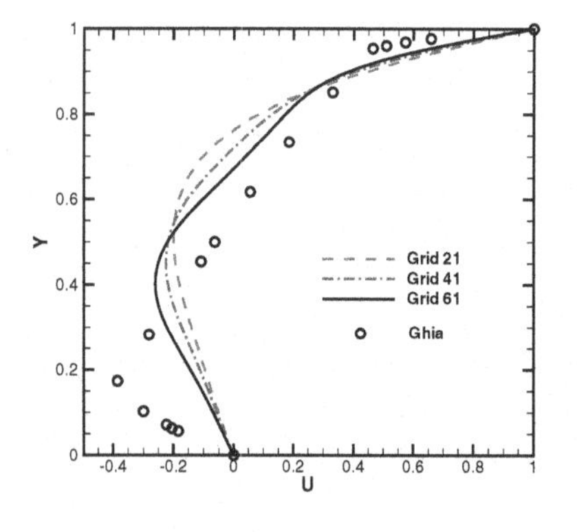

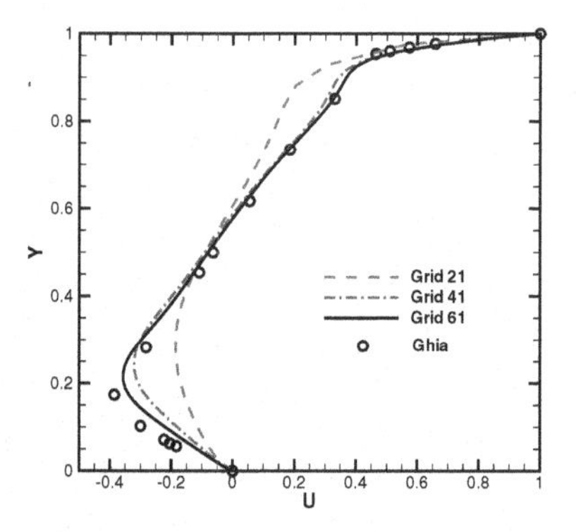

Fig. 5.3 The U velocity along the central line at $x = 0.5$ with different mesh points $(21 \times 21, 41 \times 41, 61 \times 61)$ in physical space. The left is from IMEX AP, and the right is from UGKS. The benchmark solutions are from [Ghia *et al.* (1982)].

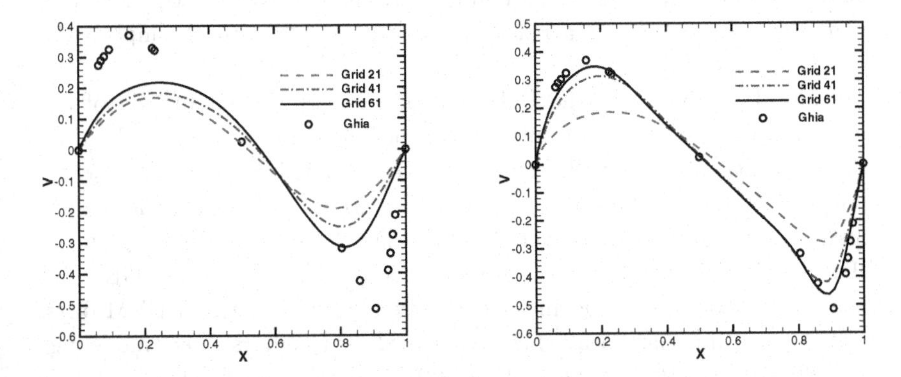

Fig. 5.4 The V velocity along the central line at $y = 0.5$ with different mesh points ($21 \times 21, 41 \times 41, 61 \times 61$) in physical space. The left is from IMEX AP, and the right is from UGKS. The benchmark solutions are from [Ghia *et al.* (1982)].

Moments and discrete velocity method vs UGKS

As discussed in Chapter 3, in the rarefied flow computations starting from the Boltzmann equation, the moment and discrete velocity methods (DVM) are two main methods currently being used and rapidly developed. The moment methods, such as the Grad, R13, and many others, are mainly concentrating on the flow description of accumulating effect of a gas system, the so-called wave or hydrodynamic properties. This can be clearly observed in the Hermit expansion of the gas distribution function in the Grad's method, and its recovering of the Euler, NS, and many other hydrodynamic type equations. However, it is also well recognized that the moment method can be hardly extended to the highly rarefied flow computation. One of the main reason is that the moment expansion doesn't take into account individual particle motion in the free transport limit. In a highly rarefied flow regime, all particles are basically independent and their accumulating moments should be independent as well. Therefore, in the rarefied regime, there is no reason to believe that the moment equations can be properly closed through any closure condition.

For the DVM method, the particle velocity space is discretized directly, and the individual particle movement is followed. Due to its particle nature, the DVM is perfectly appropriate for the highly rarefied flow computation. Many Boltzmann solvers and the DSMC fully incorporate the particle feature. But, the DVM encounters great difficulties when it is used in the continuum flow regime, because it requires the DVM to represent an accumulating effect from individual particle movement to the hydrodynamic flow behavior. As a result, it is hard for DVM to accomplish the mission.

Now if we look that the UGKS, it is somehow a combination of the moments and DVM. In the integral solution used for the flux evaluation, the initial part f_0 is actually the DVM, where the individual particle transport is followed. In the integral part of the equilibrium state, the NS gas distribution function is recovered, which has the same mechanism as the Hermit expansion to recover the NS solutions. Therefore, the UGKS can be categorized as a hybrid method between the moment and DVM ones, and the coupling has been achieved smoothly through its direct modeling.

If the lattice Boltzmann method (LBM) [Chen and Doolen (1998)] is considered as a kind of DVM method, due to its regular lattice and specific transport and collision it recovers the incompressible NS solutions, especially the velocity field, accurately. However, the LBM with the separation of particle transport and collision can be modified according to the methodology in UGKS, such as the including the collision within the transport. As a result, the solution of LBM can be much improved, especially the pressure field [Guo *et al.* (2013); Wang *et al.* (2014)].

Extension to radiative transfer with diffusive limit

In radiation transport, the governing equation is a linear kinetic equation. Similar to the Boltzmann equation, the dynamics of radiation equation is driven by a balance between photon free transport and interaction with material medium. Since the collision frequencies vary by several orders of magnitude through the optical thin or thick material, equations of this type will exhibit multiscale phenomena, such as those in the rarefied and continuum flow regimes. For the neutron and radiation transport, many AP schemes have been proposed [Larsen *et al.* (1987); Larsen and Morel (1989); Jin and Levermore (1991, 1993); Klar (1998); Jin *et al.* (2000); Naldi and Pareschi (1998)]. All these methods have advantages and drawbacks, and there is still a need for other AP schemes for the radiative and neutron transport problems.

The UGKS provides a framework for the construction of schemes covering multiple scale transport mechanism. Recently, Mieussens applied the methodology of UGKS to the radiative transfer equation [Mieussens (2013)]. While such a problem exhibits purely diffusive behavior in the optically thick (or small Knudsen) regime, he proved that UGKS is still asymptotic preserving in this regime, and captures the free transport regime as well. Moreover, he modified the scheme to include a time implicit discretization of the limit diffusion equation, and to correctly capture the solution in case of boundary layers. Contrary to many AP schemes, the UGKS-type AP method for radiative transfer is based on a standard finite volume approach, and it does neither use any decomposition of the solution, nor staggered grids. It provides a general framework for the solution of transport equation. Along the same line, recently the UGKS has been extended to solve grey radiative transfer equations [Sun *et al.* (2014)],

$$\frac{\epsilon^2}{c}\frac{\partial I}{\partial t} + \epsilon \boldsymbol{\Omega} \cdot \nabla I = \sigma \left(\frac{1}{4\pi} acT^4 - I \right),$$

$$\epsilon^2 C_v \frac{\partial T}{\partial t} = \epsilon^2 \frac{\partial U_m}{\partial t} = \sigma \left(\int I d\boldsymbol{\Omega} - acT^4 \right),$$

where I is the radiation intensity, T is material temperature, C_v is the heat capacity of the material, σ is opacity, a is radiation constant, c is the speed of light, ϵ is the Knudsen number for photon transport. Excellent results have been obtained from UGKS for the above equations.

Moving and adaptive mesh methods

Many kinetic solvers employ a uniform mesh in the particle velocity space. This kind of discretization makes the distribution function based method be very expensive for high Mach number and high Knudsen number flow computations. Owing to the wide spread of particle velocity distribution in high speed flow and narrow-kernel distribution function at high Knudsen number cases, a mesh in velocity space has to cover an extremely large domain with high resolution. When a uniform mesh is used in the velocity space, the storage and computational load will be unendurable. As pointed out in [Bird (1994)], the bottle-neck for a direct Boltzmann solver in aerospace application is the discretization of a large particle velocity space and the calculation of the complicated particle collision term. Moreover, uniform mesh is inefficient to reflect the dramatic changes of distribution function in rarefied high speed flow. In recent years, adaptive velocity space technique has been increasingly used in solving the kinetic equations [Gutnic *et al.* (2004); Mehrenberger *et al.* (2006)]. Kolobov *et al.* introduced the quadtree into the velocity space, and showed the advantages of quadtree structure in solving the collision term of the kinetic equation for the homogeneous flow [Kolobov *et al.* (2010); Kolobov and Arslanbekov (2012)]. It is believed that adaptive mesh has to be used for the Boltzmann solver in the hypersonic rarefied flow simulations. But, for an inhomogeneous flow computation the use of quadtree technique in the Boltzmann solver introduces additional difficulty to keep the momentum and energy conservation during the particle-redistribution process. Recently, many kinetic schemes with adaptive particle velocity space have been constructed [Kolobov *et al.* (2010); Kolobov and Arslanbekov (2012); Baranger *et al.* (2014); Yu (2013); Baranger *et al.* (2014)]. A quadtree technique has been introduced into the UGKS for inhomogeneous flow simulation [Chen *et al.* (2012b)], especially for the hypersonic flow computations [Yu (2013)]. At the same time, in order to simulate the moving boundary problem accurately and efficiently, a moving mesh UGKS method in the physical space has been constructed [Chen *et al.* (2012b)] along the same line as the moving mesh GKS [Jin and Xu (2007)], see Appendix D for the formulation. The related test cases will be presented in the subsequent chapters.

UGKS with full Boltzmann collision term

The UGKS presented in this chapter is mainly based on the kinetic Shakhov model. Since the UGKS is a multiscale modeling scheme, the accuracy of the scheme in different flow regimes has different sensitivity on the kinetic model equations. In the near continuum flow regime with $\Delta t \gg \tau = \mu/p$, the quality of the UGKS is mainly controlled by the flux function across a cell interface, which is similar to GKS, all kinetic models will work equally well, once the flux function takes both transport and collision into its time evolution. In the transition regime $\Delta t \geq \tau$, both the flux function and the collision term inside each control volume will contribute to the quality of the scheme. In the highly rarefied flow regime $\Delta t < \tau$, the interface flux function will automatically reduce to the particle free transport, which has no dynamic influence from the collision term, but the collision effect inside each control volume will become important. Even though the UGKS based on the Shakhov model is accurate enough for real engineering applications in all flow regimes, as shown in the following chapters, the full Boltzmann collision term can be still implemented in UGKS to improve the accuracy of the method. The implementation of the full Boltzmann collision term is only useful in the local cells where the time step is compatible with the particle collision time, i.e., $\Delta t \leq \tau$. In the cell with $\Delta t > \tau$, the direct modeling of collision effect inside each control volume will go beyond the Boltzmann modeling scale and the accumulating effect from the multiple particle collision has to be taken into account. In this case, the differences between the solutions from the kinetic collision models and accumulating solution of the full Boltzmann collision term diminish exponentially as a function of $\Delta t/\tau$.

The UGKS is to model governing equation in the scale of mesh size and time step. As shown in many studies [Sun *et al.* (2014); Liu and Zhong (2014)], from any initial non-equilibrium distribution functions, the time evolution of the gas distribution function of the full Boltzmann equation and the kinetic model equations will get close to each other after a few collisions, such as $\Delta t \simeq \tau$. Based on this observation, we can easily develop a UGKS with the full Boltzmann collision term $Q_B(f)$. With the same method for the update of the conservative flow variables,

$$\mathbf{W}_i^{n+1} = \mathbf{W}_i^n + \frac{1}{\Delta x}(\mathbf{F}_{i-1/2} - \mathbf{F}_{i+1/2}),$$

where the flux can be still obtained from the integral solution of the kinetic model equation, the update of the gas distribution function in the UGKS can be changed to

$$\begin{aligned} f_{i,k}^{n+1} = f_{i,k}^n &+ \frac{1}{\Delta x}\int_{t^n}^{t^{n+1}} u_k(f_{i-1/2,k} - f_{i+1/2,k})dt \\ &+ \Delta t\left(\eta\frac{1}{2}[Q_M(f^{n+1}) + Q_M(f^n)] + (1-\eta)Q_B(f^n)\right), \end{aligned}$$

where Q_M is the collision term of the Botzmann model equations, Q_B is the full Boltzmann collision term, and the switch function η depends on the ratio of the time step to the local particle collision time, such as $\eta = H[\frac{\Delta t}{\tau} - 1]$ and $H[x]$ is the Heaviside function. The solutions from the UGKS with the above full Boltzmann collision term and the UGKS scheme with purely Shakhov model have been compared extensively in many cases [Liu *et al.* (2014)]. The improvement from the above scheme is not so significant, because the kinetic model equation based UGKS is accurate enough in most computations. Therefore, for the UGKS, the slight improvement from the full Boltzmann collision term on the numerical solution is only limited to a small region in the transitional flow regime when the time step is less than the particle collision time. In practical engineering applications, the UGKS from the kinetic model equations is a good choice for the flow simulation covering multiple flow regimes.

5.7 Summary

In this chapter, we present the UGKS method for flow simulation in the entire Knudsen regimes. The critical ingredient in UGKS is the use of the integral solution of the kinetic model equation in the flux evaluation across a cell interface. The integral solution covers a smooth transition from the free molecular transport to the hydrodynamic NS solution. The UGKS provides a scale dependent evolution process. In comparison with many other kinetic equation solvers, the UGKS doesn't use a purely upwind technique for the flux transport. The adaptation of the upwind or free molecular transport constraints the kinetic solver to the kinetic scale only. Otherwise, a significant amount of numerical dissipation will be introduced. However, an evolution solution of the kinetic model equation, which includes the effect of particle collision, goes beyond the kinetic scale gas evolution. Theoretically, the UGKS provides a continuum spectrum of discrete governing equations from the free molecular transport to the NS solutions. The UGKS presents a general framework for simulating transport process, such as non-equilibrium flow, radiation, and neutron transport, which is especially valuable for the study of the phenomena with multiple scales. Its validity in the capturing of low speed microflows and hypersonic rarefied flows will be presented in subsequent chapters.

Chapter 6

Low Speed Microflow Studies

6.1 Introduction

Rapid advancements in micro and nano fabrication technologies over the past decades have triggered the study of microflows [Karniadakis *et al.* (2005)]. The tiny length scales usually encountered in microflow devices imply that non-equilibrium effects may turn out to be important even at normal pressures. The mean free path of atmosphere at sea level is approximately $64nm$. Small size machines, called micromachines, are being designed and built. Their typical sizes range from a few microns to a few millimeters. Consequently, the ratio of the mean free path of gas molecules to the characteristic dimensions of a MEMS device can be appreciable and cover a wide range of values, especially the regime from low transition to continuum one.

In fact, rarefied gas flows occur in many micro-electro-mechanical systems (MEMS), such as actuators, microturbines, gas chromatographs, and micro air vehicles (MAVs). A correct prediction of these flows is important to design and develop MEMS. In a modern disk drive, the read/write head floats at distances of the order of $50nm$ above the surface of the spinning platter. The prediction of the vertical force on the head is a crucial design calculation since the head will not accurately read or write if it flies too high. If the head flies too low, it can catastrophically collide against the platter. Micro-channels may have further computer applications because they are supposed to dissipate the heat generated in microchips more effectively than fans, and may be used as a more practical cooling system in integrated circuit chips. Since micro-devices are gaining popularity both in commercial applications and in scientific research, there exists a rapidly growing interest in improving the conventional design techniques related with these devices. Micro-devices are often operated in gaseous environments (typically air), and thus their performances are affected by the gas around them. The numerical study plays an important role in microflow studies.

6.2 Numerical Methods for Microflow

Due to the involvement of tiny length scales, the use of hydrodynamic scale flow physics may break down for the accurate prediction of fluid transport in micro devices. Hence, it seems problematic to adopt the Navier-Stokes model for microscale gas flows with significant effects of rarefaction. Central to these considerations, the applicability of the classical continuum hypothesis has often been seriously debated and tested in the presence of prominent micro-scale effects. Besides the validity of the constitutive relationship, another crucial factor which will effect the microflow physics is the boundary treatment. The surface-effects considerably influence a flow in the near-wall region, referred to as the Knudsen layer, which, because of the small scale of the system, represents a substantial portion of the gas volume. These effects are transmitted by the gas molecules interactions with surfaces. The structure of the Knudsen layer in the gas flow and the conditions on fundamental modeling techniques have been investigated by many researchers [Lockerby *et al.* (2005); Guo and Zheng (2008); Myong *et al.* (2005)]. When gas molecules collide with a solid boundary, those are temporarily adsorbed on the wall and are subsequently ejected. This allows a partial transfer of momentum and energy of the walls to the gas molecules. With respect to the cell size and time step resolution, if within a numerical time step the frequency of particle collisions with the wall is very large, the momentum and energy exchange is virtually complete and the no-slip/ no-jump boundary conditions may prevail for all practical considerations. However, in a less-dense system, such as the cell size is on the order of particle mean free path, deviations from such idealization are significantly more ominous. Thus, in the microflow study, both the physical size and non-continuum effect profoundly affect the mass, momentum and energy transport; and lead to additional effects like slip flow, rarefaction, compressibility, etc. In addition to exhibiting rarefied behavior, gas flows in micro-conduits also tend to exhibit significant compressibility effects even for cases with low Mach numbers, which is a significant deviation from the corresponding macro-scale counterpart.

In microflow study, the flow regimes are characterized by the Knudsen number Kn, which is defined as the ratio of molecular mean free path to a characteristic length scale. This definition of Kudsen number has different physical meaning in comparison with the number obtained from the Chapman-Enskog expansion of the Boltzmann equation, where the Knudsen number is defined as the ratio of mean free path over the length scale for the appreciable variation of macroscopic flow variables. Therefore, even with the same value of Knudesn number, different non-equilibrium effect can be presented. In the flow study, the continuum regime is in the range of $Kn < 0.001$, followed by the slip regime $0.001 < Kn < 0.1$. The Knudsen number in the transition regime is between 0.1 and 10. Even though it is commonly believed that the Navier-Stokes equations are applicable in the continuum and slip regimes, the validity of these macroscopic description depends on the physical problems and quantities to be evaluated. Even in a fully continuum flow regime, the Navier-Stokes equations cannot be claimed to describe everything properly, where the ghost effect may appear in some cases [Sone *et al.* (1990)], especially for those related to heat generated motion around sharp corner. Since experimental study of microflow is difficult due to the small physical dimension, the development of accurate numerical algorithm for microflow simulation will play an important role, especially for non-equilibrium flow with heat transfer in the low transition flow regime. The numerical challenge for flow in microdevices is that the flow transport may cover the whole flow regimes, from continuum to free molecular one. Therefore, a multiscale method with the correct capturing of both continuum and rarefied flow behavior is needed.

To the current stage, the molecular-based simulation techniques are the dominant methods for rarefied flow computation, even though these methods face statistical scattering problem in the low speed limit. The DSMC method is a particle-based simulation method for rarefied flows [Bird (1994); Belotserkovskii and Khlopkov (2010); Mohammadzadeh *et al.* (2012)]. The validity of this method has been presented in an enormous amount of research papers. Due to the particle based nature, the DSMC method cannot effectively reduce the statistical scatter encountered in microscale flows, which presents a very large noise to information ratio for flows having low speed and/or small temperature variation. Since the statistical scattering inherent in DSMC decreases with the inverse square root of the sample size, an extremely large sample size is required to reduce it to a level that is small in comparison with the small macroscopic velocity. This makes DSMC simulation of MEMS flows extremely time-consuming. Many small temperature variation phenomena can be hardly identified. In terms of accuracy, a valid DSMC simulation has strict numerical constraint to control the error from the time step, cell size, number of particles per cell, and the sample size. The numerical error in DSMC for heat conductivity is proportional to $(\Delta x/l_{mfp})^2$ and $(\Delta t/\tau)^2$, which is basically a first-order method [Rader *et al.* (2006)]. It is clear that DSMC can be hardly used to study flow in microdevices at small Knudsen number ($Kn < 0.01$). Even with so many limitations in the DSMC method for the microflow computations, the DSMC method is still considered to be the only reliable and accurate method here. In order to improve its efficiency, many attempts have been tried. The information preservation (IP) method is specially designed for low speed rarefied gas flows [Fan and Shen (2001); Sun and Boyd (2002); Masters and Ye (2007); Zhang *et al.* (2011)]. Since IP-DSMC updates the macroscopic variables for each DSMC particle, how to evolve the macroscopic variables when two DSMC particles get collision is still an active research topic. Another promising approach to improve the efficiency of the DSMC method is the low-noise DSMC through the modeling of linearized Boltzmann equation [Radtke *et al.* (2011)].

In terms of direct kinetic equation solvers, many schemes have been developed [Tcheremissine (2008); Aoki *et al.* (2001); Morinishi (2006); Beylich (2000)]. In the transport part of these Boltzmann solvers, the collisionless Boltzmann equation is usually used for the flux evaluation, i.e., the so-called Discrete Ordinate Method (DOM). As a result, similar to the DSMC method, the time step in these explicit schemes should be smaller than the particle collision time. Therefore, great difficulties will be encountered as well in these methods in the small Knudsen number limit, where the cell size may become much larger than the particle mean free path.

Some other promising solvers for microflow study are probably the moment equations, such as R13, R26, and many others [Struchtrup (2005); Gu and Emerson (2009); Cai and Li (2010); Myong (1999, 2001)]. But, their validation is only for the simple geometry and in the regimes close to the continuum, or in the low part of the transition regime. Another difficulty for these moment equations is the boundary treatment for the additionally introduced quantities. Physically, the moment of a gas distribution function is an averaged or coarse graining quantity from the kinetic scale to hydrodynamic scale, such as heat flux and stress. The dynamics of these moment equations are mostly rooted in the hydrodynamic scale approach. The philosophy underlying this direction of research is that the hydrodynamic-scale modeling can be still used to resolve kinetic scale dynamics with the inclusion of more accumulated information. There may have the possibility for success, but in practice there is more uncertainty involved due to the modeling scale differences. Theoretically, it becomes difficult that a universal moment system can be constructed in the description of vast amount of multiscale non-equilibrium flow behavior, because the starting point of all these approaches is the equilibrium one. As the Knudsen number increases, the individual particle transport will play a dominant role, such that the distribution function may become a collection of independent particle transport. But, the moment method is mainly based on their accumulating effect, where the intensive particle interaction is needed if they are associated with correlated flow behavior.

Certainly, there are many modified kinetic solvers with asymptotic property in the continuum flow regime. The continuum asymptotic limit of these schemes is mostly for the inviscid Euler solutions. Even with many AP schemes constructed recently, most of them have never been fully tested and validated with the continuum NS solution and rarefied DSMC solution.

6.3 Unified Gas Kinetic Scheme for Microflow

The microflow in microdevices is likely to cover the whole flow regimes from continuum and rarefied. With affordable computational recourses, the mesh cannot be set to be smaller than the local particle mean free path everywhere. Therefore, the ratio of the local cell size to the local particle mean free path may cover a wide range of values. With the definition of cell's Knudsen number Kn_c, this number can be varied significantly in microflows, such as $10^{-4} < Kn_c < 10$. In order to capture the corresponding flow physics, a scheme should be able to recover both the hydrodynamic and kinetic flow physics. The DSMC and many direct Boltzmann solvers are single scale numerical schemes, which require the numerical cell size to be less than the local particle mean free path. So, it will be hard to apply these methods in the microflow study.

The UGKS presented in the last chapter provides such an ideal flow solver for the multiscale modeling. The UGKS not only has the asymptotic property to capture the NS solution in the continuum regime as $Kn_c \ll 1$, but also get the exact collisionless solution as $Kn_c \gg 1$. So, the validity of the UGKS for microflow study will be fully checked in this chapter in the transition regimes. It is certainly true that the current UGKS is based on the kinetic model equations. But the numerical examples seem show that for the low speed microflows different molecular models have little influence on the flow patterns, where the complicated full Boltzmann collision term has marginal effect on the solution once the viscosity and heat conduction coefficients are correctly defined in the kinetic schemes. The collision with the solid wall plays an even more important role in microflow. The UGKS doesn't use any slip boundary condition explicitly. As presented in last chapter, the boundary condition in all flow regimes is basically implemented through the dynamic model with colliding and reflecting particles from the wall, and collisions among particles close to the wall as the cell size being larger than the local particle mean free path. The Knudsen layer next to the boundary should be captured automatically in the UGKS computation.

The focus of this chapter is to present a large number of test cases and to validate the UGKS in microflow applications. Since there is limited analytic solution in the transition regime, the validation is mainly based on the comparison between the solution of the UGKS and that of the DSMC or direct Boltzmann solvers. Another attractive property of the UGKS for the microflow simulation is that there is no statistical noise or scattering and the scheme has second order accuracy. The temperature variation associated with microdevices can be relatively small, but the second-order derivatives of the temperature may be relatively large, which can generate thermal stress slip flows. Another advantage of UGKS is the smooth transition to cover flow physics in different regimes. The UGKS has great advantage to capture the low speed and small temperature variation because the mesh points in the velocity space can be much reduced here. In order to get the DSMC solution with a reasonable amount of computational time, the magnitude of the temperature variation in some test cases in this chapter has been set to a large value, which is beyond the physical reality in a microdevice. But, for the validation purpose it is still helpful to use this kind of test cases.

In the following, we will validate the UGKS for the microflow study through different kind of test cases. The test cases cover steady and unsteady, and continuum and rarefied flows. All these test cases may be used in the future as benchmarks for any new scheme developed for non-equilibrium flow study.

6.4 Microflow Studies

In this section, we are going to present numerical tests to demonstrate the capability of UGKS for flow study in micro-scales. Most test cases involve heat. New phenomena in the transition flow regime, such as the possible reversing of flow velocity close to the boundary in the thermal creep flow, have been observed for the first time by UGKS computation, which has been confirmed later by using DSMC. The anti-Fourier heat conduction behavior in the cavity case at large Knudsen numbers has been obtained as well in the UGKS computation. In all these test cases, the fully diffusive boundary condition with accommodation coefficient 1 has been implemented. Besides cases included in this chapter, many other cases have been tried as well. The UGKS has never failed a single case for the non-equilibrium microflow study.

6.4.1 *Couette Flow*

The Couette flow is a standard simple test for the whole flow regime. It is a steady flow that is driven by the surface shear stresses of two infinite and parallel plates moving oppositely along their own planes. The Knudsen number is defined as $Kn = \lambda_{HS}/h$, where λ_{HS} is the mean free path based on hard sphere model, and h is the distance between the plates.

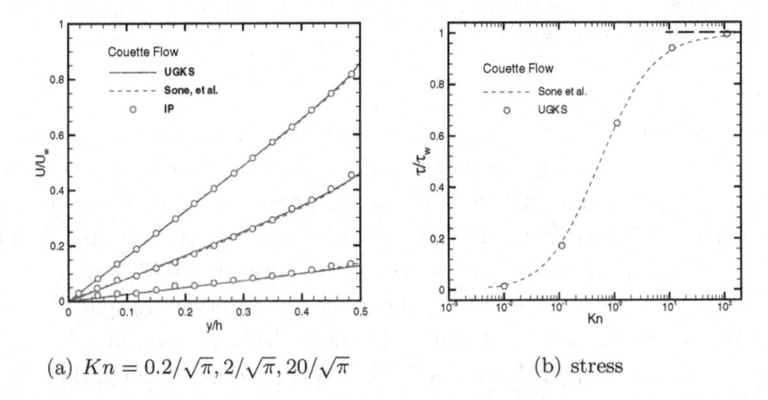

(a) $Kn = 0.2/\sqrt{\pi}, 2/\sqrt{\pi}, 20/\sqrt{\pi}$ (b) stress

Fig. 6.1 Couette flow [Huang *et al.* (2013)]. (a) Comparison of velocity profiles in the upper half channel given by the IP method [Fan and Shen (2001)], linearized Boltzmann equation [Sone *et al.* (1990)], and UGKS. (b). Relation of drag versus Knudsen number.

In the transition regime, three Knudsen numbers are considered:

$0.2/\sqrt{\pi}, 2/\sqrt{\pi}$, and $20/\sqrt{\pi}$. To resolve the flow fields well, 100 cells are employed in the current calculation for all three cases. Figure 6.1a compares the velocity profiles given by UGKS, the linearized Boltzmann equation [Sone *et al.* (1990)], and IP-DSMC results [Fan and Shen (2001)]. All solutions have excellent agreement. Figure 6.1b also compares the relation of the surface shear stress versus the Knudsen number given by various methods. The normalization factor is the collisionless solution [Fan and Shen (2001)]. Both solutions agree nicely with each other in the whole flow regime. The above test is basically an isothermal one.

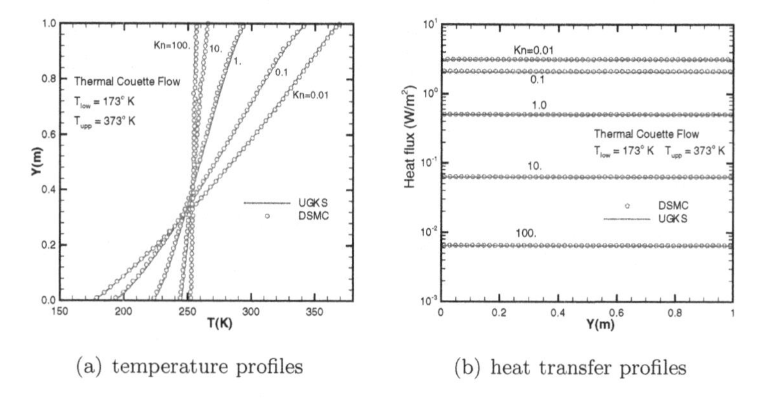

(a) temperature profiles

(b) heat transfer profiles

Fig. 6.2 Thermal Couette flow at $Kn = 0.01, 0.1, 1, 10$, and 100 calculated by DSMC and UGKS [Huang *et al.* (2013)].

Simple heat conduction problem in rarefied gas is also a valuable case to test the capability to capture thermal effect. This consists of two stationary parallel surfaces maintained at different temperatures. The same problems have been studied in [Sun and Boyd (2002); Masters and Ye (2007)]. The up and down surfaces are maintained at temperature of $173K$ and $373K$ separately with a $1m$ gap between them and the intervening space is filled with argon gas at various densities to have the corresponding Knudsen numbers $Kn = 0.01, 0.1, 1, 10$, and 100. The $1D$ computational domain is discretized with 100 cells in the physical space and 28×28 grid points in the velocity space. Figure 6.2 presents the temperature profiles and heat flux results from the unified scheme and the benchmark DSMC solution. There is an excellent agreement between UGKS and DSMC solutions in the whole range of Knudsen numbers.

6.4.2 *Pressure Driven Poiseuille Flow*

The pressure driven Poiseuille flow at Knudsen number $Kn = 0.1$ has been simulated using UGKS. As pointed out by many authors, see [Santos *et al.* (1989); Malek *et al.* (1997); Uribe and Garcia (1999); Aoki *et al.* (2002)], even for this simple case with relative small gradient and Knudsen number, the Navier-Stokes equations fail to predict qualitative correct solution in the cross stream direction. Specifically, the Navier-Stokes equations fail to reproduce a non-constant pressure profiles in the cross stream direction, which are both predicted by the kinetic theory and observed in the DSMC simulations. Furthermore, based on the Navier-Stokes equations it is not possible to correct this failure by modifying the equation of state, transport coefficients or boundary conditions. Unlike the slip phenomena, the discrepancy is not just near a boundary but throughout the system.

The set up of the pressure driven case is in the following papers [Zheng *et al.* (2002a,b)]. The simulation fluid is a hard sphere gas with reference particle mass $m = 1$ and diameter $d = 1$. At the reference density of $\rho_0 = 1.21 \times 10^{-3}$, the mean free path is $l_0 = m/(\sqrt{2}\pi\rho_0 d^2) = 186$. The distance between the thermal walls is $L_y = 10l_0$ and their temperature is set to be $T_0 = 1.0$. The reference fluid speed is $U_0 = 1 = \sqrt{2kT_0/m}$, so the Boltzmann constant is taken as $k = 1/2$. The reference sound speed is $c_0 = \sqrt{\gamma kT_0/m} = 0.91$ with $\gamma = 5/3$ for a monatomic gas. The reference pressure is $p_0 = \rho_0 kT_0/m = 6.05 \times 10^{-4}$. The pressure gradient is chosen so that the flow will be subsonic and laminar. The length of the channel is about 30 mean free paths. Specifically, $dp/dx = 1.08 \times 10^{-7}$ ($p_+ = 3p_0/2, p_- = p_0/2, L_x = 30l_0$) is applied, the Knudsen number is $Kn = l_0/L_y = 0.1$, and the Reynolds number is of order one. In all calculations, the cell size is the half of the mean free path of the initial data.

Fig. 6.3 shows the flow distribution along the centerline of the channel as well as the distributions in the cross stream direction, where the solid lines are from the UGKS and the circles are the DSMC solutions. Since the Maxwell boundary condition with accommodation coefficient 1 is implemented in the UGKS, the slip boundary boundary condition appears automatically. In this pressure driven Poiseuille flow case, reasonable agreement can be obtained between UGKS and DSMC solutions.

This case has $Kn = 0.1$, which is in the upper slip flow regime. It is commonly agreed that the Navier-Stokes equations can be still applied here. Actually, with the implementation of slip boundary condition, the NS

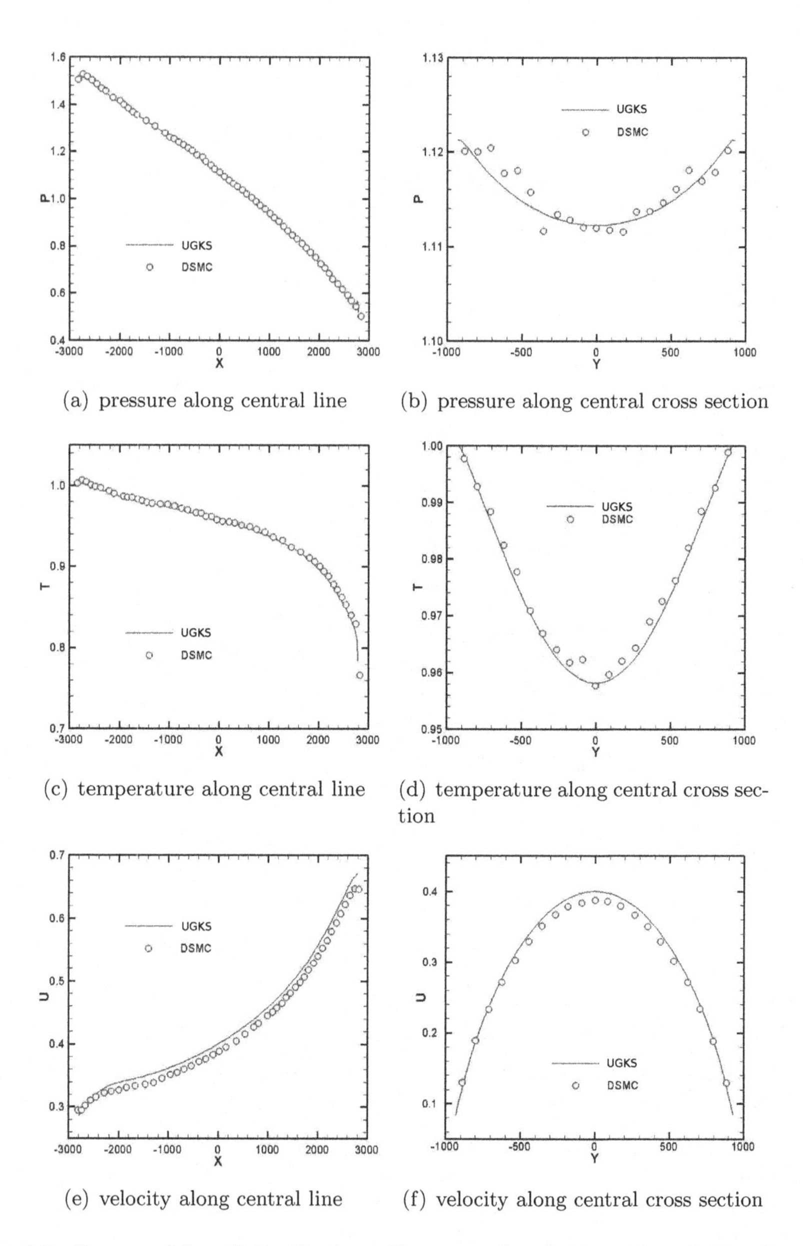

(a) pressure along central line

(b) pressure along central cross section

(c) temperature along central line

(d) temperature along central cross section

(e) velocity along central line

(f) velocity along central cross section

Fig. 6.3 Pressure driven Poiseuille flow. The comparison between the solutions from DSMC and UGKS.

solutions along the centerline of the channel are similar to the results from UGKS. However, in the cross-stream direction at the middle of the channel the flow distribution calculated by GKS for the NS solution is shown in Fig. 6.4. This figure clearly shows the inconsistency in pressure distribution between the NS and DSMC solutions. For example, the Navier-Stokes equations with slip boundary condition predict the maximum pressure at the center. But, the DSMC gives the opposite conclusion. This phenomenon is irrelevant with the slip boundary condition or constitutive relationship. So, the claim of the applicability of the NS equations in the slip regime should be careful. It depends on the flow variables to be evaluated. The NS solution here is obtained using GKS, which is presented in Chapter 4. The main difference of GKS from UGKS is that the distribution function in GKS is reconstructed using the Chapman-Enskog expansion. This test clearly shows that the NS distribution function through the Chapman-Enskog expansion is different from the real solution, even at $Kn = 0.1$.

6.4.3 *Slider Air Bearing Problem*

The slider air bearing problem arises from the hard disk industry. The demand for higher recording density requires the spacing between the read/write head and the disk to be in nano-scale, where the gas is rarefied and non-equilibrium phenomena may occur. New technologies have also been developed over the years to further increase the recording density, such as the heat-assisted magnetic recording (HAMR) that introduced a heat source into the slider system. Both the nano-scale flying height of the slider and the heat source could affect the force on the slider, which is considered crucial in the slider design.

The schematic of the two-dimensional slider air bearing problem is shown in Fig. 6.5(a). A stationary wall with length L_1 and temperature T_0 is inclined above a moving wall with velocity U, temperature T_0 and length L. Both walls are assumed to be fully diffusive. Inlet and outlet boundaries are connected to the environment and are assumed to have inlet and outlet pressure, with $p_0 = 1\text{atm}$ and $T_0 = 273\text{K}$ of the ambient air.

The system is characterized by the length/height ratio L/h_0, the inclination angle α, the Mach number Ma and the Knudsen number Kn. The Mach number and Knudsen number are defined as follows,

$$\mathrm{Ma} = \frac{U}{\sqrt{\gamma R T_0}}, \qquad \mathrm{Kn} = \frac{l}{h_0}, \tag{6.1}$$

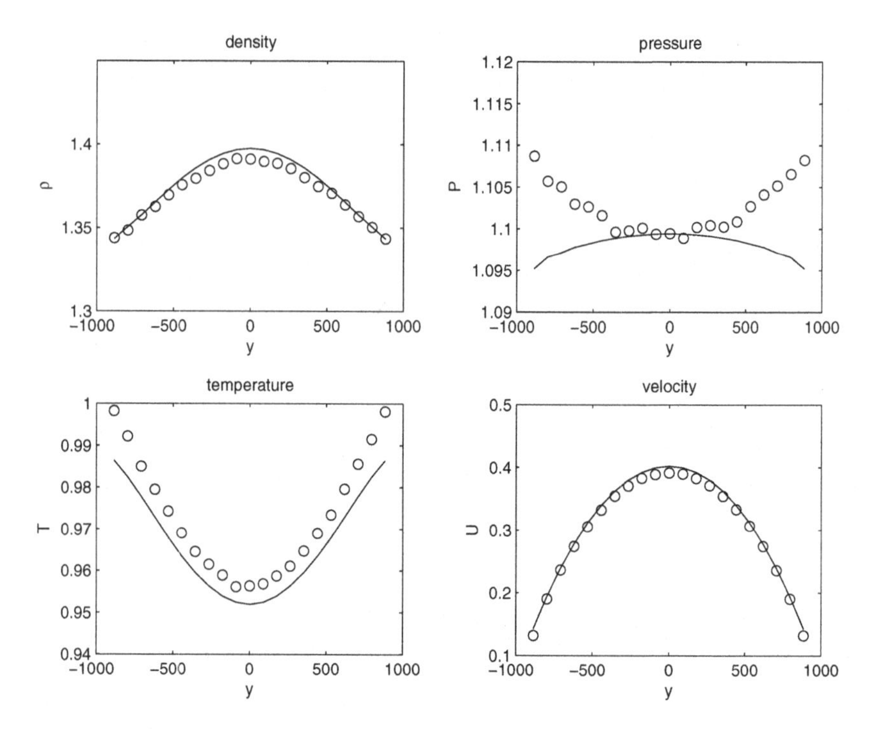

Fig. 6.4 Flow distribution in the cross-stream direction in the pressure driven case. Circles: DSMC results [Zheng *et al.* (2002b)], solid lines: GKS solutions [Xu (2001)]. The Navier-Stokes equations have opposite curvature in comparison with the DSMC solution in the pressure distribution.

where γ is the ratio of specific heat, R is the gas constant and l is the mean free path.

The load carrying capacity w on the upper surface is defined as [Alexander *et al.* (1994)]

$$w = \frac{1}{L}\int_0^L \frac{p(x) - p_0}{p_0} dx, \tag{6.2}$$

where $p(x)$ could be the gas pressure distribution along the upper surface or the distribution of force exerted on the upper surface per unit area. The former is calculated by taking the corresponding moment of the distribution function, and the latter is calculated from the momentum transfer of particles striking the wall.

To verify UGKS, simulations are carried out to compare with the DSMC solution in [Alexander *et al.* (1994)]. The length/height ratio is $L/h_0 = 100$ and the inclination angle is $\alpha = 0.01$rad. The gas is assumed to be argon with $\mu \sim T^{0.5}$. In our simulation, 5×350 grids are used in all cases. Fig. 6.5(b), 6.5(c), 6.5(d) show the comparison of pressure distribution for several different Mach numbers and Knudsen numbers. The load capacity is also computed and compared in Table 6.1.

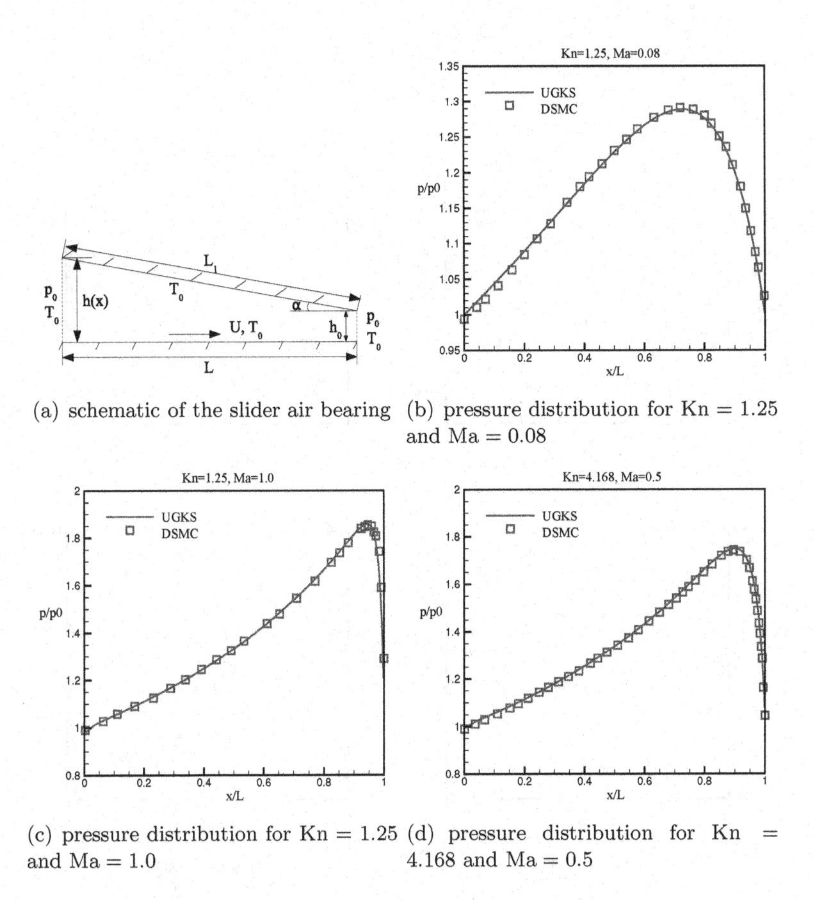

(a) schematic of the slider air bearing (b) pressure distribution for Kn = 1.25 and Ma = 0.08

(c) pressure distribution for Kn = 1.25 and Ma = 1.0 (d) pressure distribution for Kn = 4.168 and Ma = 0.5

Fig. 6.5 Slider air bearing simulations [Wang and Xu (2014)].

6.4.4 *Unsteady Rayleigh Flow*

The Rayleigh flow is an unsteady flow problem in which a plate below a gas at rest suddenly acquires a constant parallel velocity and a constant temperature. The set up of this test follows the work in [Sun (2003)]. The initial argon gas is at rest with a temperature of $273K$. When $t > 0$,

Table 6.1 Comparison of the load capacity computed from pressure and force.

Load	Kn = 1.25 Ma = 0.08	Kn = 1.25 Ma = 1.0	Kn = 4.168 Ma = 0.5
DSMC (pressure)	0.175	0.370	0.357
DSMC (force)	0.174	0.371	0.355
UGKS (pressure)	0.174	0.299	0.329
UGKS (force)	0.174	0.308	0.330

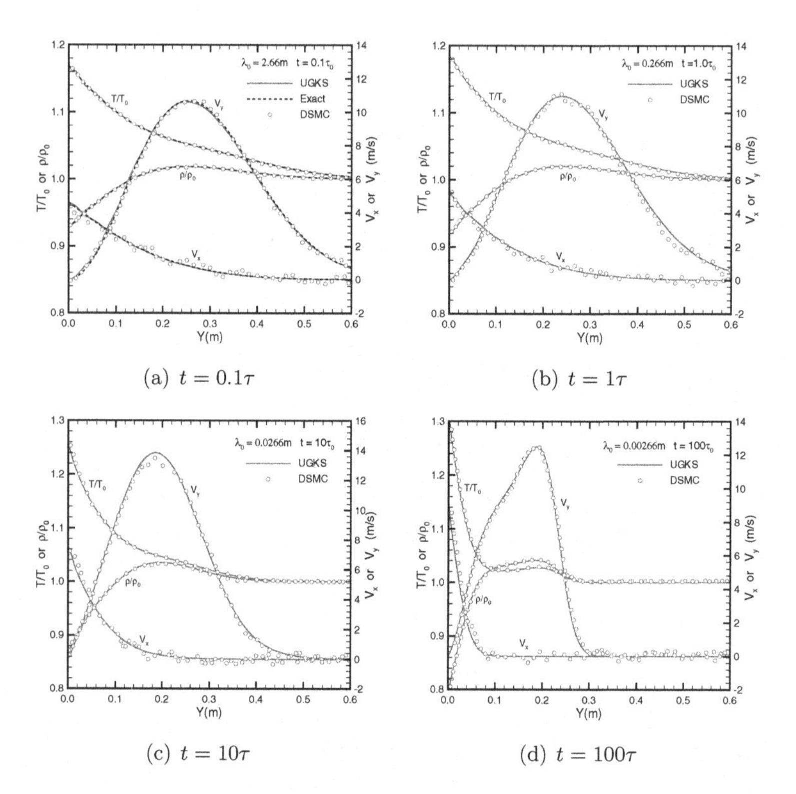

(a) $t = 0.1\tau$ (b) $t = 1\tau$ (c) $t = 10\tau$ (d) $t = 100\tau$

Fig. 6.6 Rayleigh problem at times $t = 0.1\tau, \tau, 10\tau$ and 100τ, which cover the free molecular, transition, and continuum regimes. The exact solution at 0.1τ means the solution from collisionless Boltzmann equation [Huang *et al.* (2013)].

the plate obtains a constant velocity $10m/s$ and a constant temperature $373K$. There is an analytical solution to the Rayleigh flow for times much less than the mean collision time $\tau_0 = \lambda_m/\nu_m$, where λ_m is the particle mean free path, and ν_m is the mean molecular speed with $\nu_m = \sqrt{8RT/\pi}$ [Bird (1994)]. In order to simulate the solutions of the Rayleigh problem at different output time, with a fixed mesh point in space, such as 100 cells in the vertical direction into the gas, the mesh size has to be changed because the wave propagating distance into the initial rest gas will be different. Figure 6.6a shows the simulated results at 0.1τ from the unified scheme and the DSMC solution along with the analytical solution of the collisionless Boltzmann equation at early times. All three solutions agree with each other very well. In comparison with DSMC method, the unified scheme has no statistical scattering in the velocity V_x profile. At time $t = 0.1\tau$, the unified scheme recovers the exact collisionless Boltzmann solution. As the time increases to $t = 1\tau$ and 10τ in the transition regime, the unified solutions and the DSMC solutions are shown in Figs. 6.6b and 6.6c. The small wiggles in V_x and V_y of the DSMC solution are absent in the unified results. The largest discrepancy appears in the peak V_y value at $t = 10\tau$. This discrepancy is most likely due to the low resolution in the DSMC solution. As time goes to $t = 100\tau$, the flow goes to the continuum flow regime. Figure 6.6d presents the results from the unified scheme and the DSMC. Continuing in this limit, the unified scheme will recover the Navier-Stokes solution accurately. The excellent agreement between unified and DSMC solutions validates the unified scheme in the unsteady flow computation, at least in this simple case.

The main difference between the unified scheme and the traditional DOM method is the different way to evaluate the cell interface flux. For the unified scheme, the integral solution (5.31) is used for the determination of the interface gas distribution function, which covers both hydrodynamic and kinetic scales gas evolution physics. For the DOM methods, only the f_0 term in (5.31) is kept, such as using the collisionless Boltzmann solution for the interface flux. As a result, the DOM methods lack the hydrodynamic part and will have problem in cases where the numerical cell size is much larger than the particle mean free path. For the same Rayleigh problem, as time increases, the wave starting from the wall will propagate and effect a large domain. With the same number of cells to resolve the flow in a large domain, such as 100 or 200 cells, the ratio of the cell size over the particle mean free path will become large. Then, the hydrodynamic effect will gradually become important with the increase of cell size relative to the particle mean free path. So, theoretically there will have deviation between the solutions of the unified scheme and the DOM method. Figures 6.7a and 6.7b present the solutions of the unified and DOM at $t = 100\tau$ with 100 and 200 mesh points in a physical space. It is clear the the numerical solution of DOM depends sensitively on the mesh size, but the unified scheme gives the

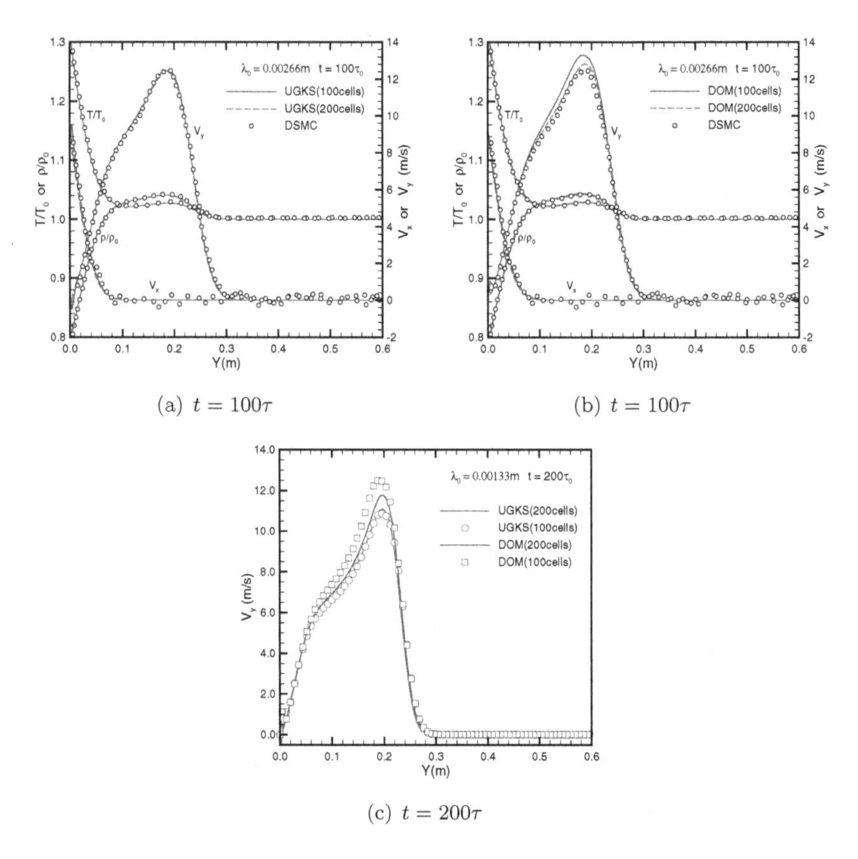

(a) $t = 100\tau$

(b) $t = 100\tau$

(c) $t = 200\tau$

Fig. 6.7 Rayleigh problem at times $t = 100\tau$ and 200τ. The comparison of the Unified and DOM solutions with different physical mesh points [Huang *et al.* (2013)].

same result even with the change of the ratio of the cell size over particle mean free path. From the DOM solutions, it is clear that the fine mesh size solution is more accurate than the coarse mesh solution because a small mesh size is close to its modeling assumption. To get the solution at $t = 200\tau$, an even large numerical mesh size is needed in order to cover the whole wave propagating domain. The difference between unified and DOM solutions becomes more obvious, see Fig. 6.7c. Therefore, if we go further into the continuum flow regime with the mesh size being much larger than the particle mean free path, the DOM solution will deteriorate due to its particle free transport mechanism across a cell interface. For the unified scheme, a smooth transition from kinetic to the hydrodynamic description can be obtained with the increasing of the mesh size.

6.4.5 *Thermal Transpiration*

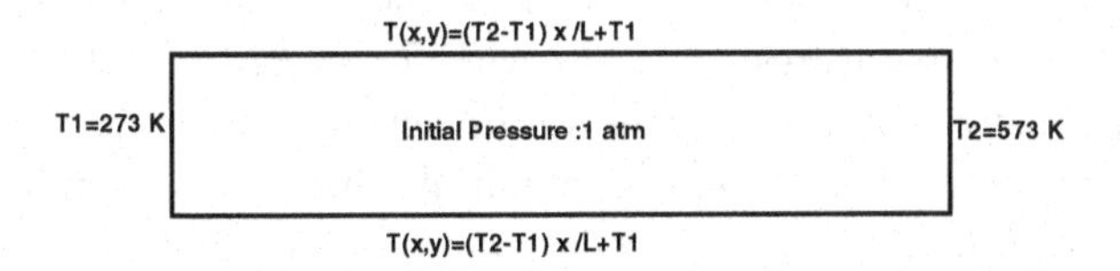

Fig. 6.8 Set-up of thermal creep flow with closed walls. Channel length to width ratio is 5.

Thermal transpiration is important physical phenomena in microdevices due to temperature variations. Consider a system wherein two cavities are maintained at the same initial pressure but dissimilar temperatures and joined by a tube of some length. If the width of the tube is large in comparison with the mean free path of the gas, the Navier-Stokes equations have a uniform pressure solution and a purely diffusion solution for the temperature field without fluid velocity. As the tube becomes narrow, such that the flow goes to transition and free molecular one, then molecules will creep through the tube from the cold to the hot reservoir. If the reservoir is sealed, the result will be a static pressure gradient. If they are open, then the result will be a continuous transport or pumping the gas from the cold to the hot reservoir. The detailed explanation can be found in [Karniadakis *et al.* (2005); Masters and Ye (2007)]. In the transition regime, the creep flow may eventually be balanced by reverse Poiseuille flow driven by the induced pressure gradient. Many experimental observations of thermal transpiration have been reported [Annis (1972); Knudsen (1950)] as well as practical application of the effect in Knudsen pumps and micropropulsion systems [Vargo *et al.* (1999); Alexeenko *et al.* (2005); Han (2010)]. From a modeling standpoint, a number of different techniques have been proposed, including various solutions of the linearized Boltzmann transport equation (BTE) [Sharipov (1999)], near continuum slip models [Karniadakis *et al.* (2005); Bielenberg and Brenner (2006)], and DSMC simulations [Alexeenko *et al.* (2005); Han (2010)]. Linearized BTE methods are suitable for problems with small thermal gradients, i.e., weakly non-equilibrium, but are likely inadequate for the complex geometries and large thermal gradients that may be encountered in micro- and nanoscale systems. Near continuum models are only applicable for a small range of flow conditions.

The first problem to be considered is the same case as that presented in [Masters and Ye (2007)]. There is a sealed 2D microchannel with a rectangular cross section and geometry suitable for MEMS applications as shown in Fig. 6.8. The two ends of the channel are maintained at two different temperatures $T_1 < T_2$ with $T_1 = 273K$ and $T_2 = 573K$. The temperature of the side walls varies linearly along the length of the channel and the working argon gas is initially in thermal equilibrium with the walls, i.e., $T(x, y) = (T_2 - T_1)x/L + T_1$ and at a uniform pressure of one atmosphere, i.e., $P(x, y) = P = 1atm$. Therefore, the mean free path of the gas is about $64nm$. The channel is discretized using a fixed grid points of 200 cells along the length and 40 along the width. With the change of channel width, the cell size will be changed as well. Again, the particle velocity space is discretized with a fixed grid points of 28×28 in the UGKS. The wall accommodating coefficient is equal to 1.

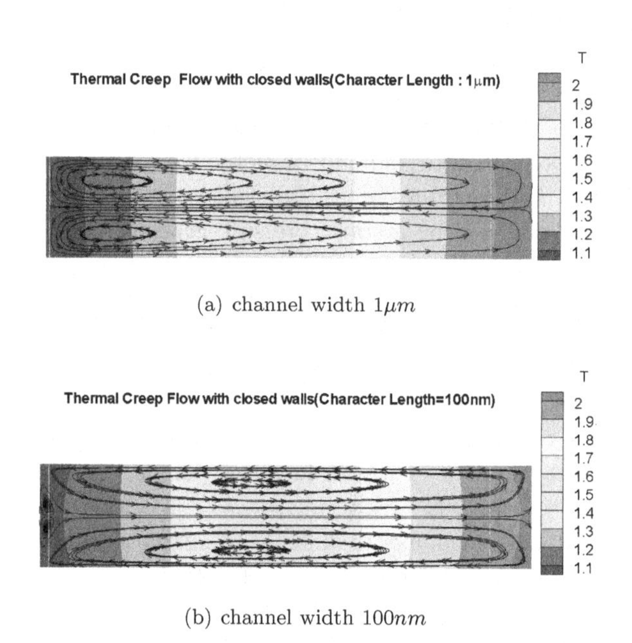

(a) channel width $1\mu m$

(b) channel width $100nm$

Fig. 6.9 Thermal creep flow with closed walls. Streamline distributions with channel widths $1\mu m$ and $100nm$ [Huang *et al.* (2013)].

Figure 6.9 presents streamline distributions inside the tubes calculated by the unified scheme at different channel widths $1\mu m$ and $100nm$. At $h = 1\mu m$, the corresponding Knudsen number is $Kn = 0.064$, and the flow is moving from low temperature region to the high temperature region along the boundary, and the flow returns in the central region. However, at $h = 100nm$ with $Kn = 0.64$, the flow direction is reversed, even though the flow velocity is very small. This phenomena of reversing gas velocity along the wall surface were first observed by the UGKS calculation, which were confirmed later by intensive DSMC calculation.

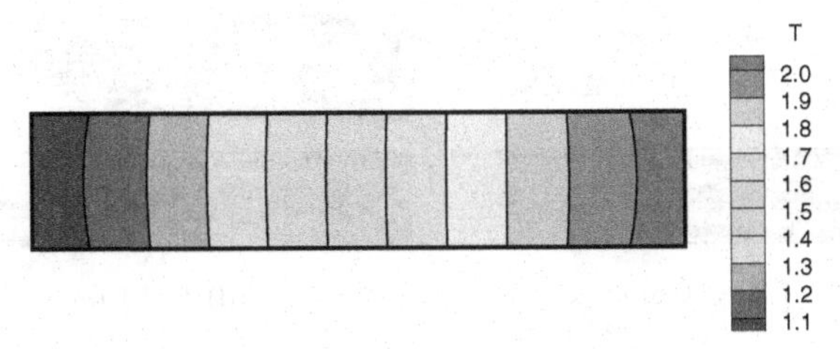

(a) temperature contours. Solid line: UGKS, dash-dot line: DSMC.

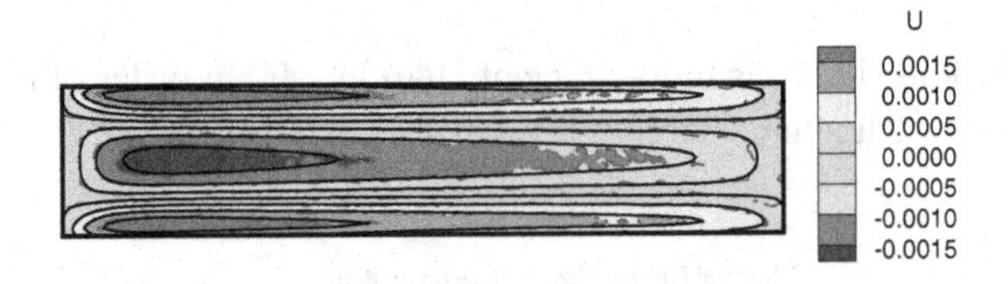

(b) velocity contours. Solid line: UGKS, dash-dot line: DSMC.

Fig. 6.10 Temperature and velocity contours for the thermal creep flow at channel width $1\mu m$ calculated by unified and DSMC method [Huang *et al.* (2013)].

At width $h = 1\mu m$, the comparison of temperature and velocity contours between UGKS and DSMC solutions is shown in Fig. 6.10, where perfect match has been obtained, especially for the temperature distribution.

With the changing of channel widths and the Knudsen numbers, the flow patterns inside the channel become complicated. In order to validate the observations from the UGKS, intensive DSMC calculations have been conducted. Figure 6.11 shows the streamline and temperature distributions from the UGKS and DSMC computations at different channel widths. To our surprising, the flow patterns from UGKS and DSMC match very well in all cases, even though the solutions of DSMC are a little bit more noise than UGKS ones. These results fully confirm the reliability and accuracy of

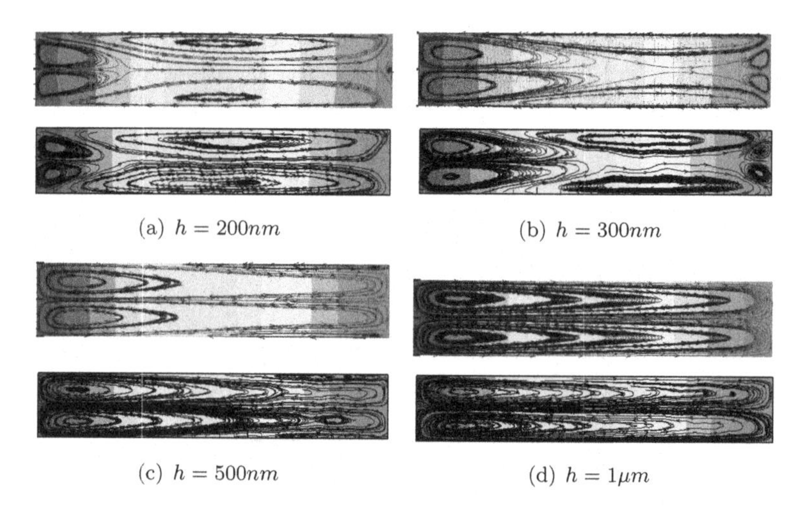

(a) $h = 200nm$ (b) $h = 300nm$

(c) $h = 500nm$ (d) $h = 1\mu m$

Fig. 6.11 The comparison of streamline distributions from UGKS and DSMC solutions at different channel widths. In each sub-figure, up: UGKS, down: DSMC. (a) $h = 200nm$; (b) $h = 300nm$;(c) $h = 500nm$; (d) $h = 1\mu m$.

the UGKS. The UGKS is more efficient than DSMC in order of magnitude, even in cases with such a large temperature variation.

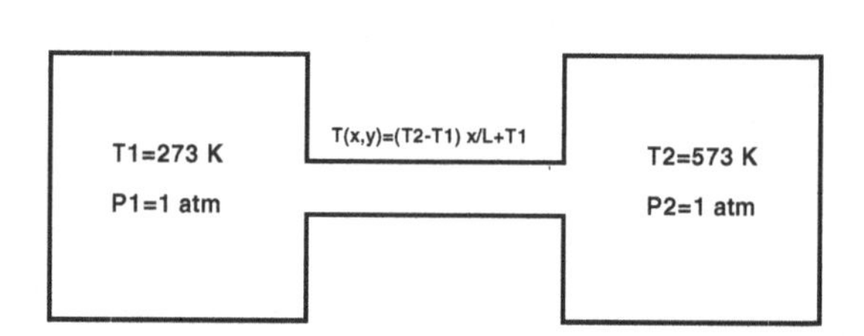

Fig. 6.12 Micro-channel connected with two cavities.

In order to compare the solution differences between the closed tube and a tube connected with two cavities, we test the cases of the same tube with widths $1\mu m$ and $100nm$, which are connected with two cavities with the same initial pressure and two temperatures T_1 and T_2 separately, see Fig. 6.12 for the set up. Different from the closed tube case, initially the gas will flow from the low temperature cavity to high temperature one. As

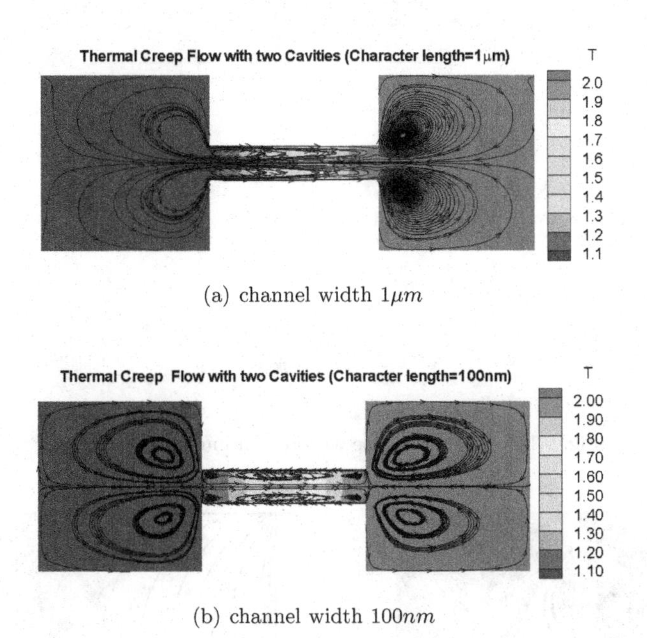

(a) channel width $1\mu m$

(b) channel width $100nm$

Fig. 6.13 Streamlines for the thermal creep flow with cavities [Huang *et al.* (2013)].

the pressure piles up inside the high temperature cavity, a back-flow will be formed inside the tube. To the steady state solution, the final mass flux through the tube will be zero, but the pressure gradient is kept along the tube. Figure 6.13 shows the streamlines for these cases.

6.4.6 *Flow Induced by Temperature Discontinuity*

In a sealed 2D domain, there are temperature discontinuities at the boundaries. The schematic of the boundary temperature distribution is shown in Fig. 6.14.

For UGKS calculation, the whole physical domain is covered by 80×80 mesh points, and the velocity space has 28×28 mesh points. The initial uniform pressure inside the domain is $1atm$, and the length of the square is set to make the corresponds Knudsen number $Kn = 0.1$. The results of the UGKS and DSMC solutions are presented for the temperature and streamlines in Fig. 6.15a, and temperature contours in Fig. 6.15b. Reasonable agreements have been obtained between these two solutions.

In order to show the computational efficiency of UGKS, we have tested

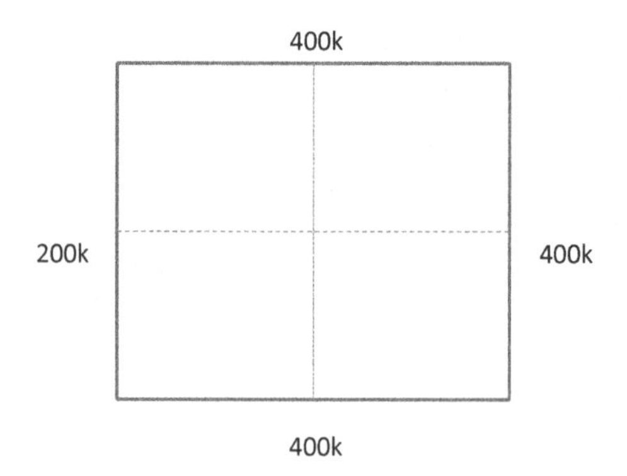

Fig. 6.14 Schematic case with discontinuous temperature.

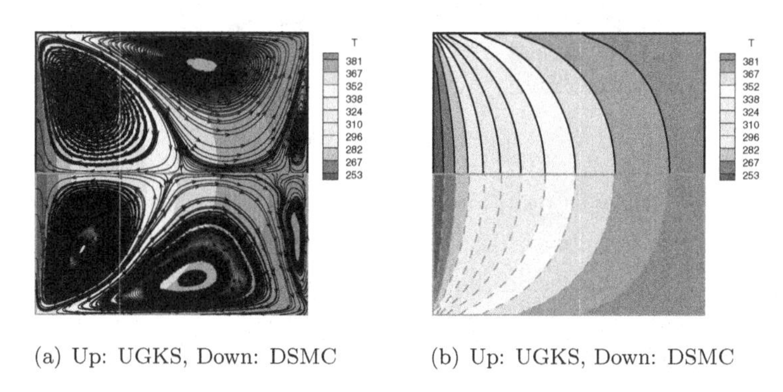

(a) Up: UGKS, Down: DSMC (b) Up: UGKS, Down: DSMC

Fig. 6.15 Comparison between UGKS and DSMC solutions for the case with temperature difference $200k$ at boundaries. (a): streamline, (b): temperature contours [Huang *et al.* (2013)].

this case with two temperature differences from $200k$ to $400k$, and from $200k$ to $202k$ at the boundaries. Defining a maximum error at each computational step,

$$\epsilon_n = \text{Max}_{i,j}(|\mathbf{W}^{n+1} - \mathbf{W}^n|),$$

in the whole physical domain, the condition $\epsilon_n \leq 10^{-7}$ is used for the convergence. The Table 6.2 presents the mesh points, temperature difference, total steps for convergence, and total CPU time in the computation with the machine Intel (R) Core TM I5-2500, CPU3.3GHz. Since it is very hard to compare the efficiency of different schemes using different machines, the above table gives a good reference for others to figure out the relative speed

Table 6.2 Maximum error $\epsilon \leq 10^{-7}$ based on machine: Intel (R) Core TM I5-2500, CPU @ 3.30GHz 3.30GHz [Huang *et al.* (2013)].

Physical space	Velocity space	ΔT	Total steps	CPU time (min)
80×80	28×28	200	5667	40.2
80×80	25×25	200	5667	39.5
80×80	20×20	200	6661	22.5
60×60	28×28	200	5159	18.0
60×60	25×25	200	5159	17.8
60×60	20×20	200	5159	9.8
40×40	28×28	200	3703	5.9
40×40	25×25	200	3703	5.7
40×40	20×20	200	3703	4.2
80×80	28×28	2	3148	19.2
80×80	25×25	2	3148	18.8
80×80	20×20	2	3148	10.5
60×60	28×28	2	2775	9.5
60×60	25×25	2	2775	9.4
60×60	20×20	2	2775	5.2
40×40	28×28	2	1466	2.4
40×40	25×25	2	1585	2.3
40×40	20×20	2	1485	1.2

of UGKS in comparison with their in-house codes.

6.4.7 *Thermal Creep Flow Instability*

Thermal creep flow is induced in a channel due to the temperature gradient along the wall boundary in the transition flow regime. The flow is induced in the direction of the temperature gradient from the cold to hot regions. Sometimes, this phenomenon was explained by Maxwell in 1879 in terms of the "slip" gas/solid boundary condition in the presence of appreciable temperature gradients along the interface. This property of rarefied gas flow is important in the design of pumping techniques, where the flow can be generated without any moving mechanical components. Many experimental and numerical techniques have been used in the study of thermal creep flows. It is surprising that the flow patterns due to the thermal creep effect are very complicated. We can category the flow patterns according to the Knudsen number and temperature gradients. The flow separation at the boundary can be clearly observed. This kind of flow motion provides new flow instability phenomena. To understand its mechanism will be important in the study of transitional flows.

Consider a gas inside a confined channel with temperature gradient along the wall direction. Since the particles from the hotter region impart more parallel momentum to the wall than those from colder region, as a result a shear stress is exerted on the wall. As a reaction force, a gas will be pushed to move from colder to hotter regions. The resulting stationary velocity of the gas is the "thermal creep" velocity, and is parallel to the wall. The flow is directly proportional to the temperature gradient. Even with zero-gravity, there exists convective flow patterns here. Many experimental observations of thermal transpiration have been reported [Annis (1972); Knudsen (1950)] as well as practical application of the effect in Knudsen pumps and micropropulsion systems [Vargo *et al.* (1999); Alexeenko *et al.* (2005); Han (2010); Aoki *et al.* (2009)].

In 1879, James Clerk Maxwell published a paper on the viscous stresses arising in rarefied gases [Maxwell (1879)], where a famous velocity slip boundary condition was proposed. This boundary condition was successful in predicting two prior experimental observations: (a) that a rarefied gas could slide over a surface, and (b) that inequalities in temperature could give rise to a force tending to make the gas slide over a surface from colder to hotter regions (which had been discovered by Reynolds, and was known as "thermal transpiration" - now more commonly known as "thermal creep"). Maxwell related the tangential gas velocity slip to the tangential shear stress, and the heat flux. At a planar wall, Maxwell gave

$$U_s = \frac{2-\sigma}{\sigma}\lambda_{mfp}\frac{\partial U_x}{\partial n} - \frac{3}{4}\frac{\mu}{\rho T}\frac{\partial T}{\partial x}, \tag{6.3}$$

where σ is the momentum accommodation coefficient ($\sigma = 1$ here), μ is the gas viscosity at the wall, λ_{mfp} is the molecular mean free path, n is the coordinate normal to the wall, x is the coordinate tangential to the wall, U_x is the x-component of the gas velocity, U_s is the x-component of slip velocity, and ρ and T are the density and temperature of the gas at the wall respectively. Because of its simplicity, equation (6.3) is remembered as Maxwell's main theoretical result. In practical applications, most surface geometries may have curvature and/or rotational motion, a generalized Maxwell boundary condition with a complete expression for the tangential shear stress can be constructed [Lockerby *et al.* (2004); Guo *et al.* (2014)].

The purpose of the following calculation is to present the possible flow patterns in a 2D closed tube with temperature gradients along the boundary and show the thermal flow instability. The complicated flow patterns induced have close relation to the temperature gradient and flow Knudsen numbers. Contrary to the general accepted flow patterns, such as the flow moving from the cold to hot region along the boundary, the real flow patterns in a closed tube is much more complicated. In the following, we are going to category the flow patterns and define this phenomena as a new kind of instability: thermal creep instability. Contrary to the Rayleigh-Bernard instability, the thermal creep instability does not need any gravity field and the temperature gradient can generate it. Also, as the Knudsen number increases in the transition regime, the Maxwell boundary condition is inadequate in the description of what happening at boundary.

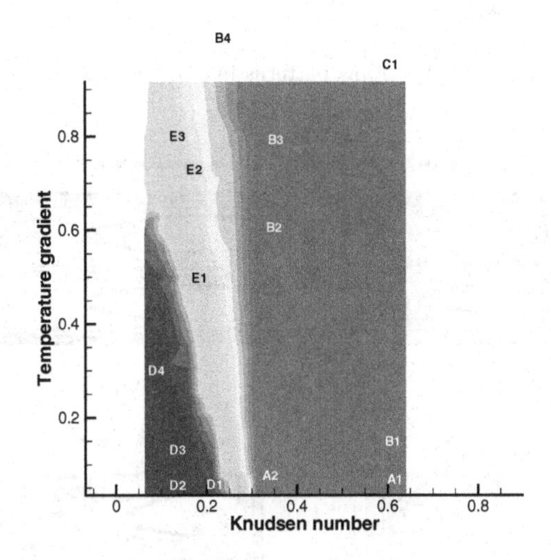

Fig. 6.16 Phase diagram for the topological flow patterns.

The result reported here is pertained to a two-dimensional cartesian cavity, with the length L in the x-direction and height H in the y-direction with a ratio $L/H = 2$. The two ends of the channel are maintained at two different temperatures $T_1 < T_2$ with different temperature to have a temperature gradient $(T_2 - T_1)/L$. The Knudsen number is defined as $Kn = \lambda/H$, where λ is the particle mean free path. Since the flow patterns only depend on the Knudsen number and the temperature gradients, we have done over 300 calculations with a variety of chosen parameters [Yu (2013)]. Figure 6.16 presents a phase diagram for the cases tested. Different zones correspond to different flow patterns. As indicated by the numbers in the Fig. 6.16, the associated flow patterns in different zones are shown in Figs. 6.17-6.21.

It will be interesting to analyze the thermal creep instability, like the Rayleigh-Bernard instability. Unfortunately, for the above thermal creep instability in the transition flow regime there is no any valid macroscopic

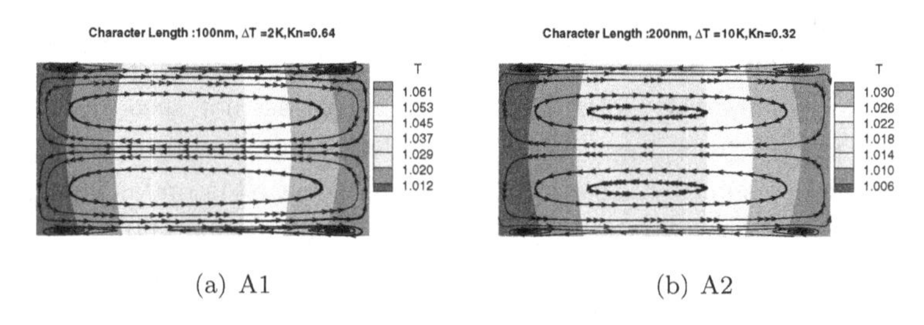

(a) A1 (b) A2

Fig. 6.17 Flow patterns in domain A.

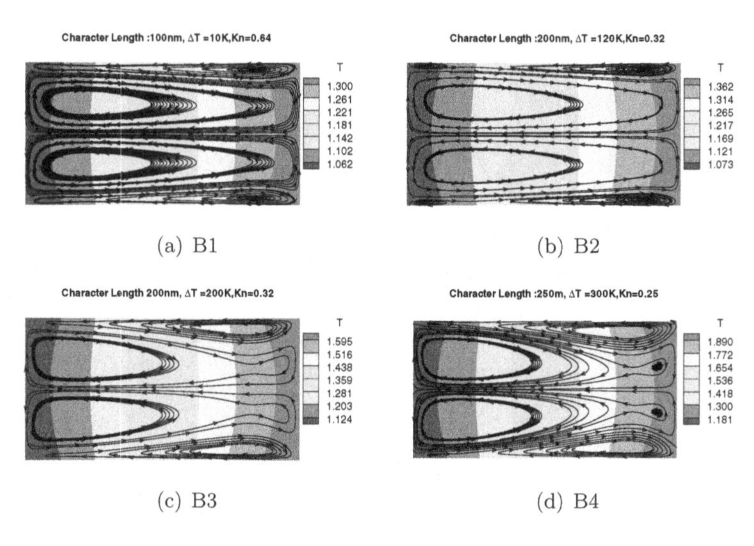

(a) B1 (b) B2

(c) B3 (d) B4

Fig. 6.18 Flow patterns in domain B.

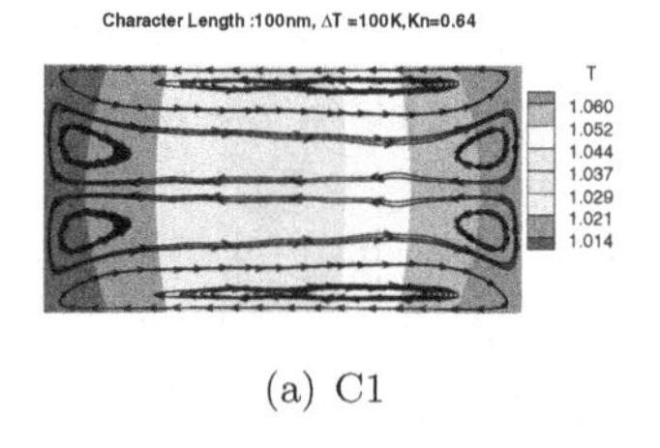

(a) C1

Fig. 6.19 Flow patterns in domain C.

governing equation. The direct analysis of the instability based on the kinetic equation will be an interesting research topic.

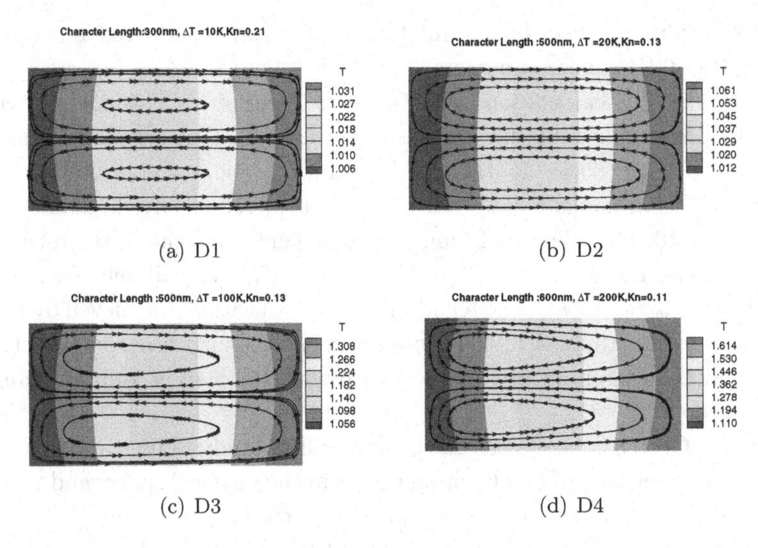

(a) D1 (b) D2

(c) D3 (d) D4

Fig. 6.20 Flow patterns in domain D.

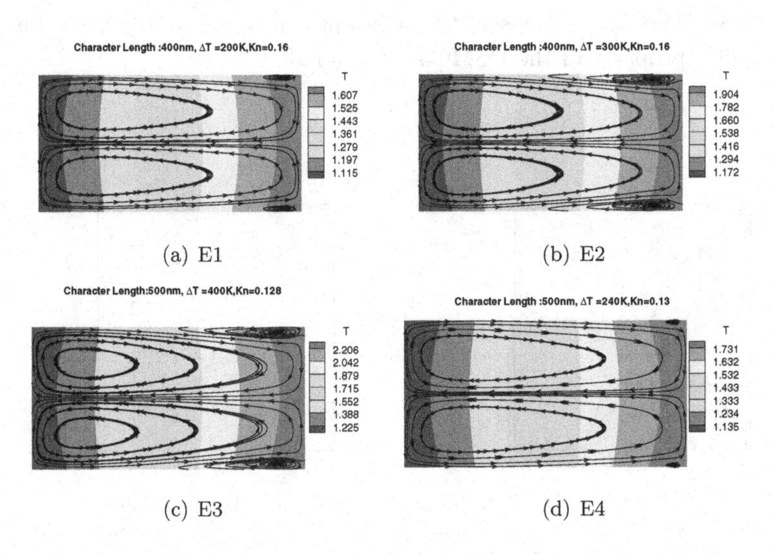

(a) E1 (b) E2

(c) E3 (d) E4

Fig. 6.21 Flow patterns in domain E.

6.4.8 *Cavity Flow*

John, Gu, and Emerson in [John *et al.* (2011)] studied non-equilibrium heat transfer in a cavity flow using DSMC method at three different Knudsen numbers $Kn = 10, 1.0$, and 0.075. In order to get a reliable non-equilibrium flow solution, an intensive computation has been conducted on a Blue Gene/P (BGP) supercomputer.

For all flow calculations, the gaseous medium is assumed to consist of monatomic molecules corresponding to that of argon with mass, $m = 6.63 \times 10^{-26} kg$. In the DSMC solution, the variable hard sphere (VHS) collision model has been used, with a reference particle diameter of $d = 4.17 \times 10^{-10} m$. The wall temperature is kept the same as the reference temperature, i.e. $T_w = T_0 = 273K$. In the study, the wall velocity is kept fixed, i.e., $U_w = 50m/s$. The Knudsen number variation is achieved by varying the density. Maxwell model is used to represent surface accommodation, where in the current study only the case with full wall accommodation is presented.

The UGKS has been used in the cavity flow study. The computational domain is composed of 61×61 mesh points in the physical space, and 28×28 mesh points in the particle velocity space with the Gauss Hermit quadrature integrations. In order to match with the DSMC VHS model, the collision time taken in the unified scheme is $\tau = \mu/p$, where $\mu = \mu_{ref}(T/T_{ref})^{\omega}$ and $\omega = 0.81$. The reference viscosity coefficient can be calculated based on the molecular property of the DSMC simulation.

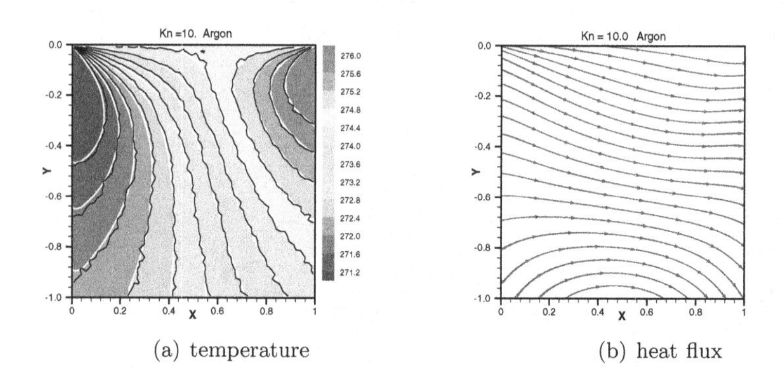

(a) temperature (b) heat flux

Fig. 6.22 Cavity at $Kn = 10$ by unified scheme. (a) temperature contours, black lines: DSMC, white lines and background: UGKS, (b) heat flux, black line: DSMC, red-dash line: UGKS [Huang *et al.* (2012)].

Thermal patterns at different Knudsen numbers in the cavity are illustrated in Figs. 6.22 - 6.24, which show plots of temperature contours and heat flux at $Kn = 10, 1$, and 0.075. Both DSMC solution and the results

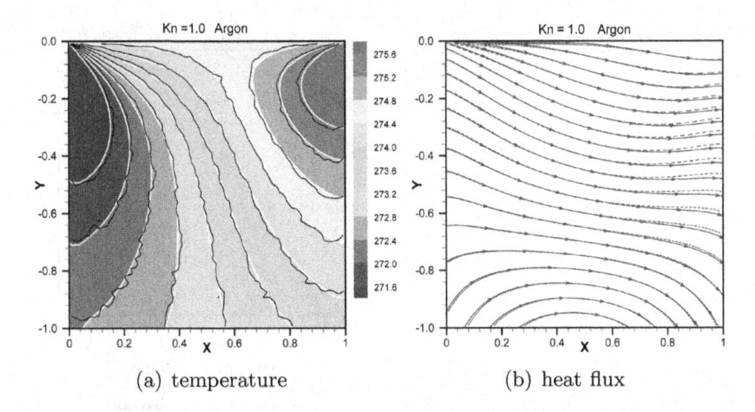

Fig. 6.23 Cavity at $Kn = 1.0$ by unified scheme. (a) temperature contours, black lines: DSMC, white lines and background: UGKS, (b) heat flux, black line: DSMC, red-dash line: UGKS [Huang *et al.* (2012)].

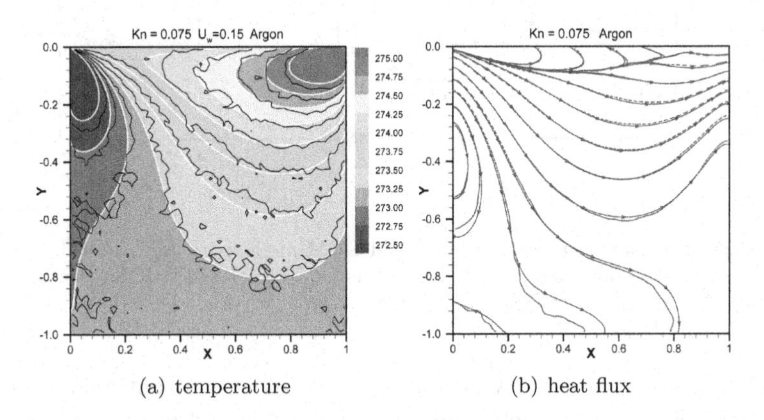

Fig. 6.24 Cavity at $Kn = 0.075$ by unified scheme. (a) temperature contours, black lines: DSMC, white lines and background: UGKS, (b) heat flux, black line: DSMC, red-dash line: UGKS [Huang *et al.* (2012)].

from UGKS are presented. The good agreements between DSMC and unified solutions have been obtained for all flow variables at different Knudsen numbers. Same as DSMC solution, from the heat flux plots the direction of heat flow is found to be mainly from the cold to the hot region. The gaseous heat transfer direction denotes a counter-gradient heat flux, which implies that thermal energy transfer doesn't always follow the gradient transport

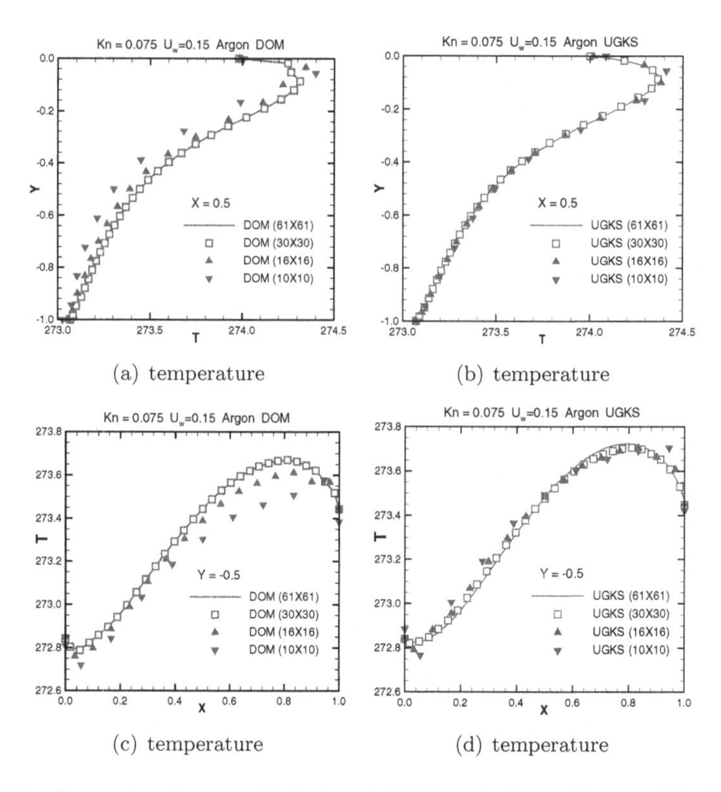

(a) temperature (b) temperature

(c) temperature (d) temperature

Fig. 6.25 Comparison between Unified and DOM methods at $Kn = 0.075$ with the reduction of mesh points. (a) and (c): DOM method, (b) and (d): UGKS [Huang *et al.* (2012)].

mechanism of Fourier's law of continuum flow. This is contradicting with the NS solutions. The non-equilibrium expansion and compression of gas flow effect the heat transport significantly.

The merit of UGKS is the coupled treatment of particle free transport and collision. The DOM solves the free transport and collision of a kinetic equation separately, which is a limiting case by setting $\tau \to \infty$ in the integral solution of UGKS, where a free transport mechanism is used for the evaluation of interface fluxes. Fig. 6.25 shows the performance of UGKS and DOM method by reducing the mesh points in the physical space from 61×61 to 10×10 in the cavity simulation at $Kn = 0.075$. As shown in the temperature distributions along the vertical and horizontal symmetric lines, the UGKS solution is not too sensitive to the increasing of physical mesh spacing. Even with 10×10 mesh points, the solution can be still well

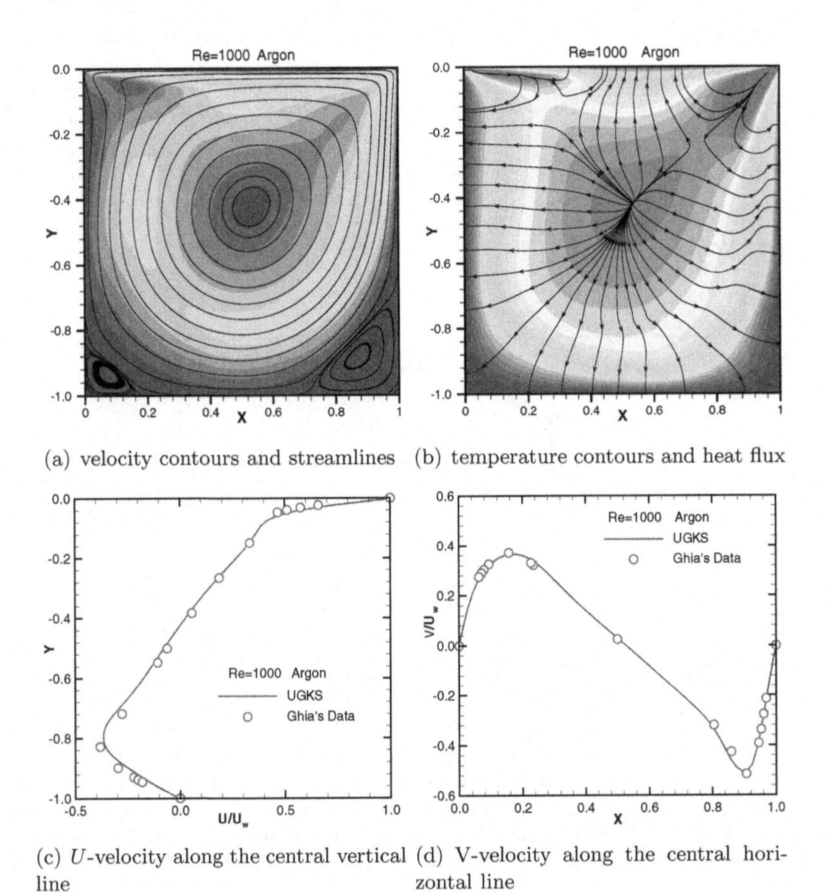

(a) velocity contours and streamlines (b) temperature contours and heat flux

(c) U-velocity along the central vertical line (d) V-velocity along the central horizontal line

Fig. 6.26 Cavity simulation using UGKS at $Kn = 5.42 \times 10^{-4}$ and $Re = 1000$. circles: NS solution [Ghia *et al.* (1982)], line: UGKS [Huang *et al.* (2012)].

captured. However, for the DOM method, the solution deteriorates quickly. The solution deviation between the UGKS and DOM will become more significant in the high Reynolds number continuum flow computations.

In order to further validate UGKS in the continuum flow regime, we continuously reduce the Knudsen numbers to the order of $Kn = 10^{-4}$ and increase the Reynolds numbers to 1000. In the continuum flow regime, at the low speed limit there is a well-defined incompressible NS solutions [Ghia *et al.* (1982)]. For the UGKS, in the continuum flow limit we can much reduce the velocity space mesh points. With the same of 61×61 mesh points in the physical space, the velocity space reduces to 16×16 mesh

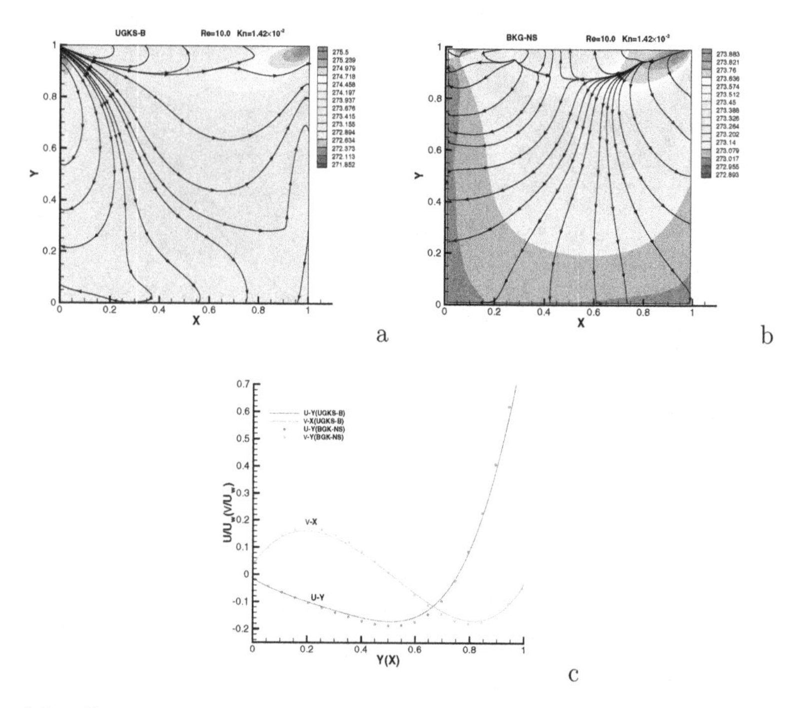

Fig. 6.27 Cavity simulation using UGKS-B (UGKS with full Boltzmann collision term in modeling) and BGK-NS (an earlier name of GKS) at $Kn = 1.42 \times 10^{-2}$ and $Re = 10$. (a) temperature contour and heat flow: UGKS-B, (b) temperature contour and heat flow: BGK-NS, (c) U-velocity along the central vertical line and V-velocity along the central horizontal line, circles: BGK-NS, line:UGKS-B [Liu *et al.* (2014)].

points. Figure 6.26 presents the velocity contours, stream lines, temperature contours, heat flux, and velocity profiles along the symmetric lines. Different from the flow behavior in the transition regime, the heat flux now becomes consistent with Fourier's law, where heat flows from high temperature region to low temperature one. Also, the velocity profiles match with the incompressible NS solution very well, even though both methods are based on totally different physical models. This case clearly shows that the UGKS goes back to GKS for the Navier-Stokes solutions in the continuum regime, where the GKS solutions are shown in [Su *et al.* (1999); Xu and He (2003); Guo *et al.* (2008)].

Since the UGKS provides such an important tool to cover the whole transition regime, in order to get a more clear picture about the transition of the heat flux at small Knudsen number, the solutions at Reynolds numbers

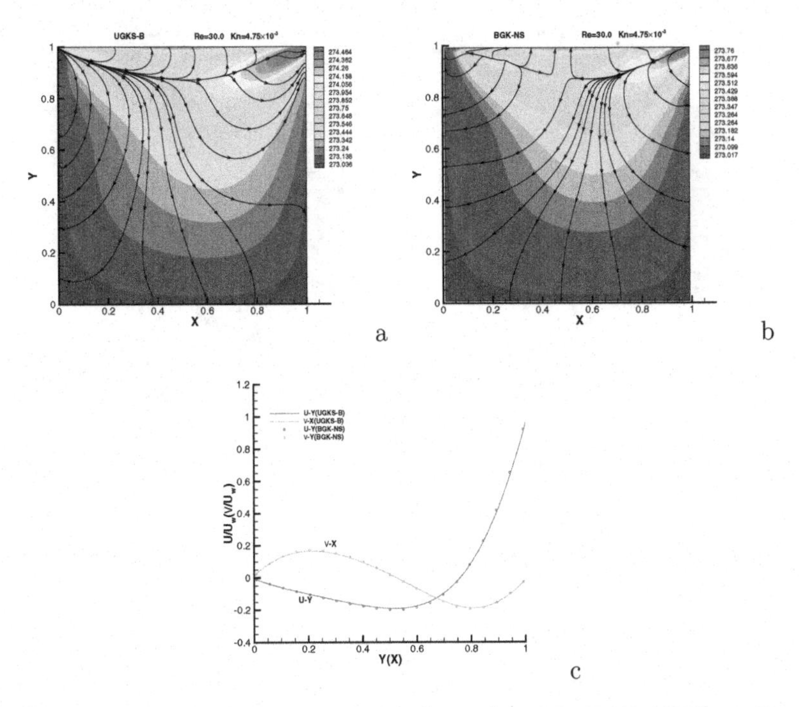

Fig. 6.28 Cavity simulation using UGKS-B (UGKS) and BGK-NS (GKS) at $Kn = 4.75 \times 10^{-3}$ and $Re = 30$. (a) temperature contour and heat flow: UGKS-B, (b) temperature contour and heat flow: BGK-NS, (c) U-velocity along the central vertical line and V-velocity along the central horizontal line, circles: BGK-NS, line:UGKS-B [Liu *et al.* (2014)].

$Re = 10, 30$ and 50 from both UGKS and GKS are shown in Figs. 6.27-6.29. The current UGKS-B used for the above calculations implements the full Boltzmann collision term and kinetic models in their valid scale, but there are no difference in solutions between UGKS-B and UGKS with kinetic model equation only in the current cavity study [Liu *et al.* (2014)]. Since GKS is solving the NS equations, see Chapter 4, the heat flux is always opposite of the temperature gradient. However, the UGKS provides a Knudsen number dependent heat flow, but the velocity profiles from both UGKS and GKS are close to each other. At least for the cavity flow here, it is surprising that at Reynolds numbers 10, 30 and even 50, which are not too small, the validity of the NS equations for gas dynamics with heat is questionable. The flow phenomena in the transition regime are basically unexplored before due to the lack of reliable numerical tool. The UGKS

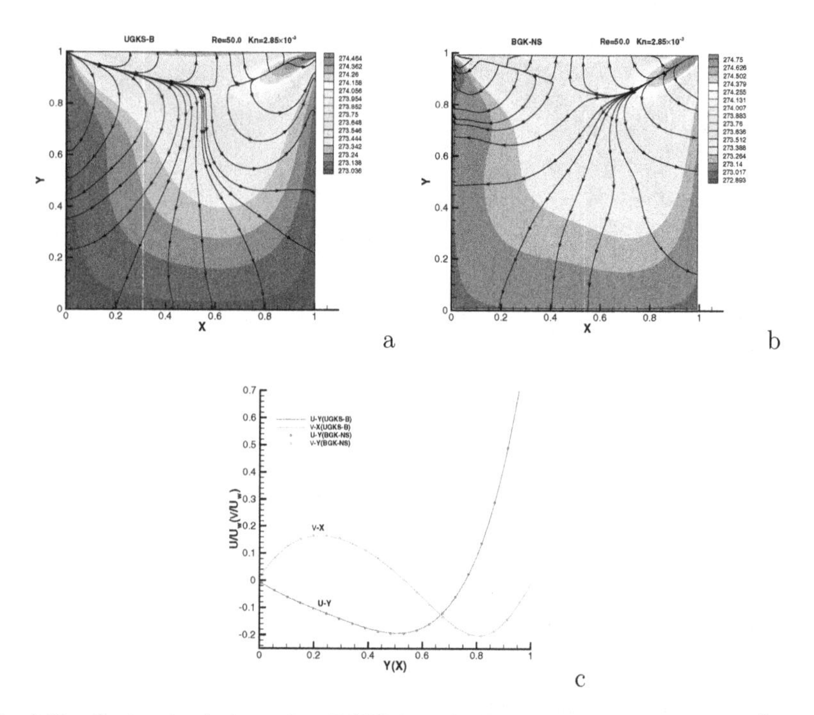

Fig. 6.29 Cavity simulation using UGKS-B and BGK-NS at $Kn = 2.85 \times 10^{-3}$ and $Re = 50$. (a) temperature contour and heat flow: UGKS-B, (b) temperature contour and heat flow: BGK-NS, (c) U-velocity along the central vertical line and V-velocity along the central horizontal line, circles: BGK-NS, line:UGKS-B [Liu *et al.* (2014)].

provides us a powerful method here. The cavity case with supersonic speed is also studied recently using the UGKS [Venugopal and Girimaji (2014)].

As the temperature variation becomes small, a discrete UGKS (DUGKS), which has the similar structure as the Lattice Boltzmann Method (LBM) [Chen and Doolen (1998)], has been developed recently and used for the isothermal flow computations [Guo *et al.* (2013)]. The DUGKS seems present more accurate solutions than LBM even with the same number discrete velocity points in the incompressible limit [Wang *et al.* (2014)].

In the following, we extend the above 2D cavity simulation to 3D case at Knudsen number $Kn = 0.1$. The set-up of the 3D calculation is the same as the above 2D case, but with a change of the upper-wall velocity to Mach number $M = 0.8$ for the convenience of DSMC solution. Figures 6.30 and 6.31 present the 3D temperature contours and flow distribution in different cut-planes of the 3D result. There is excellent match between

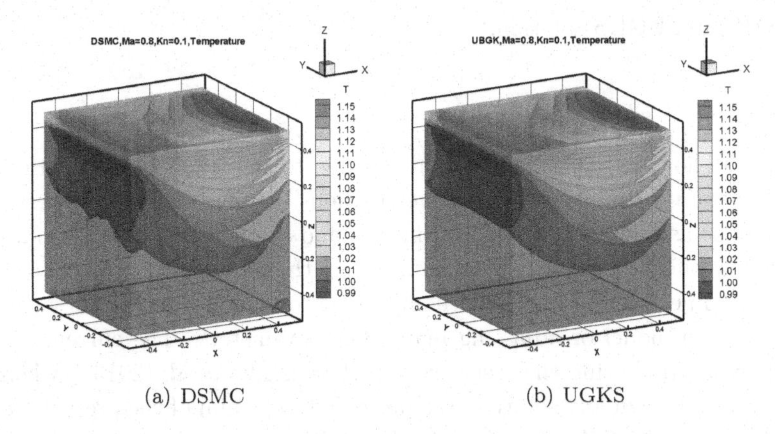

(a) DSMC (b) UGKS

Fig. 6.30 Temperature distribution in 3D cavity simulation at Kn=0.1 and M=0.8 [Huang *et al.* (2012)].

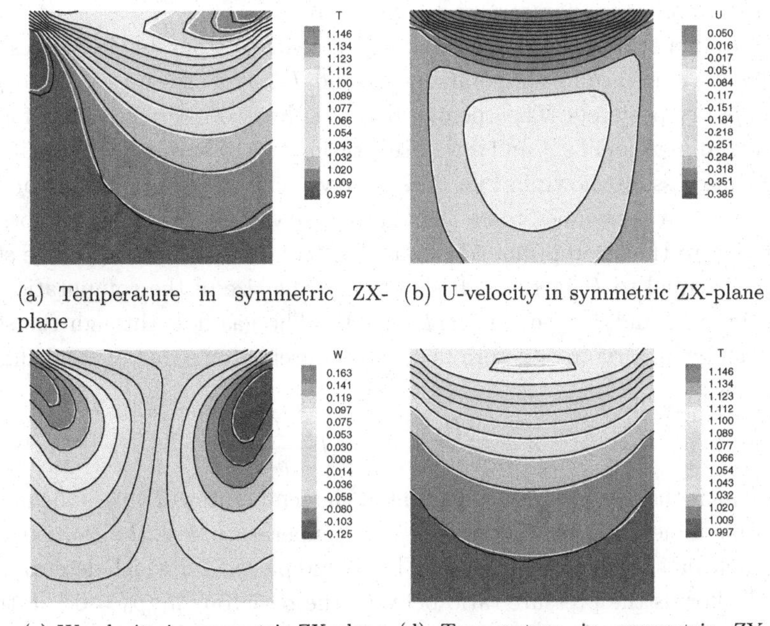

(a) Temperature in symmetric ZX-plane (b) U-velocity in symmetric ZX-plane

(c) W-velocity in symmetric ZX-plane (d) Temperature in symmetric ZY-plane

Fig. 6.31 Flow distributions in different cut-planes in 3D simulation. Black lines: DSMC, white lines: UGKS [Huang *et al.* (2012)].

DSMC and UGKS solution.

6.4.9 *Slit Flow*

The slit flow attracts many attentions due to its practical application in the vacuum technology, space investigations, like the re-entry hypersonic flights or astrophysical research, micro/nano engineering. In addition, the slit flow offers the possibility of comparing theory and experiment, or different numerical approaches without having to assume a particular gas-wall interaction model or adjusting many flow parameters. The current study is basically following the test cases in [Rovenskaya *et al.* (2013)], where a direct full Boltzmann solution was presented. The aim of the present study is to compare UGKS solution with the full Boltzmann solution. For this purpose the two-dimensional flow through a slit, placed between two reservoirs, is simulated. The pressure ratio between two reservoirs is 10, and different rarefaction parameters are considered.

The simulating gas is monatomic argon, which is separated in two reservoirs with equilibrium temperatures T_0 and T_1, and pressures p_0, and p_1 ($p_1 < p_0$) respectively. The opening between the two reservoirs has a form of a two-dimensional slit and a wall separating two reservoirs is of infinitesimal thickness, as shown in Fig. 6.32. The velocity distribution function in each reservoir is assumed to be the Maxwellian distribution function corresponding to the appropriate reservoir pressures and temperature. The slit height is equal to H in the y-direction, and the size of the computational domain is L, and the ratio is $H/L = 0.01$. The gas flow through the slit between two reservoirs is determined by the rarefaction parameter δ defined as follows

$$\delta = \frac{p_0 H}{\mu_0 v_0}, \quad v_0 = \sqrt{\frac{2kT_0}{m}},$$

where μ_0 is the gas viscosity coefficient at the temperature T_0, v_0 is the most probable molecular velocity at the same temperature, k is the Boltzmann constant, m is the molecular mass. The second parameter which determines the slit flow is the pressure ratio between the reservoirs $p_1/p_0 = 0.1$ in the current study. The Knudsen number is on the order of $Kn = 1/\delta$.

The numerical calculations are carried out for the rarefaction parameter ranging from $\delta = 1, 10$ (transition regime) and $\delta = 100$ (hydrodynamic regime). The UGKS uses a total 88,400 mesh points in the physical space, which is about 1/4 of the spatial mesh points used by the full Boltzmann solver in [Rovenskaya *et al.* (2013)], the velocity space is 28×28 mesh

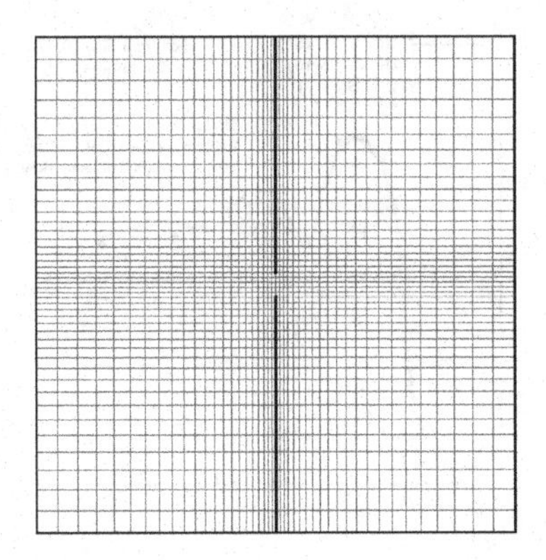

Fig. 6.32 Gas flow through a slit. Mesh distribution around slit.

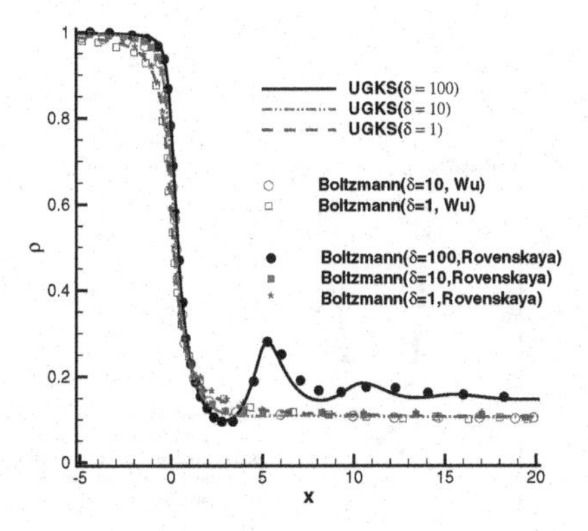

Fig. 6.33 Density along central line. Comparison of the solutions from the full Boltzmann equation of Rovenskaya *et al.*, the full Boltzmann equation of Wu *et al.*, and the UGKS results [Yu (2013)].

points. The axial distributions of the number density, x-component of the bulk velocity and the temperature for $\delta = 1, 10$, and 100 obtained from the UGKS and the reference full Boltzmann solutions are shown in Figures 6.33-6.35. In all three cases, the density variations are qualitatively the same. The axial velocity U_x is close to zero at the distance $\backsim -5H$ and increases considerably around the slit. The decreasing of the velocity is not as fast as for the density. The flow acceleration in the slit is more important for the smaller pressure ratios. The temperature variation along the axis depends essentially on the pressure ratio. It is seen that the flow distributions of the full Boltzmann solution of Rovenskaya *et al.* are different from those

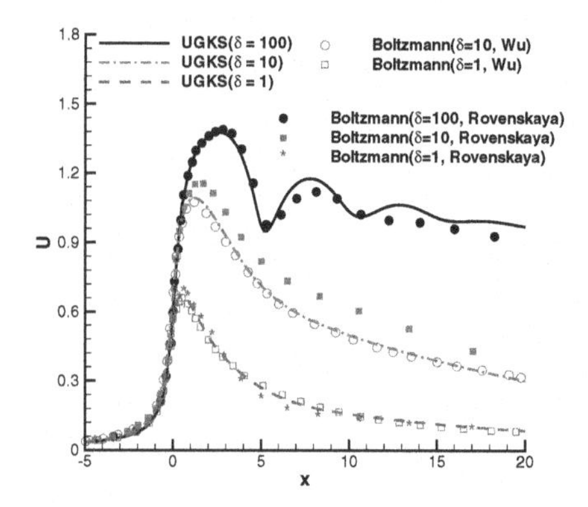

Fig. 6.34 Velocity along the central line. Comparison of the solutions from the full Boltzmann equation of Rovenskaya *et al.*, the full Boltzmann equation of Wu *et al.*, and the UGKS results [Yu (2013)].

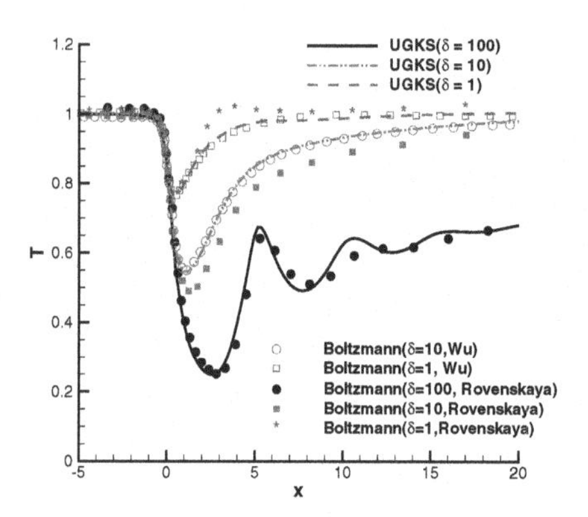

Fig. 6.35 Temperature along the central line. Comparison of the solutions from the full Boltzmann equation of Rovenskaya *et al.*, the full Boltzmann equation of Wu *et al.*, and the UGKS results [Yu (2013)].

obtained by the UGKS, especially for the velocity and temperature at $\delta = 10$.

In order to figure out the reason for the solution differences from UGKS and the above full Boltzmann solver of [Rovenskaya *et al.* (2013)], another full Boltzmann solver developed at University of Strathclyde is used to test the same case [Wu (2013); Wu *et al.* (2013, 2014)]. The results from the new full Boltzmann solver is presented in Figs. 6.33-6.35 as well. Surprisingly, there is a perfect match between UGKS and Wu's Boltzmann solution.

Based on these results, we can realize that even for the same equation different numerical treatments may present different solutions. At least in the slit case, the UGKS is as accurate as the full Boltzmann solver. Also, we believe that at $\delta = 100$ (hydrodynamic regime), the UGKS solution should be more accurate than that of the full Boltzmann one.

6.4.10 *Sound Wave Propagation*

The wave propagation problem in different sound wave frequency is studied using UGKS [Wang and Xu (2012)]. At a sufficient small Knudsen number, the sound propagation in gas can be adequately described by the Navier-Stokes equations. However, as Knudsen number increases to the transition regime, the sound wave parameters, i.e., phase speed and attenuation coefficient, are known to deviate from the classical prediction. Most existing hydrodynamic equations fail to describe the ultrasound propagation. This is not surprising because the scale for the hydrodynamic equations is in the diffusive scale due to the variation or expansion of local equilibrium states, and this modeling scale is much larger than the kinetic scale which is in the particle mean free path and particle collision time. At high frequency, the period of the sound wave propagation can easily come to the kinetic scale, which is comparable with the particle collision time. In order to investigate the high frequency sound wave propagation, many researchers turned attention to the kinetic equations by means of theories based on the expansion of Boltzmann equation. Wang Chang and Uhlenbeck utilized the Super-Burnett equations [Wang Chang and Uhlenbeck (1970)], which were then extended by Pekeris *et al.* up to 483 moments [Pekeris *et al.* (1962)]. However, the success of these theories cannot be extended to high Knudsen number flow regime. A remarkable success that performs well for a wide range of Knudsen numbers was the work of [Sirovich and Thurber (1965)], and also [Buckner and Ferziger (1966)]. Sirvoich and Thurber used Gross-Jackson model and analyzed the dispersion relation, where Buckner and Ferziger solved the half-space problem by means of elementary solutions, with diffusely-reflecting boundary. Besides the Gross-Jackson model, another popular kinetic model used for the study of sound wave is the BGK model. In [Thomas and Siewert (1979)] and [Loyalka and Cheng (1979)], they adopted the BGK model and solved the problem in half-space together with diffusely-reflecting boundary. Their results agreed with each other. As reported in [Loyalka and Cheng (1979)], they obtained better agreement with experimental measurements than the BGK results in [Sirovich and Thurber (1965)] and [Buckner and Ferziger (1966)]. Another successful method in simulating ultrasound wave propagation is the Direct Simulation Monte Carlo (DSMC) method [Hadjiconstantinou and Garcia (2001)], and the DSMC method is basically solving the Boltzmann equation as well.

The UGKS is used to perform numerical simulation of sound propagation in monatomic gases for the whole Knudsen regime with hard-sphere molecule. Comparison is made with experimental results of Greenspan [Greenspan (1956)] and Meyer and Sessler [Meyer and Sessler (1957)], and the DSMC results in [Hadjiconstantinou and Garcia (2001)]. Good agreement is found between UGKS solutions and the experimental measurements.

Figure 6.36 shows the schematic of the simulation geometry. The monatomic argon gas is enclosed between two walls separated by a distance L. The left wall is the transducer which is imposed by a periodical velocity $U(t) = U_0 \cos \omega t$ and the particles are diffusely reflected from the wall. On the other hand, the right wall is a resting receiver and is assumed to have a specular boundary condition, which leads to total reflection of the propagating waves. The flow field is assumed to be one-dimensional.

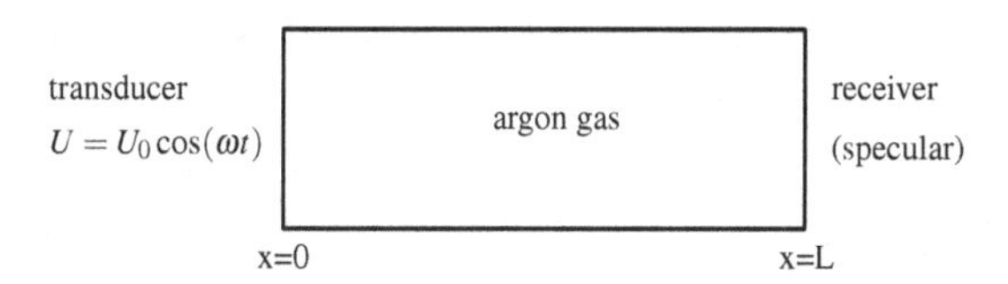

Fig. 6.36 Schematic of the simulation geometry.

The UGKS results are compared with the original experimental data presented in [Greenspan (1956)] and [Meyer and Sessler (1957)] in Fig. 6.37 and Fig. 6.38, where k is wave number and m is attenuation coefficient. It can be seen that the UGKS results have excellent agreement with the experimental data.

6.4.11 *Crookes Radiometer*

As an application of the UGKS with moving meshes, the Crookes radiometer problem has been studied recently [Chen *et al.* (2012a)]. The Crookes radiometer [Crookes (1874)] consists of a sealed glass bulb which is evacuated to a partial vacuum and a set of vanes mounted inside the bulb.

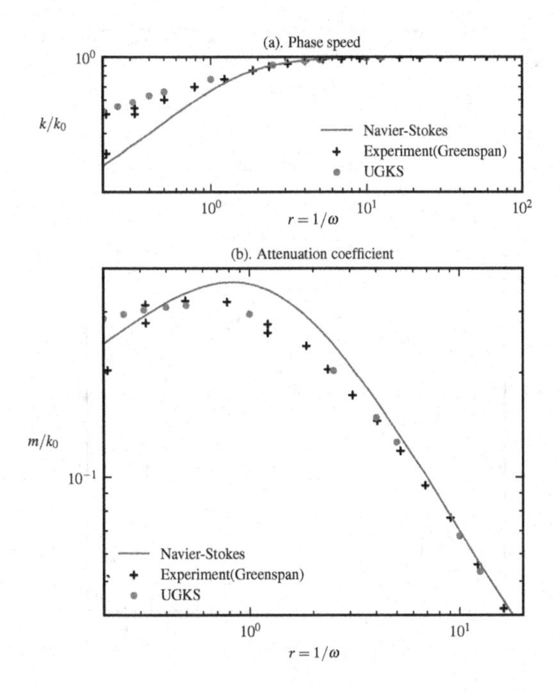

Fig. 6.37 Results comparison between UGKS and experimental data of Greenspan [Wang and Xu (2012)].

The vanes are blackened on one side, and kept glossy on the other side. When exposed to light, the radiometer rotates. This interesting apparatus has attracted many renowned scientists' interests, such as [Reynolds (1876, 1879)], [Maxwell (1879)] and [Einstein (1924)] etc. The driving force of Crookes radiometer, named radiometric force, had been identified. In the range of pressure for the maximum radiometric force [Westphal (1920)], both forces generated from the area and edge effect are important [Selden *et al.* (2009a)]. The area effect corresponds to the pressure difference between the two surfaces of the vanes in free molecular flow. And the edge effect corresponds to the well known thermal transpiration near the edge of the vanes. In 1986, Binnig *et al.* proposed the conception of atomic force microscope which employs microcantilever as a probe [Binnig *et al.* (1986)]. This work encouraged the investigations of the Knudsen force in the transitional regime. For example, Passian *et al.* studied the radiometric phenomena experimentally and analytically in application to microcantilevers

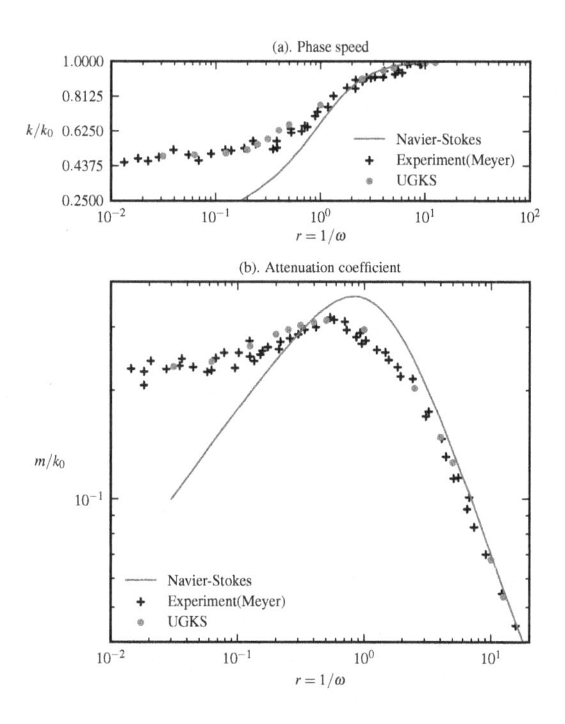

Fig. 6.38 Results comparison between UGKS and experimental data of Meyer [Wang and Xu (2012)].

[Passian *et al.* (2002, 2003a,b)]. In recent years, great efforts have been made to get a better understanding of the radiometric force [Scandurra *et al.* (2007); Han *et al.* (2010); Taguchi and Aoki (2011); Zhu and Ye (2010); Nabeth *et al.* (2011); Anikin (2011a)], and many theories have been advanced to depict the radiometric force in the transition regime. However, as shown by Selden *et al.* [Selden *et al.* (2009b)], most of these theories are valid only in a small range of Knudsen number. It is hard to identify any preferred theory here.

Although plenty of studies have been made to explain the Crookes radiometer, there are few works targeting on the study of dynamical process of an unsteady freely rotating motion. After the discovery of the peak of radiometric force by Westphal [Westphal (1920)], Ota *et al.* [Ota *et al.* (2001)] reported that, in the transition regime, there was a shift between the maximum torque and maximum rotational speed in terms of Knudsen numbers in their experiment. This phenomenon indicates that the situation of a rotating vane differs from a static one. For a moving fan, Anikin simulated the radiometric force of a 2D vane in a non-inertia rotational reference frame which is fixed on the vane [Anikin (2011b)]. This is the first time that a moving vane with a given constant angular velocity is considered numerically. He found that the total radiometric force is a linear function of the angular velocity of the vane. However, this methodology requires a constant rotation of the reference frame. As a result, it doesn't provide the acceleration dynamical process of the fan.

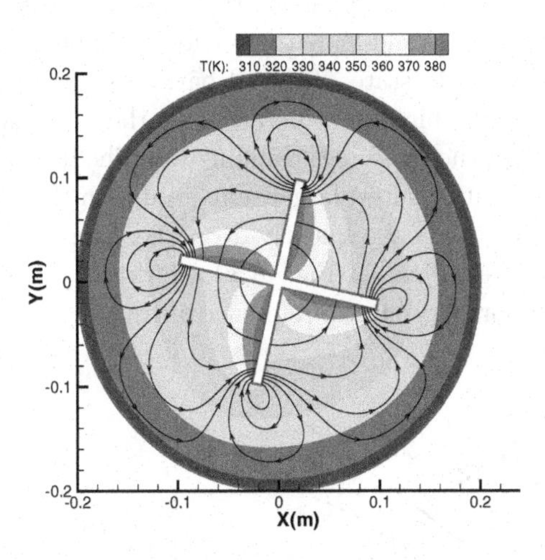

Fig. 6.39 The contours of Temperature field and the streamlines in laboratory inertia reference of frame at Kn = 0.1 for an idling moving fan [Chen *et al.* (2012a)].

The UGKS with a moving mesh was used to study the dynamic process of a rotating fan [Chen *et al.* (2012a)]. A cross shaped fan mounted in the center of a computational domain. The outside boundary is a circle with a radius of $0.2m$, which is twice of the length of the vane. The thickness of the vane is $0.01m$. The rotational moment of inertia of the fan is $4.9 \times 10^{-9} kg \cdot m^2$. And the fan can rotate freely with a mechanical rotational resistance. The gas simulated here is argon with $\Pr = 2/3$, and viscosity coefficient $\mu \sim T^{0.68}$. In the set up of the simulation, the outside boundary is a solid fixed wall with a constant temperature of $300K$. The temperature at one side of each vane has a value of $350K$, and the other side keeps at $400K$. The temperature distribution at the lateral side of the vanes changes continuously from $400K$ to $350K$. The topic of thermal variations on the radiometric forces is discussed in [Lereu *et al.* (2004)] and [Nabeth *et al.* (2011)]. The diffusive boundary condition with fully accommodation coefficient is used at all solid boundaries.

The Knudsen number is defined in the current study by the molecular mean free path over the length of the vane. And it varies from 0.001 to 10 in the current study which covers a wide range of flow regime from continuum flow to nearly free molecular one. The numerical experiments include two kinds of fan movements.

The following simulation is an idling case with a freely rotational movement with the neglect of mechanical friction of the rotor. Figure 6.39 shows a typical temperature field at Kn = 0.1 for an idling moving fan after settling down to a steady state. The red parts present the high temperature areas. Owing to the higher temperature on the black surfaces induced by radiation heating, the molecules colliding with the hot black side leave with an increased velocity, exert a larger momentum on the black side than that on the white side, and consequently drive the vanes rotating counterclockwise. All the streamlines in Fig. 6.39 are drawn from the absolute flow velocity in a laboratory inertia reference of frame.

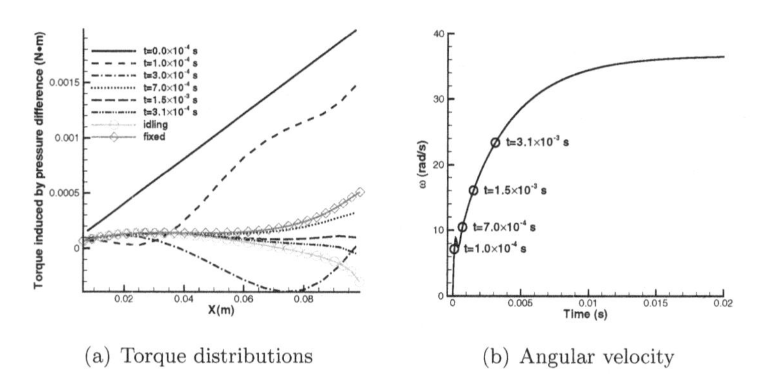

(a) Torque distributions (b) Angular velocity

Fig. 6.40 (a). The time series of the torque distributions along the long arm for Kn = 0.1, (b). The time-dependent angular velocity of a starting moving fan [Chen *et al.* (2012a)].

Figure 6.40 shows a time-dependent variation of the pressure torque along the long arm in the initial starting movement phase of a rotating vane. The extent of the long arm is from 0.005 to 0.1. As shown in the figure, the initial pressure torque distribution is a straight line due to the constant pressure difference between the hot and cold sides along the long arm. Owing to the great pressure difference, the gases respond and get to a quasi steady state quickly. At $t = 7 \times 10^{-4}s$, the angular velocity ω is still small, and the torque distribution is similar to that of a fixed fan, which is represented by a diamond line in Fig. 6.40. As the fan rotates and gets a high angular velocity, the fan's tip velocity gets much higher than that in the center region. Then, the vane tip starts to push the surrounding gas away more efficiently. As a result, the pressure on the leading edge (cold side) near the tip becomes higher than that on the hot side. As shown by the line at $t = 3.1 \times 10^{-3}s$, the pressure difference reverses its sign from positive to negative near the tip of the vane. Therefore, the pressure torque near the tip will resist the fan's rotation. So, the acceleration of the angular velocity decreases gradually. When the positive part gets balanced with the negative part, which is represented by circle line in Fig. 6.40, the fan settles to a steady state rotation. By examining all torques on the vane, the pressure torque generated near the center of the fan along the long arm is the only propulsive part. All other ones act oppositely. In this case, the force near the tip within one mean free path provides resistance for the fan movement. It is basically balanced by inner contributions of the vane. Therefore, the role played by the edge effect near the tip is different in the idling and in the fixed fan case. This dynamical torque evolution has been identified for the first time using the current UGKS method. More detailed analysis of physical mechanism for the Crookes radiometer can be found in [Chen *et al.* (2012a)].

6.4.12 *Knudsen Forces on Heated Microbeams*

The case is to study the Knudsen force of a heated microbeam in rarefied environment. This kind of study has important applications related to the experimental and analytical works by many others [Passian *et al.* (2002, 2003a,b); Gotsmann and Durig (2005); Scandurra *et al.* (2007); Selden *et al.* (2009a,b)]. The Knudsen force is negligible in both continuum and free-molecule regimes, it only happens in the transition regime. But, the mechanism is not clearly known, especially about the force distribution in whole transition regimes. In a recent paper, Zhu and Ye presented a 2D DSMC simulation for the Knudsen numbers from 0.05 to free molecular one, and explained the original of Knudsen forces [Zhu and Ye (2010)]. Here, the UGKS will be used to simulate the same case with a wide flow regime from free molecular to the continuum one. The simulation results may provide useful information for the theoretical analysis and macroscopic model construction.

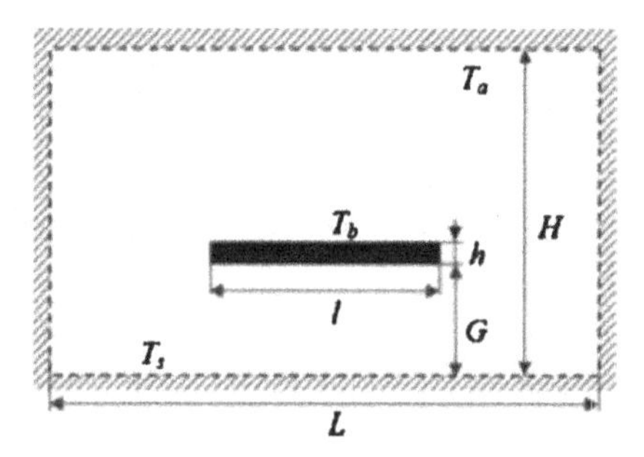

Fig. 6.41 Schematic illustration of the model problem [Zhu and Ye (2010)] (courtesy of W.J. Ye).

The set up of the simulation is shown in Fig. 6.41. The 2D model employed in this study consists of a hot rectangular beam with a thickness of h and a width of l, which is encompassed by a rectangular enclosure with a width of L and a height of H. The beam is placed at a fixed distance, G, away from the bottom wall of the enclosure which represents the substrate. The temperatures of the beam and the substrate are denoted as T_b and T_s, respectively. The top and the side walls of the enclosure are assumed to have the same temperature denoted as T_a. When the temperature of the beam, T_b, is different from T_s and T_a, the vertical linear momentum fluxes imparted to the top and bottom surfaces of the beam are unbalanced, which gives rise to a net force, the Knudsen force, along the same direction.

Argon is filled between the beam and the chamber, of which the temperatures are $500K$ and $300K$, respectively. Variable hard sphere molecules and Maxwell gas-surface interaction model are employed. All walls are assumed to be fully diffuse. The Knudsen numbers are calculated based on the average temperature of the beam and the chamber, and the gap G. The DSMC solutions in Zhu and Ye [Zhu and Ye (2010)] are only in the transition and free molecular regimes. At the Knudsen numbers of 0.11 and 1.1, the DSMC and UGKS solutions are shown in Fig. 6.42. Similar flow patterns at corresponding Knudsen numbers have been obtained using different schemes. At small Knudsen number $Kn = 0.11$, the UGKS solution is much smoother than that of DSMC.

Knudsen forces per unit length at various Knudsen numbers are calculated and plotted in Fig. 6.43 from both DSMC and UGKS simulations. Since it is hard for DSMC to go to continuum flow regimes [Zhu and Ye (2010)], only UGKS calculations are conducted at smaller Knudsen

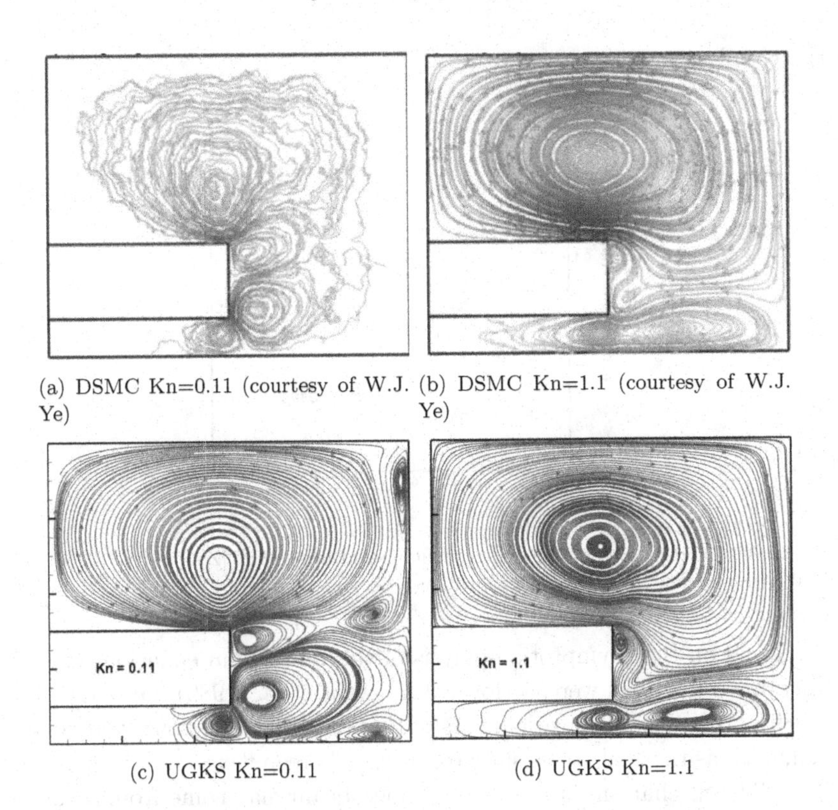

(a) DSMC Kn=0.11 (courtesy of W.J. Ye)

(b) DSMC Kn=1.1 (courtesy of W.J. Ye)

(c) UGKS Kn=0.11

(d) UGKS Kn=1.1

Fig. 6.42 Comparison of flow structures at Kn = 0.11 and 1.1 from DSMC [Zhu and Ye (2010)] (a,b) and UGKS (c,d).

numbers, i.e., $Kn \leqslant 0.05$. The direction and the general trend of the simulated Knudsen force are qualitatively consistent with the measured data within a large range of the Knudsen number [Passian *et al.* (2002, 2003a,b)], i.e., the force is inversely proportional to the Knudsen number in the low-pressure range and proportional to the Knudsen number in the high-pressure range. The maxima of the Knudsen force occurs between $Kn = 0.3$ and $Kn = 0.4$, which is not too far from 0.6, the measured data in Passian *et al.*'s experiment. Another important feature in the simulated Knudsen force is the reverse in its direction near $Kn = 1$. From $Kn = 1.5$ to $Kn = 30$, the direction of the force is pointing toward the gap instead of pointing away from the gap. This phenomenon was not observed in Passian *et al.*'s experiment, probably due to the small magnitude of the force within this range. But, they are observed from both DSMC and UGKS simulations. At the same time, the UGKS provides more detailed force information as the Knudsen number goes to zero.

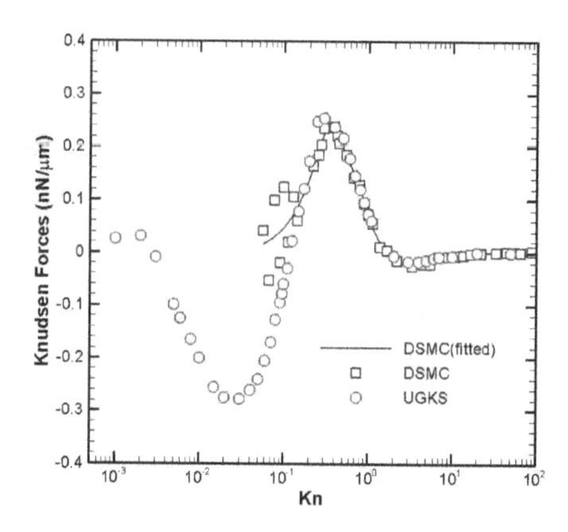

Fig. 6.43 Knudsen force versus Knudsen numbers from DSMC [Zhu and Ye (2010)] and UGKS calculations (courtesy of J.C. Huang).

Based on the asymptotic analysis of the Boltzmann equation, systematical investigations were performed in [Aoki *et al.* (1989); Ohwada *et al.* (1889); Sone (2007)]. Several types of thermally induced flows were identified, such as thermal stress slip flow, thermal creep flow, and thermal edge flow. The mechanism of the current microbeam may come from a combination of the above thermal flow mechanism. Figure 6.44 presents a few flow patterns covering a wide range of flow regimes, which may be useful for the future theoretical analysis.

6.5 Summary

In this chapter, the UGKS has been applied to many low speed microflow simulations. The numerical examples cover the whole range of flow regimes from free molecular to the continuum one. The capacity of the UGKS in capturing non-equilibrium flow behavior has been fully explored in this study. The success of the current method is mainly due to its multiscale modeling in recovering the flow physics for a wide range of Knudsen numbers. The interface flux in UGKS is capable to cover both hydrodynamic and kinetic scale physics with a variation of local Kn. In the low speed limit, the physical solution is not so sensitive to the particle scattering law, and the kinetic model equations can do a perfect job here. Since the full

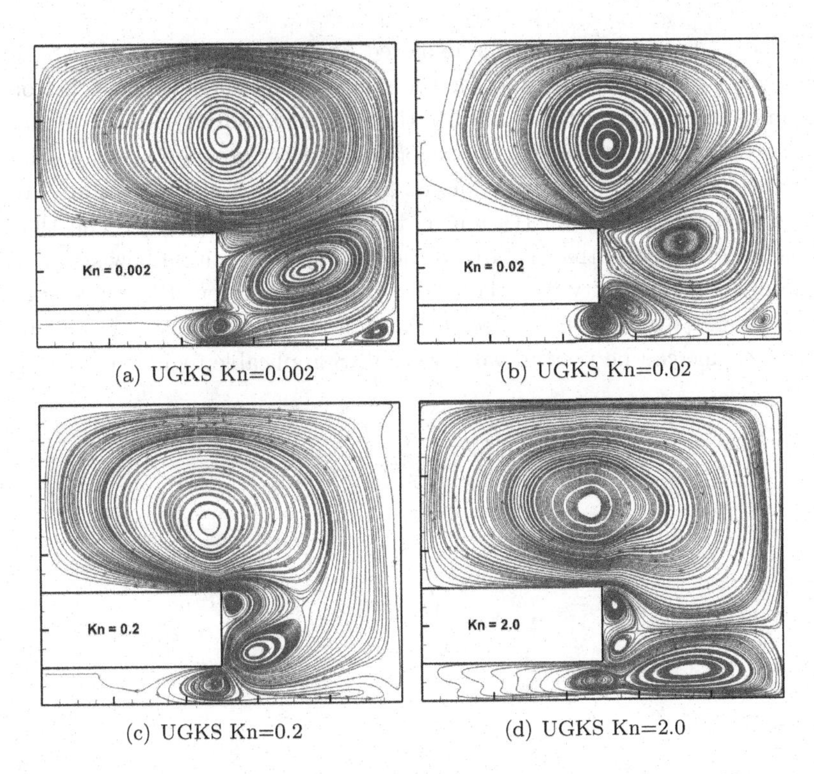

(a) UGKS Kn=0.002
(b) UGKS Kn=0.02
(c) UGKS Kn=0.2
(d) UGKS Kn=2.0

Fig. 6.44 Flow structures in a wide range of flow regimes from UGKS (courtesy of J.C. Huang).

Boltzmann equation is constructed in the kinetic scale, when the numerical time step used in UGKS is a few times of the local particle collision time, the importance of the full Boltzmann collision term diminishes due to the insensitivity of the solution to the particle collision model after multiple particle collisions. In comparison with the particle-based DSMC method, there is no statistical scattering in UGKS, and the non-equilibrium flow motion due to the small temperature variation can be easily captured. For the low speed microflow, the non-equilibrium effect can be captured reasonably well using a limited number of grid points in the particle velocity space. Therefore, there is a reasonable efficiency for the UGKS here. These tests clearly demonstrate that the UGKS is a valuable method for low speed microflow study, especially for the cases with the co-existence of continuum and rarefied flow regimes and with small temperature variations.

At end, we can confidently declare that the UGKS can be faithfully used as a numerical experimental tool. The transition from the equilibrium to the non-equilibrium can be explored quantitatively using UGKS. The UGKS is a valuable method for the study of non-equilibrium thermodynamics and can be used to validate the extended hydrodynamic equations. For example, the applicable regime of Onsager reciprocal relation has been recently tested quantitatively for complicated temperature-pressure driven flow in a complex geometry under different flow conditions. The UGKS has better capacity than DSMC in the low speed transition flow study. For example, in the cavity flow the UGKS can easily present the whole transition process from the free molecular $Kn = 10$ to continuum one $Kn = 10^{-5}$. The scheme can be used to validate the NS applicable regime as well.

Chapter 7

High Speed Flow Studies

7.1 Introduction

Design of fight vehicles in near space environment is an interesting research topic. Such a design needs to incorporate wind tunnel testing, flight experiments, modeling, and computer simulation. Computer simulation is particularly attractive due to its relatively low cost and its ability to deliver data that cannot be measured or observed in a laboratory testing. Its effectiveness relies upon the quality of the underlying numerical methods and physical modeling for the particular application. The design of the vehicles requires accurate prediction of near field and surface properties, such as heat flux, shear stress, and pressure, along the entire vehicle surface and throughout all possible fight trajectories. In this application, the characteristic length scale and particle mean free path may vary significantly throughout the whole computational process, especially when the application is in the near space transitional regime between fully rarefied and continuum one. Within a single computation, there may have both continuum, where the NS solutions are applicable, and highly non-equilibrium flow regime, where the kinetic solvers, such as DSMC, are needed. But, unfortunately there is no a validated method in the transition regime yet. Although there are techniques to extend continuum flow solvers to the transition regime by implementing slip boundary conditions, the reliability of the approach has not been fully confirmed. Same as the kinetic solvers, it is hard to extend them to the near continuum flow regime. At the current stage, we do need a multiscale method to accomplish the task.

In high speed flight in near space, both continuum and rarefied flow can co-exist around a flight vehicle. With a reasonable number of mesh points, the ratio between mesh size to the local particle mean free path depends on the flight altitude and the location of the flow field. For this kind of flow, the application of DSMC alone to the entire flow field with the mean free path resolution everywhere would be computationally expensive, especially at relative low altitude, such as the flight below $60km$. On the other hand, hydrodynamic scale description may not be accurate for the non-equilibrium region. A common practice is to use the expensive, kinetic description in required regions, while using the macroscopic description throughout the rest of the flow field. However, this methodology will introduce uncertainties, such as the location of interface between different regimes, and the validity of both approaches in the buffer zone, especially for the unsteady flow problem.

The UGKS presented in this book provides a suitable tool to simulate such a multi-physics flow. To capture such a complicated flow dynamics solution does need a numerical algorithm with available modeling scale to cope with the significant but continuous changes of the ratio between the mesh size and local particle mean free path. In this chapter, we are going to present some test cases related to the high speed rarefied flow by using unified gas-kinetic scheme. The power of UGKS is that a continuous spectrum of governing equations can be obtained automatically from the kinetic free molecular motion to the NS solutions. The UGKS provides a useful tool for the flow study in the near space, such as altitude between $30km$ to $80km$, due to its high-order accuracy and efficiency in comparison with DSMC method in the low transition regime. Actually, many interesting flow phenomena in this regime have not yet been fully explored, such as the laminar to turbulence transition and rarefied to continuum flow transition, due to the lack of a valid governing equation and a trustable numerical method.

7.2 Physical Modeling and Numerical Difficulties

The last chapter presents the applications of the UGKS to the low speed microflow simulations. This chapter will concentrate on high-speed flows, where specific numerical difficulties will be encountered in such an application.

The UGKS is a direct multi-scale modeling method with the update of both macroscopic conservative flow variables and microscopic gas distribution function. The novelty of the approach is the coupled treatment of particle transport and collision with a scale dependent weight in the evaluation of flux, which cover both kinetic to hydrodynamic flow evolution. The time-dependent flow transport across a cell interface combines the contribution from two limiting solutions, and the real solution controlling the local evolution depends on the ratio of local time step over the particle collision time. The time step used in UGKS is determined by the CFL condition. In the continuum and near continuum flow regimes, the flux is mainly contributed from hydrodynamic scale evolution solution due to the intensive particle collisions $\Delta t/\tau \gg 1$. On the other hands, the molecular free transport mechanism will play an important role in the highly rarefied region, due to inadequate particle collisions. In the transition regime, both kinetic and hydrodynamic scale physics will contribute to the local time evolution of a non-equilibrium flow, such as those inside a high Mach number shock layer. The governing equation underlying UGKS depends on the mesh size and time step, which are the modeling scale. This is completely different from the philosophy of numerical PDEs, where the grid-independent solution is the target of the simulation. Here, as cell size and time step change, the modeling physics will change as well, which corresponds to different flow behavior in different regime.

For the high speed rarefied flow computations, the DSMC method is a direct physical modeling method, which has no rivals for practical computations. Mathematically, the DSMC and the Boltzmann equation have the similar modeling mechanism in their constructions, and both capture the dynamics in the mean free path scale. Therefore, both direct Boltzmann solver and the DSMC method require the mesh size to be compatible with the local particle mean free path. Instead of directly discretizing the kinetic equation, the UGKS uses a time evolution solution of the kinetic equation in its algorithm construction. The UGKS is not a direct solver of the kinetic equation, but uses the kinetic solution to model the flow physics in the mesh size scale. The dynamics to be recovered depend on the ratio of the mesh size to the particle mean free path. It should not be surprising about the unified scheme if better results can be obtained than many previous attempts which discretize the kinetic equation directly. It should not be surprising either if in some test cases the UGKS may be more efficient than the DSMC method, because the direct modeling in UGKS covers flow physics in different regimes through the coupling of particle transport and collision, instead of using operators splitting approach.

Due to the direct modeling and flexibility in incorporating the collision term in UGKS, the method can achieve the same accuracy as DSMC in the capturing of the high-speed non-equilibrium flow. Even with the Shakhov collision model, as shown in this chapter, the simulation results of UGKS are accurate enough for practical engineering applications. However, like many other kinetic equation solvers, the major barrier encountered by UGKS in the high speed application is the wide spread of particle velocity space. If a uniform mesh is adopted, many grid points have to be used in the velocity space. Therefore, the UGKS can hardly become a competitive method in comparison with particle based DSMC in the high speed and high temperature limit. The use of adaptive mesh in the velocity space can greatly release the large memory burden in UGKS. Due to high-order accuracy and multiscale nature, the UGKS can use relatively large cell size in the physical space, which, on the other hand, enhances the efficiency of the scheme.

The physics related to high speed flow is complicated due to the high temperature effect. For example, the rotation, vibration, ionization, and chemical reaction will emerge with the increasing of the temperature. The modeling of all these physical phenomena involve all kind of knowledge from thermodynamics, statistical mechanics, kinetic theory, and even quantum mechanics. The results presented in this chapter is the simple cases, where ideal monatomic gas under different rarefied condition is studied. There is space to further develop the scheme to enlarge its applicable regime. Based on the performance of the scheme, it is optimistic that the UGKS will eventually become a powerful engineering design tool for the flight vehicles in the non-equilibrium flight environment, especially in the near space. The UGKS may be the only method which could present accurate solution in the whole transition regime from rarefied all the way to the continuum one.

7.3 Supersonic Flow Studies

In high speed rarefied flow study, the Knudsen number is usually defined as the particle mean free path over the vehicle length scale. Since the characteristic length is somewhat arbitrary, a length defined by $L = \rho/(\partial\rho/\partial x)$ may be more appropriate to describe the flow behavior. Traditionally, the flow is categorized into different regimes, where $Kn < 0.01$ corresponds to the Navier Stokes flows, but they may fail in localized regions such as very close to leading edges or inside shock wave due to highly non-equilibrium distributions. For large Knudsen number, such as $Kn > 10$, the free molecular flow is valid. The regime between continuum and free molecule flow is called transitional. In practical simulations, the Knudsen number varies widely over the locations around a vehicle at high speed flight. Except pointed out explicitly, the computations in the following are based on the UGKS with Shakhov collision model for the evaluation of a time dependent gas distribution function at a cell interface and the collision term integration inside each cell. The results from UGKS with the full Boltzmann collision term are presented in the shock structure calculation as well.

7.3.1 *Shock Structures*

One of the simplest and most fundamental non-equilibrium gas dynamic phenomena that can be used for the model validation is the internal structure of a normal shock wave. There are mainly two reasons for this. First, the shock wave represents a flow condition that is far from thermodynamic equilibrium at high Mach numbers. Second, shock wave phenomenon is unique in that it allows one to separate the differential equations of fluid motion from the boundary conditions that would be required to complete a well-posed problem. The boundary conditions for a shock wave are simply determined by the Rankine-Hugoniot relations. Thus, in the study of shock structure, one is able to isolate effect due to the differential equations themselves. Since 1950s, the computation of shock structure has played an important and critical role in validating theories and numerical schemes for the study of non-equilibrium flow.

In the following, we will apply the UGKS to the shock structure calculations. The shock strength covers from low Mach number cases, where the NS equations are still valid, to the high Mach number ones, where the highly non-equilibrium effect appears. Besides the density and temperature distributions inside the shock layer, the stress and heat flux will also be presented in some cases. The solutions from the Boltzmann equation [Ohwada (1993)], DSMC results [Bird (1970)], and possible experimental measurements [Alsmeyer (1976); Steinhilper (1972)], will be used to validate the UGKS.

First we present test cases on the shock structure for a monatomic gas with the non-equilibrium limited to the translational energy mode. Comparisons of UGKS results are made with the solution of the full Boltzmann equation which were obtained by Ohwada for the hard sphere molecules at Mach numbers 1.2, 2 and 3 [Ohwada (1993)]. For the hard sphere molecules, the viscosity coefficient $\tau \sim \mu \sim T^{0.5}$, where the x-coordinate is normalized by $\sqrt{\pi} l_0/2$ and l_0 is the mean free path of the gas molecules at the upstream condition. The UGKS used in this computation includes two schemes,

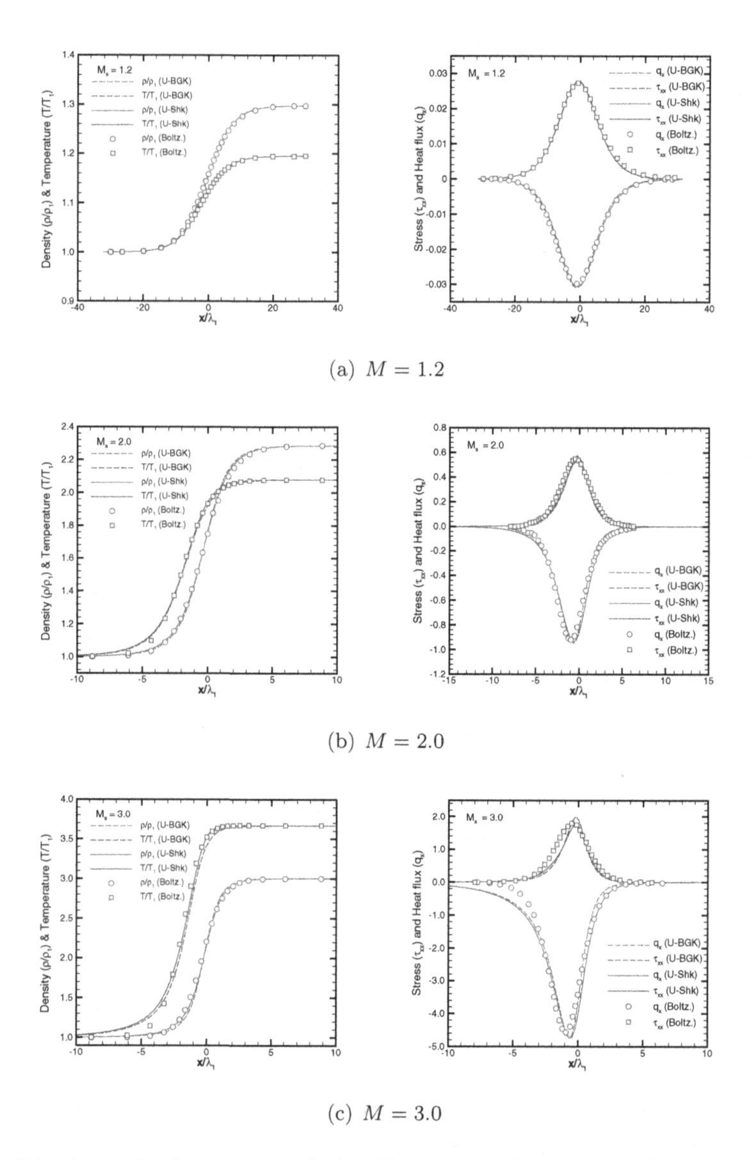

(a) $M = 1.2$

(b) $M = 2.0$

(c) $M = 3.0$

Fig. 7.1 Argon shock structures calculated by the unified schemes and the direct Boltzmann solver [Xu and Huang (2011)].

which are based on the U-BGK model (U-BGK), where the Prandtl number is fixed through the modification of the updated macroscopic flow variable [Xu and Huang (2010)], and the Shakhov model (U-Shk)[Xu and Huang

(2011)], for the evaluation of the interface gas distribution function and the collision term treatment inside each cell. Figure 7.1 shows the density, temperature, stress and heat flux inside a shock layer at different Mach numbers by both unified schemes. Comparisons of the unified schemes are made with the solutions of the Boltzmann equation. For all the Mach numbers presented, the results from the direct Boltzmann solver and the current unified schemes have good agreement. At Mach number 1.2, the local Knudsen number, which is defined as the local mean free path over the density variation scale $l/(\rho/\partial_x \rho)$, is less than 0.02. As expected, both the Boltzmann and the unified solutions are very close to the standard Navier-Stokes solution.

At Mach 3, there is slight discrepancy in stress and heat flux between the UGKS (based on Shakhov model) and the full Boltzmann solution. However, based on the full Boltzmann collision term the UGKS has been developed as well [Liu *et al.* (2014)]. Fig. 7.2 shows the result from the UGKS-B and the Ohwada's Boltzmann solution.

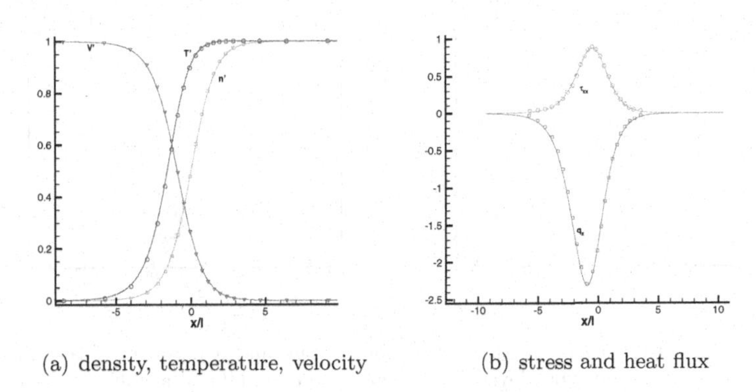

(a) density, temperature, velocity

(b) stress and heat flux

Fig. 7.2 Mach 3 shock structure. Solid lines are result from UGKS-B and the symbols are the results from the Ohwada's finite difference method [Liu *et al.* (2014)].

Figure 7.3 shows the argon shock structures for a viscosity coefficient of $\mu \sim T^{0.72}$ at Mach numbers 8.0 and 9.0 by both UGKS methods, where the experimental measurements are used as benchmark solutions [Alsmeyer (1976); Steinhilper (1972)]. As shown in this figure, the solutions from the unified schemes match with the experimental data very well in the density distributions. Again, at high Mach numbers the U-BGK presents a thinner shock thickness and narrower temperature-density distance than that of U-Shk.

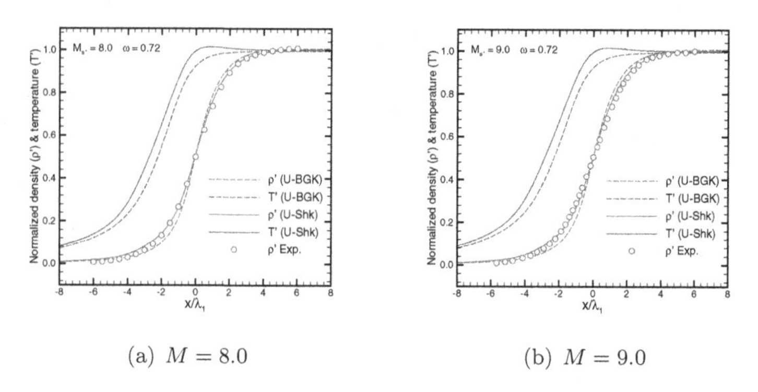

(a) $M = 8.0$ (b) $M = 9.0$

Fig. 7.3 Argon shock structures calculated by the unified schemes and the comparison with the experimental measurements [Xu and Huang (2011)].

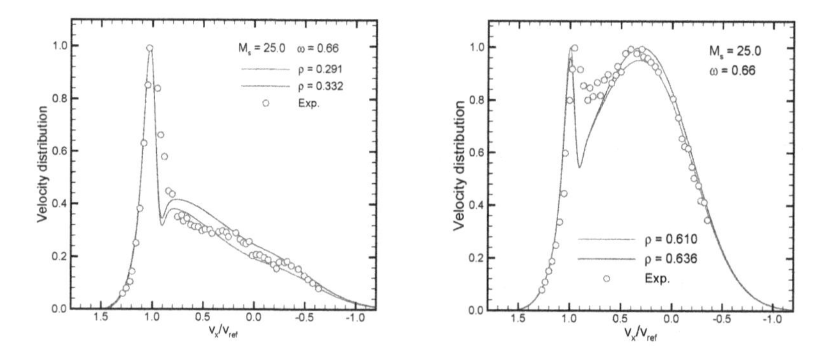

Fig. 7.4 Gas distribution functions from both U-Shk and experiment measurements inside $M = 25.0$ helium shock layer [Pham-Van-Diep *et al.* (1989)] at different locations with normalized densities $(0.291, 0.332)$ (left) and $(0.610, 0.636)$ (right) [Xu and Huang (2011)].

Figure 7.4 presents the helium shock structure calculations at Mach number 25. Instead of the shock structure, we present the gas distribution functions directly at different locations inside the shock layer [Pham-Van-Diep *et al.* (1989); Erwin *et al.* (1991)]. With the normalized density $\hat{\rho} = (\rho - \rho_1)/(\rho_2 - \rho_1)$, the locations for the distribution function presentation are at $\hat{\rho} = 0.291, 0.332$ and $\hat{\rho} = 0.610, 0.636$. In figure 7.4, the direct simulation results from U-Shk are compared with the experimental data in [Pham-Van-Diep *et al.* (1989)]. It seems that the distribution functions are not too sensitive to the locations inside the shock layer. From figure 7.4, the validity of the unified scheme for the description of non-equilibrium flow has been proved.

In the above simulation, the viscosity coefficient used in the unified scheme is $\mu \sim T^{\omega}$, which is mainly to match with the VHS model of the DSMC simulations. Practically, there is no any additional cost for the unified scheme to use any other viscosity law. A more accurate viscosity law for argon gas is the Sutherland's law. The dimensionless viscosity coefficient for Sutherland's law will depend on the incoming temperature. For example, with the appropriate choice of incoming temperature, the shock thickness of Alsmeyer's measurements [Alsmeyer (1976)] at different Mach numbers can be perfectly matched from UGKS solutions.

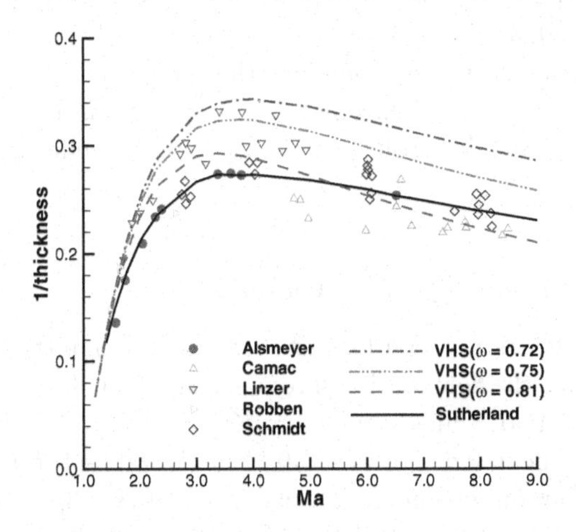

Fig. 7.5 Reciprocal shock thickness vs. Mach numbers for argon gas from experimental measurements and UGKS calculations. The UGKS uses different viscosity, such as VHS and Sutherland's law [Yu (2013)].

Figure 7.5 presents the numerical solutions calculated by UGKS with different viscosity laws. The reciprocal shock thickness vs. Mach numbers from 1 to 9 for argon gas with different ω in VHS viscosity model are compared with experimental measurements. This clearly shows that the VHS viscosity model is not adequate to describe the shock structure at all Mach numbers. At the same time, with the changing of the incoming temperature, and with the adoption of the Sutherland's viscosity law, Alsmeyer's experimental shock thickness at different Mach numbers can be perfectly recovered. Therefore, with the confidence on the Sutherland's law and the UGKS solution, the information about the experimental set up can

Table 7.1 Incoming gas temperature at different Mach numbers for recovering Alsmeyer's experimental shock thickness.

M	3.0	4.0	5.0	6.0	7.0	8.0	9.0
T_{in}	120k	110k	100k	95k	90k	82k	77k

be extracted. In order to have a perfect match between the numerical and experimental solutions, a change of incoming temperature at different Mach numbers seems necessary. The required incoming temperatures are shown in Table 7.1. So, based on the numerical solutions, it is possible to figure out the experimental conditions, which were not provided in Alsmeyer's paper. Certainly, all these conclusions depend on the reliability of the numerical solutions [Yu (2013)].

The shock structure tests validate the UGKS in the non-equilibrium flow computations. In the following, the computations are mostly based on the UGKS with Shakhov collision model, if not specifically stated about the model used.

7.3.2 *Flow around Circular Cylinder*

In order to further test the performance of the UGKS in the high speed rarefied flow regime, the flow passing through a circular cylinder of argon gas at Mach number 5 and Knudsen numbers $Kn = 0.1$ and 1.0 relative to cylinder radius are calculated. For $Kn = 0.1$, the DSMC setup is the following. The incoming argon gas has a velocity $U_\infty = 1538.73m/s$ with temperature $T_\infty = 273K$, molecule number density $n = 1.2944 \times 10^{21}/m^3$, and the viscosity coefficient at upstream condition $\mu_\infty = 2.117 \times 10^{-5}Ns/m^2$. The cylinder has a cold wall with a constant temperature $T_w = 273K$, with diffusive reflection boundary condition. For the $Kn = 1.0$, the only change is the incoming molecule number density, which is reduced to $n = 1.2944 \times 10^{20}/m^3$. The DSMC solution is provided by Quanhua Sun using their in-house parallelized DSMC code at Institute of Mechanics in Beijing. In DSMC simulations for both Knudsen numbers, 15000 mesh points are used in the physical space. Theoretically, at $M = 5$ much less mesh points can be used for a valid DSMC solution, especially at $Kn = 1.0$. The reason for using such an amount of mesh points in DSMC simulation is to get an accurate solution. Therefore, the following comparison between the solutions of DSMC and UGKS is not for the efficiency purpose, but for the validation of the accuracy of the unified scheme only.

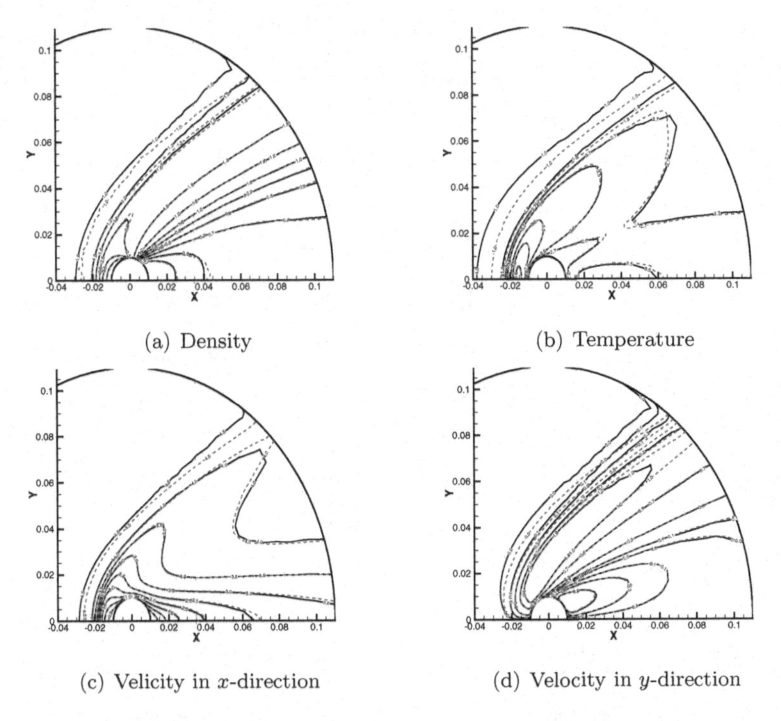

(a) Density

(b) Temperature

(c) Velicity in x-direction

(d) Velocity in y-direction

Fig. 7.6 Flow distributions around a cylinder at $M = 5$ and $Kn = 0.1$. Black solid line: UGKS, dash line: DSMC [Huang *et al.* (2012)].

For the calculations with UGKS, a mesh with $50 \times 64 = 3200$ points in physical space is used, and 93×93 in velocity space. The particle collision time is determined according to the VHS viscosity law with $\omega = 0.81$. In the current case, the outer boundary condition for the UGKS is based on the distribution function extrapolation.

At $Kn = 0.1$, the comparison between UGKS and DSMC solutions are shown in Figs. 7.6 - 7.8. Figure 7.6 presents the density, temperature, U-velocity, and V-velocity contours outside the circular cylinder. Most contour lines from UGKS match with DSMC very well. Figure 7.7 shows the density, pressure, temperature, and velocity distributions along the symmetric axis in front of the stagnation point. Since the cylinder wall has a low temperature, the density piles up sharply close to the cylinder surface. Figure 7.8 presents pressure, heat flux, and wall stress along the cylinder surface from the stagnation point to the trailing edge. Perfect match has been obtained between UGKS and DSMC solutions. Figures 7.9 and 7.10 show the same flow distributions in front of the cylinder and on

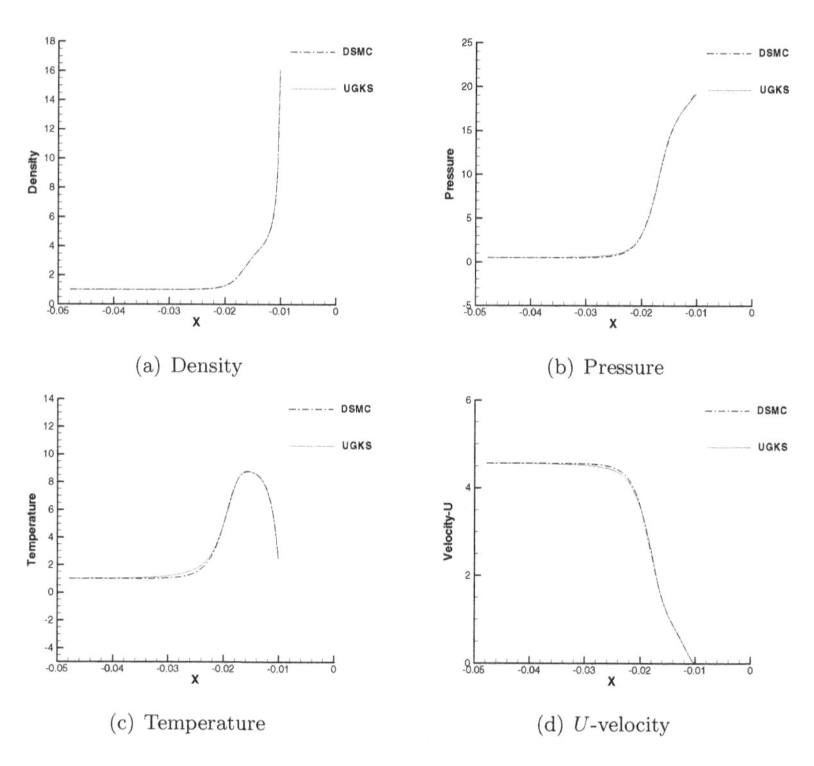

(a) Density (b) Pressure

(c) Temperature (d) U-velocity

Fig. 7.7 Flow variable distributions along central symmetric line in front of the stagnation point for case $M = 5$ and $Kn = 0.1$. Solid line: UGKS, dash-dot line: DSMC [Huang *et al.* (2012)].

the surface of the cylinder at $Kn = 1.0$. Perfect match has been obtained as well. In the current case, all solutions obtained by DSMC and UGKS take similar computational time. With the single machine with 6 cores, the explicit UGKS for unsteady flow computation needs about one day to get convergent solutions. Certainly, the efficiency of UGKS can be much improved if a local time step is used here for the steady state solution. In comparison with the low speed flow simulation, the efficiency of the UGKS for high speed flow over DSMC is not so obvious. One of the main reason is that the UGKS needs many mesh points in the particle velocity space. So, for high speed flow the bottleneck for the further development of the UGKS is the numerical integration in the velocity space. So, an adaptive mesh in velocity space is needed. Besides the local time stepping, other convergence acceleration techniques for CFD can be implemented in UGKS as well, such as multigrid, LU-SGS, at least for the update of the

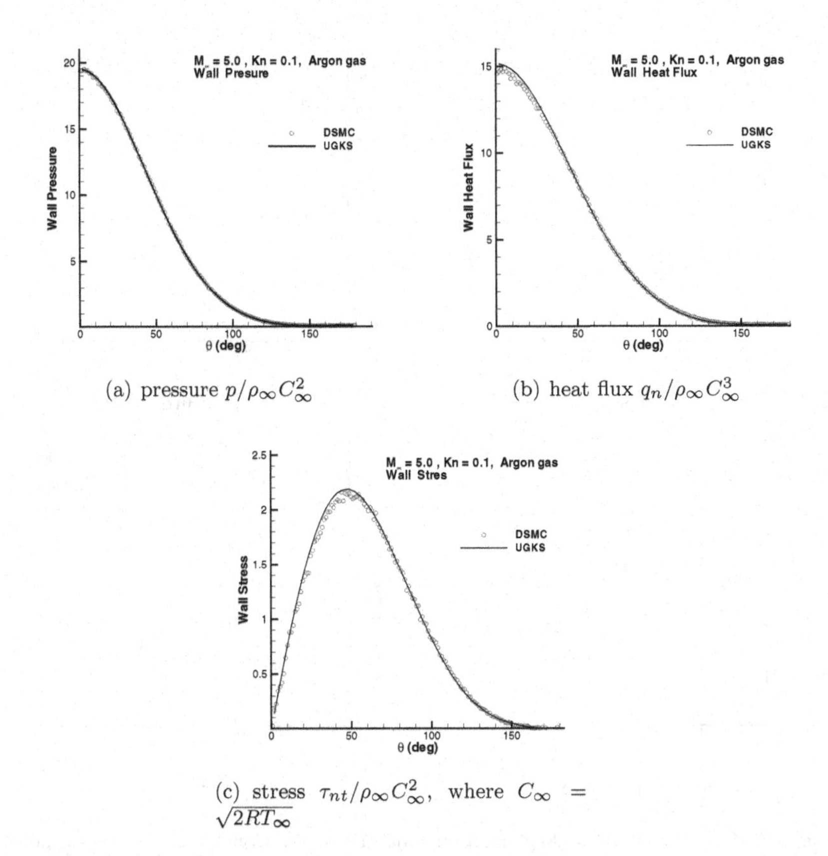

(a) pressure $p/\rho_\infty C_\infty^2$

(b) heat flux $q_n/\rho_\infty C_\infty^3$

(c) stress $\tau_{nt}/\rho_\infty C_\infty^2$, where $C_\infty = \sqrt{2RT_\infty}$

Fig. 7.8 Pressure, heat flux, and stress distributions along cylinder surface for $M = 5$ and $Kn = 0.1$. Solid line: UGKS, circles: DSMC [Huang *et al.* (2012)].

conservative flow variables [Xu *et al.* (2005)]. If succeed, the efficiency of UGKS will be improved significantly. For the steady state calculation, a recent study shows that the implicit UGKS can use a local time step of hundreds of CFL number.

In the whole transition regimes, there are a few experimental measurements. For the circular cylinder case, there are experimental data of drag coefficients at different Mach numbers [Maslach and Schaaf (1963); Metcalf *et al.* (1965)]. The drag coefficients of a cylinder are calculated by adaptive UGKS (AUGKS) for a wide range Knudsen numbers [Chen *et al.* (2012b)], where an adapted particle velocity space is used. The results are compared with the experimental data [Maslach and Schaaf (1963)]. Monatomic gas

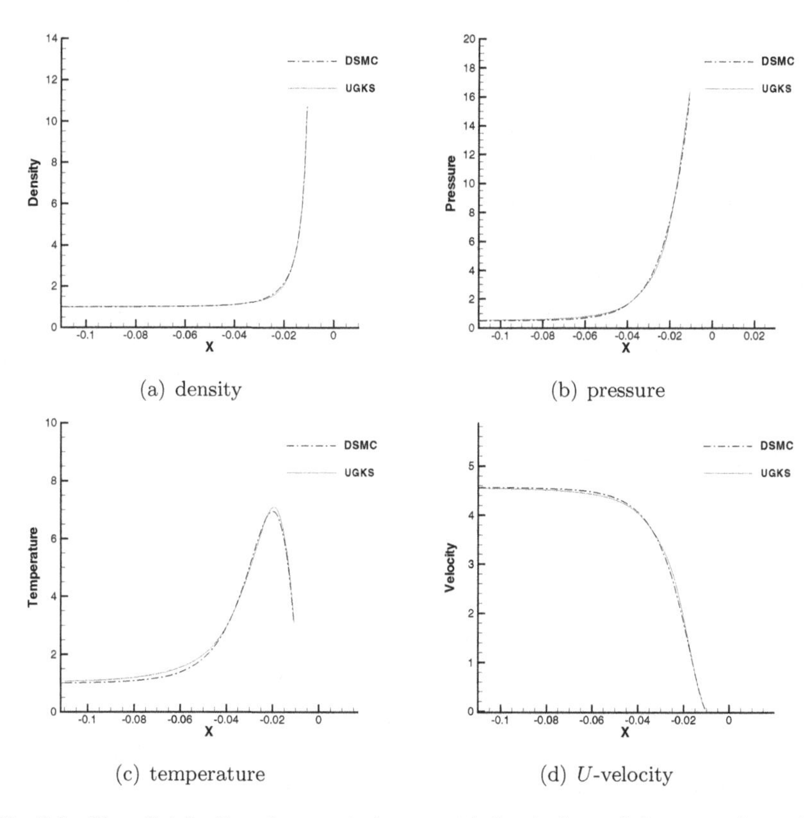

(a) density (b) pressure

(c) temperature (d) U-velocity

Fig. 7.9 Flow distribution along central symmetric line in front of the stagnation point for $M = 5$ and $Kn = 1.0$. Solid line: UGKS, dash-dot line: DSMC [Huang *et al.* (2012)].

is chosen as the medium with Prandtl number $\Pr = 2/3$, the ratio of specific heats $\gamma = 5/3$. The Mach number of incoming flow is 1.96. And the range of Knudsen numbers is from 0.001 to 100. The diameter of cylinder is denoted by D, and the diameter of whole computational domain is $11D$. The wall boundary condition is diffusive reflection. The mesh used in physical space is 100×60, and an adaptive velocity mesh is used with the initial quadtree mesh of 16×16. The wall temperature T_w is updated dynamically through the heat transfer from the flow (denoted by Q). Specifically, given the thermal capacity of the cylinder C, the temperature change is

$$\frac{\partial T_w}{\partial t} = \frac{Q}{C}, \tag{7.1}$$

where Q is obtained by the AUGKS, and C is a given appropriate value by numerical experiments. Actually, the value of C only determines how fast

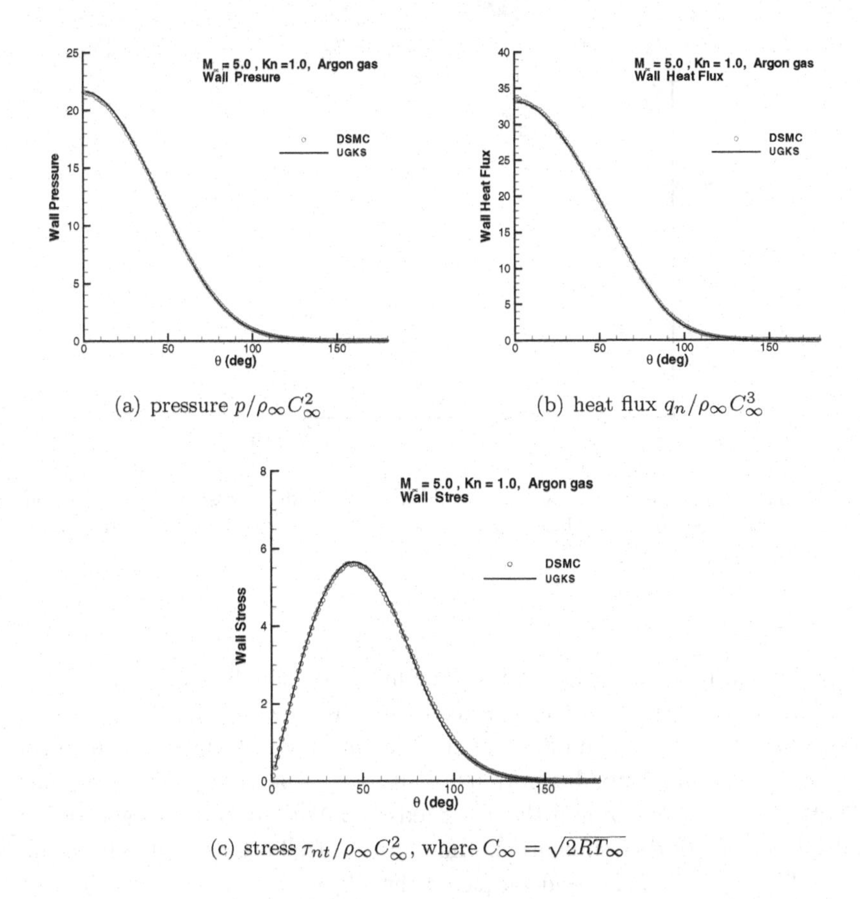

(a) pressure $p/\rho_\infty C_\infty^2$

(b) heat flux $q_n/\rho_\infty C_\infty^3$

(c) stress $\tau_{nt}/\rho_\infty C_\infty^2$, where $C_\infty = \sqrt{2RT_\infty}$

Fig. 7.10 Pressure, heat flux, and stress distributions along cylinder surface for $M = 5$ and $Kn = 1.0$. Solid line: UGKS, circles: DSMC [Huang *et al.* (2012)].

the T_w goes to a fixed value, but doesn't affect the final result. The final T_w is determined only by the heat flux Q computed by AUGKS, which will go to zero when the temperature of the cylinder rises. Figure 7.11 shows numerical drag coefficients vs. experimental measurements. As shown in the figure, the UGKS can basically capture the drag coefficients accurately in the whole flow regimes with reasonable amount of mesh points.

7.3.3 *Moving Ellipse*

A freely moving ellipse rests initially in a flow with velocity of $1538.73m/s$, temperature of $273K$, and dynamic viscosity of $2.117 \times 10^{-5} m^2/s$. The center of ellipse locates at (0,0) and the angle of attack of the ellipse is

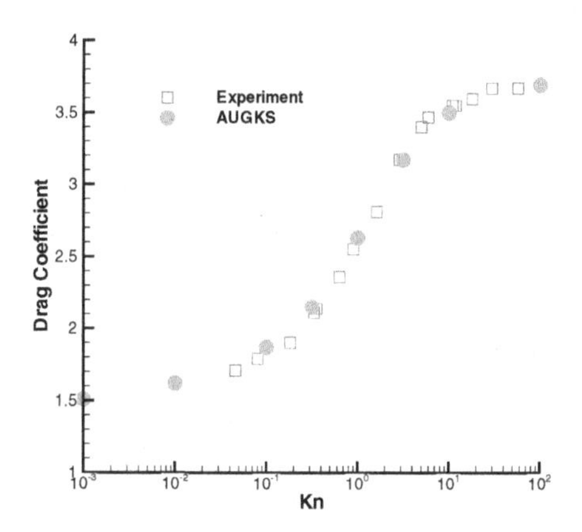

Fig. 7.11 Drag coefficient vs Knudsen number for the cylinder case at $M = 1.96$ from experiment measurements [Maslach and Schaaf (1963)] and UGKS calculations [Chen *et al.* (2012b)].

$-45°$ when the calculation starts. The incoming flow has a Mach number 5. Three cases with different upstream densities, i.e., 8.582×10^{-3} kg$/m^3$, 8.582×10^{-5} kg$/m^5$, and 8.582×10^{-6} kg$/m^3$, are calculated. The corresponding Knudsen numbers are 0.001, 0.1, 1, respectively. The long axis of the ellipse is $0.02m$ and the short axis is $0.01m$. In order to get visible displacement during simulation, the density of ellipse is relatively small, say, $1kg/m^3$. The force and torque on the ellipse are calculated during the flight, which determine the ellipse's flight trajectory and its rotation.

Figures 7.12 and 7.13 show the density and temperature field at $t = 1.28 \times 10^{-3}s$. At Kn $= 0.001$ case, similar to shock capturing scheme for continuum flow, the AUGKS presents a sharp shock front. The gas distribution functions are very close to Maxwell distribution in the whole computational domain expect at the shock front and close to the boundary. As the Knudsen number increases, the shock thickness gets broaden with a resolved shock structure. For example, at $Kn = 0.1$, the shock thickness is comparable with the size of the ellipse. The distribution functions inside the shock show non-equilibrium properties, which are shown in Fig. 7.14. At $Kn = 1$, the dissipative shock layer extends almost to the whole upstream region. As suggested by the Mott-Smith approximation [Mott-Smith (1951)], a bimodal type distribution function shows up in the shock region, see figure 7.15.

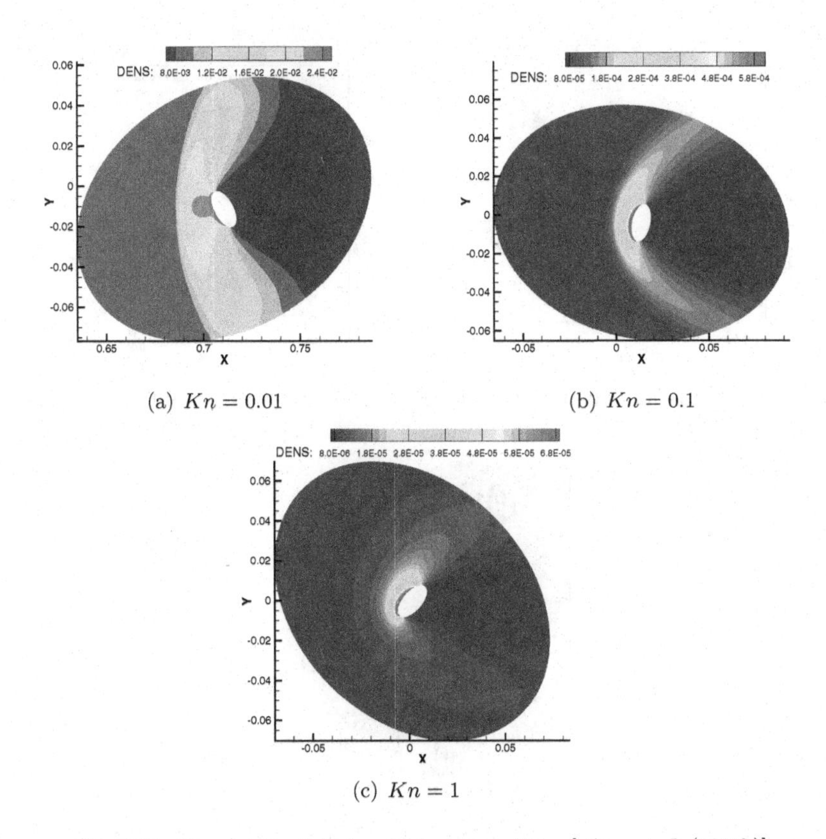

(a) $Kn = 0.01$

(b) $Kn = 0.1$

(c) $Kn = 1$

Fig. 7.12 The density field for shock driven ellipse [Chen *et al.* (2012b)].

7.3.4 *Continuum Flow Expansion into Rarefied Environment*

Here we consider a flow with the co-existence of both continuum and rarefied regimes in a single computation [Chen *et al.* (2012b)]. This is a multi-scale and challenging problem for any single scale-based numerical scheme. But, this is exactly the place where the UGKS can be faithfully applied to get accurate solution in different regimes. Otherwise, a hybrid approach with domain decomposition may be adopted. But, for the unsteady case, like this one, the hybrid approach can be hardly applied due to the absence of a valid interface.

The nozzle is designed to have a rectangle shape with a round head. Figure 7.16 gives the nozzle's configuration and the initial position and attitude. The mass of the nozzle is $1.57 \times 10^{-2} kg$. Both the dense gas inside

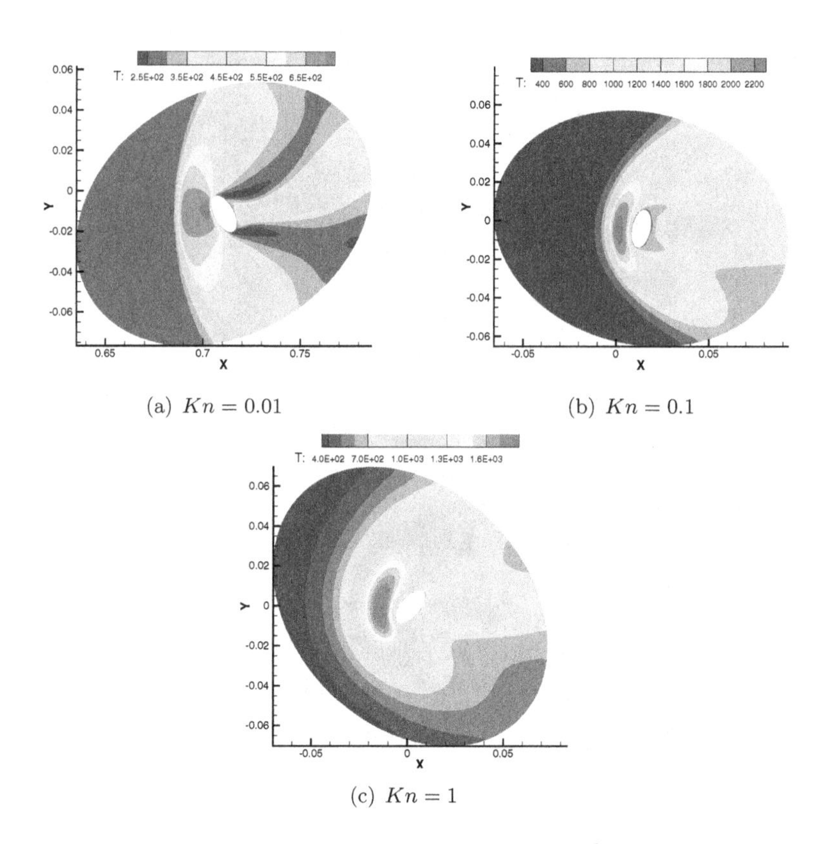

(a) $Kn = 0.01$

(b) $Kn = 0.1$

(c) $Kn = 1$

Fig. 7.13 The temperature field for shock driven ellipse [Chen *et al.* (2012b)].

the nozzle and the dilute gas outside the nozzle are stationary initially. The temperature is $273K$ for all gas and solid walls. Density inside the nozzle is $\rho_{in} = 8.582 \times 10^{-2} kg/m^3$, and outside $\rho_{out} = 8.582 \times 10^{-6} kg/m^3$, with a viscosity coefficient $2.117 \times 10^{-5} m^2/s$. Even with the density ratio of 10^4, the same cell size is used inside and outside the nozzle in the current calculation. The Knudsen number of the external gas is 0.05, based on the diameter of head. The whole expansion consists of three stages. Initially, the gas exhausting from the nozzle. In this stage, the gas behaves as a free expansion. Since the external gas is so dilute, the dense gas expansion meets almost no resistance from environment gas. The stream lines become radial, see Fig. 7.17. Associated with the intensive expansion, the temperature of gas drops quickly to almost $10K$. The distribution function becomes a narrow-kernel shape in the velocity space. After free expansion, a jet appears behind the nozzle. Owing to the reduced density difference outside,

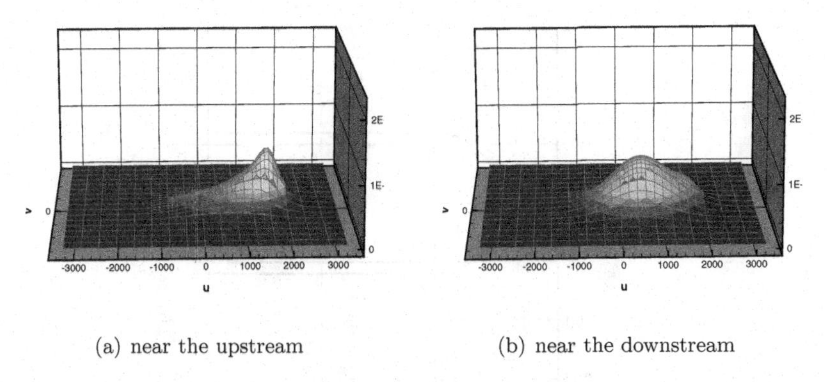

(a) near the upstream (b) near the downstream

Fig. 7.14 The velocity distribution function inside the shock for Kn = 0.1 [Chen *et al.* (2012b)].

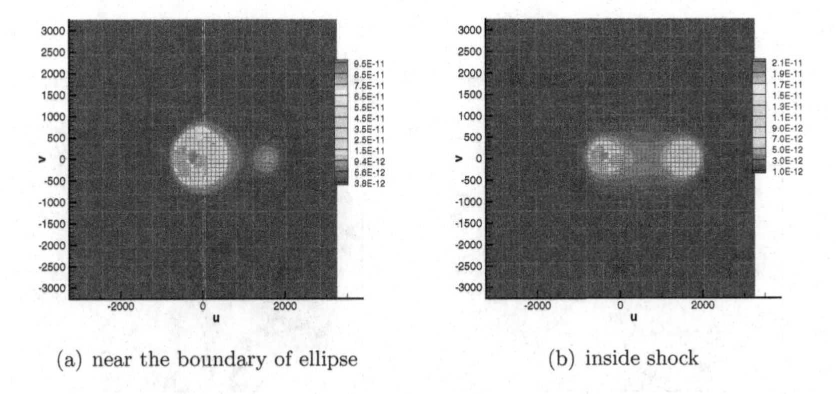

(a) near the boundary of ellipse (b) inside shock

Fig. 7.15 The velocity distribution function through the shock at $Kn = 1$ [Chen *et al.* (2012b)].

the follow-up gas out of the nozzle is confined in a narrow region. Figure 7.18 shows the flow field. At the same time, due to the re-action force from the expansion, the nozzle gets accelerated and then moves by its own inertia. Eventually, the expansion stage terminates. Figure 7.19 shows that the pressure inside the nozzle is lower than the surroundings. The streamline around the nozzle presents a typical characteristic flow at low speed. Because the environment density is very low, the deceleration is very weak. This stage will sustain a long time until the nozzle gets stopped.

Figure 7.20 shows the quadtree mesh and contour of particle distribution

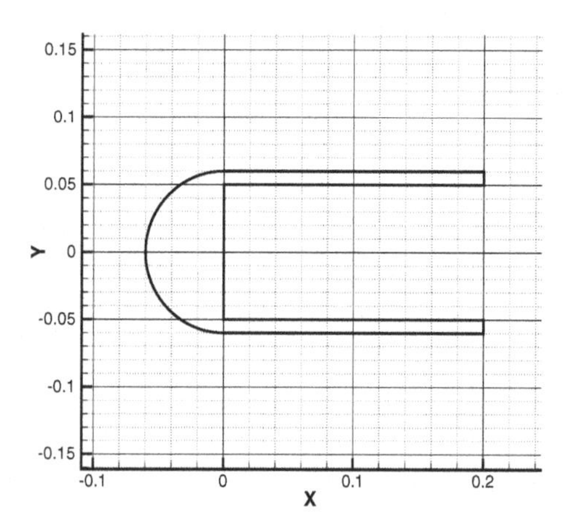

Fig. 7.16 The configuration of nozzle.

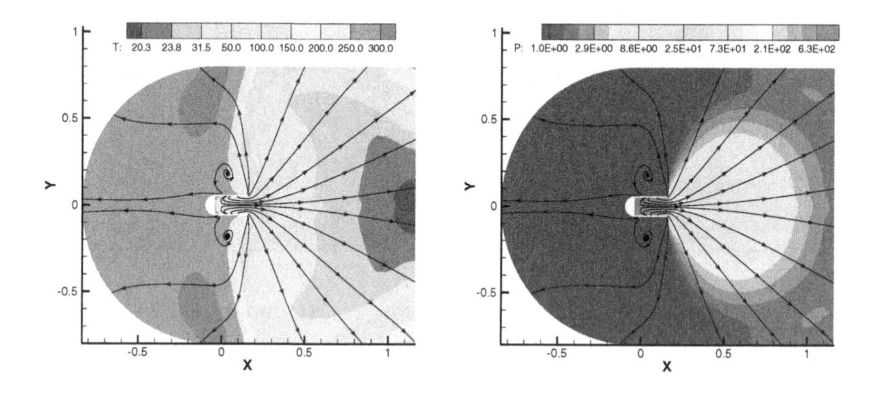

Fig. 7.17 The flow field during expansion stage [Chen *et al.* (2012b)].

function at a location $(1.2, 0)$ during the expansion stage. At first the distribution function keeps a Maxwell distribution (Fig. 7.20(a)). Then the fast particles arrive, and distort the distribution function a little bit (Fig. 7.20(b)). The corresponding macroscopic picture shows that a shock wave sweeps this location. The followed expansion makes the temperature drop significantly to a very low value. As shown in figure 7.20(c), the kernel of distribution function becomes much more concentrated than the initial one.

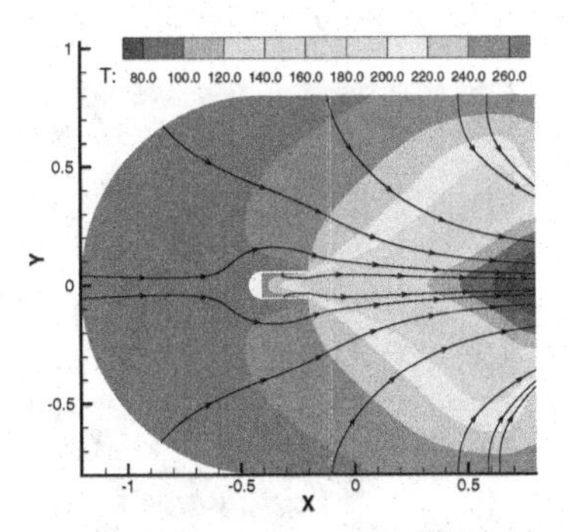

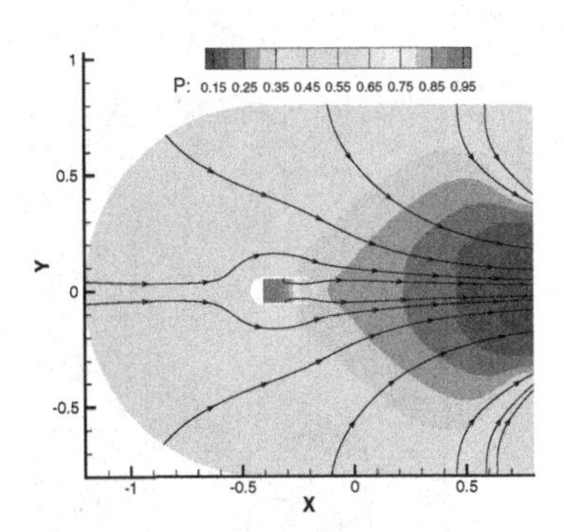

Fig. 7.18 The jet-like flow field [Chen *et al.* (2012b)].

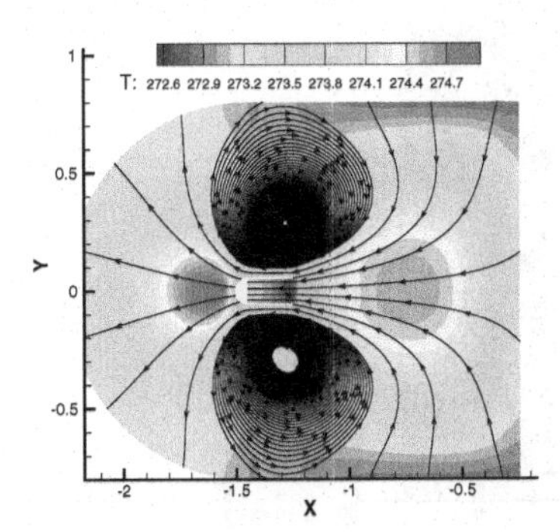

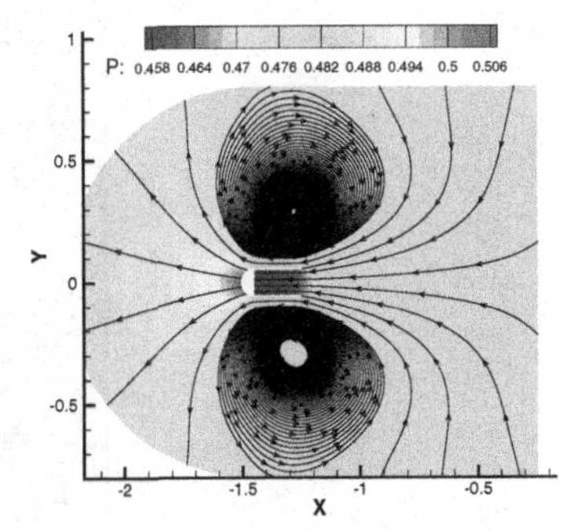

Fig. 7.19 The flow field during deceleration stage [Chen *et al.* (2012b)].

For the AUGKS, the mesh is concentrated in the area where particles accumulate, as shown in this example. For the shock driven ellipse, the AUGKS is about four times faster than the UGKS when a uniform mesh in the velocity space is used. The UGKS requires 64×64 velocity elements to achieve the same accuracy, where the AUGKS only uses 760 velocity elements. The additional storage and load for quadtree structure are less than 20%. For the nozzle case, the extreme low temperature limits the finest mesh size in the velocity space. According to velocity mesh in Fig. 7.20, for the same accuracy a uniform mesh with 256×256 grid points is necessary. This number is much larger than 1300 velocity elements used in the AUGKS calculation. In this case, the route of mesh refinement may not be an optimized one. As shown, the mesh does not follow the density of particles precisely, and the mesh depends on the refining history. This

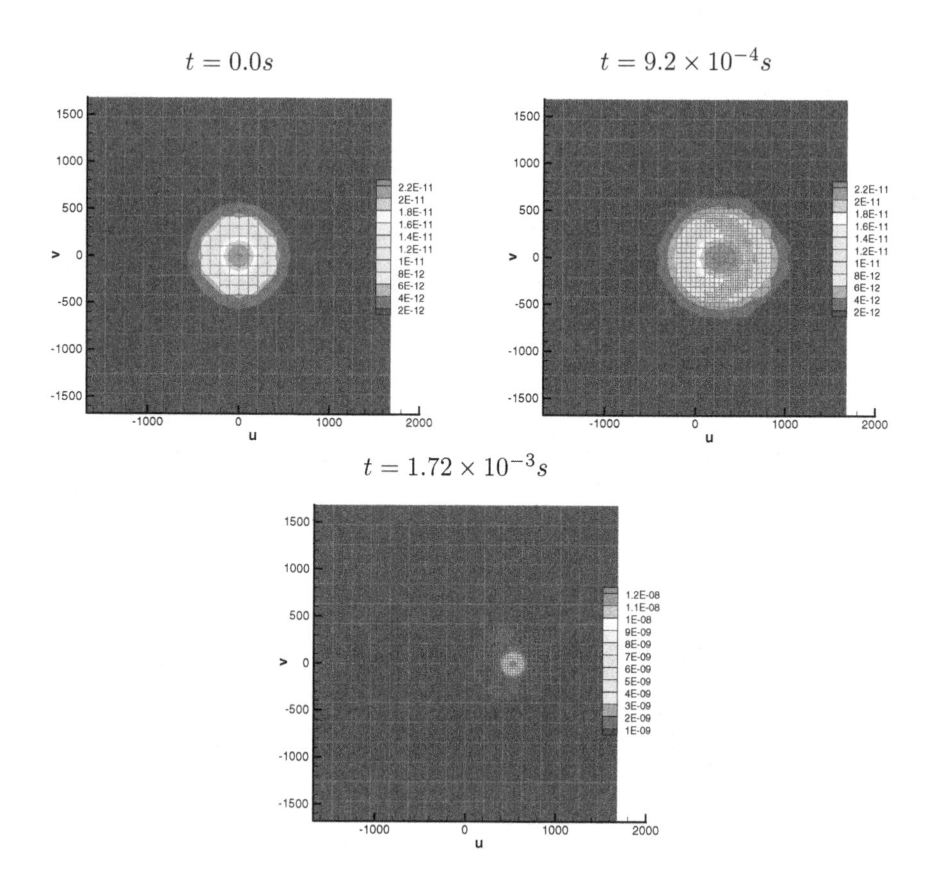

Fig. 7.20 The typical distribution function at location (1.2, 0) [Chen *et al.* (2012b)].

kind of optimization problem can be partially solved by a global monitor function for the mesh adaptation, which is a research topic for the further development of UGKS.

7.4 Hypersonic Flow Studies

It is challenging for any kinetic equation solver with discrete particle velocity space to simulate hypersonic flow. Theoretically, the particle velocity domain is infinite, finite bounds must be chosen such that only a negligible fraction of molecules lies outside the computational domain. For hypersonic flow, due to the wide range of the velocity change, it is difficult to set a proper bound. A small fraction of molecules with very high velocities

Table 7.2 Computational conditions for DSMC and UGKS.

Kn_∞	$n_\infty[1/m^3]$	$\rho_\infty[kg/m^3]$	$T_\infty[K]$	$T_w[K]$	$U_\infty[m/s]$	Ma_∞
0.01	1.294×10^{22}	8.585×10^{-4}	273	273	3077.587	10
					6155.174	20
0.1	1.294×10^{21}	8.585×10^{-5}	273	273	3077.587	10
					6155.174	20
1	1.294×10^{20}	8.585×10^{-6}	273	273	3077.587	10
					6155.174	20

relative to the bulk of the gas can have a significant effect on the macroscopic flow properties. As pointed out in [Bird (1994)], if 100 grid points are employed for each direction in the velocity space, the Boltzmann solver for a three-dimensional unsteady flow computation may require 10^{14} grid points. Therefore, the use of adaptive mesh in the velocity space must be done for the kinetic solver in the hypersonic case. If the grid in the particle velocity space can be automatically adapted to where it is needed, such as the particles in the DSMC method, it is possible that a kinetic solver can become a practical method for hypersonic rarefied flow study.

In order to demonstrate the capacity of the UGKS with adaptive velocity method in the hypersonic flow computation, the classic hypersonic rarefied flow of argon gas around a circular cylinder will be considered at hypersonic speed. The adaptive method used in this section is presented in [Yu (2013)]. The UGKS results will be compared with two DSMC solutions, one is obtained from the DS2V software of G.A. Bird with Cartesian mesh, and the other is calculated by Q.H. Sun with a DSMC code with curvilinear mesh.

The inflow boundary conditions are Mach numbers $Ma = 10$ and 20, and the free stream density of the flow is related to the Knudsen numbers of $Kn = 0.01$, 0.1, and 1.0, from the near continuum regime through the transitional to the rarefied regime. The computational conditions are listed in Table 7.2. The Knudsen number is defined based on the cylinder radius, using the hard-sphere model for particle mean free path λ,

$$\lambda = \frac{1}{\sqrt{2}\pi d^2 n}, \tag{7.2}$$

where d is the molecular diameter and n is the number density. The density ρ_∞ is evaluated by

$$\rho_\infty = nm, \tag{7.3}$$

in which m is the molecular mass. The Mach number is defined as the ratio of the inflow velocity with the sound speed,

$$Ma = \frac{U_\infty}{C}, \tag{7.4}$$

where $C = \sqrt{\gamma R T_\infty}$ is the sound speed, γ is the ratio of specific heat and $R = k/m$ is the gas constant and k is the Boltzmann constant. For the argon molecular, d is equal to 4.17×10^{-10} meter, m is equal to $6.63 \times 10^{-26} kg$, and γ is equal to $5/3$. The reference temperature T_∞ and the temperature at the cylinder surface T_w are set as a constant: $T_\infty = T_w = 273K$.

The viscosity is calculated using variable hard sphere (VHS) model in DSMC simulations,

$$\mu = \mu_{ref} \left(\frac{T}{T_{ref}} \right)^{\omega},$$

$$\mu_{ref} = \frac{15\sqrt{\pi m k T_\infty}}{2\pi d^2 (5 - 2\omega)(7 - 2\omega)},$$

where μ is the dynamic viscosity coefficient, T is the temperature, ω is the VHS temperature exponent and $\omega = 0.81$.

For the UGKS method, the reference density and temperature are set as ρ_∞ and T_∞ respectively, and reference velocity C_∞ is calculated by $C_\infty = \sqrt{2RT_\infty}$. The initial condition is set with the inflow free stream parameters, while the initial velocity distribution function in phase space is given by a Maxwellian.

The computational domain and mesh are shown in Fig. 7.21, in which the radius of the cylinder is set as $R = 0.01m$, and the computational domain is divided by body-fitted mesh for DSMC (Sun) and quadrilateral mesh for UGKS. The mesh size for DSMC method is much less than the mean free path, especially for the small Knudsen number cases, while there is no such limit on the mesh size for UGKS method.

The local tree in the velocity space for each cell of the physical space is generated based on the Maxwellian function, and the domain in phase space for $Ma = 10$ and $Ma = 20$ are estimated as $(u, v) \in [-20, 20] \times [-20, 20]$ and $(u, v) \in [-35, 35] \times [-35, 35]$, respectively, and the tree levels L are set as $3 \sim 6$ and $3 \sim 7$, respectively, while C_{min} and C_{max} are set as $C_{min} = 1.0 \times 10^{-6}$ and $C_{max} = 1.0 \times 10^{-3}$ [Yu (2013)]. Two-point Gauss-Legendre quadrature rule is adopted at the leaf cells of local trees. For these wide domains, the equal-spaced Newton-Cotes rule is no longer suitable, and the massive computer memory requirements far exceed the capacity of current personal computers. The hypersonic inflow and outflow boundary conditions are automatically determined by the angle between the direction of velocity in fluid field and the direction of free stream inflow. Fully diffusive boundary condition is used on the cylinder surface. In the following, all the DSMC solutions are presented in terms of non-dimensional variables.

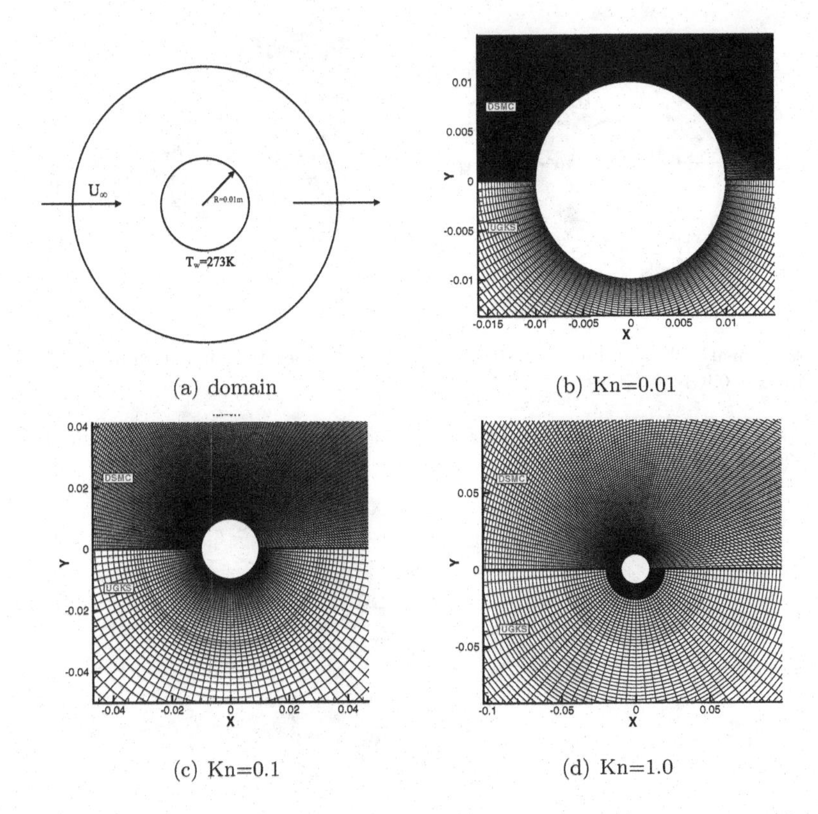

(a) domain (b) Kn=0.01

(c) Kn=0.1 (d) Kn=1.0

Fig. 7.21 Computational domain and mesh for UGKS and DSMC [Yu (2013)].

7.4.1 *Flow Fields*

The comparison between DSMC and UGKS solutions for the flow field properties are illustrated in Figs. 7.22–7.27, including the contours for density, velocity and average temperature, and velocity stream lines.

The velocity streamlines from both DSMC and UGKS are almost symmetrical in all cases, since the flow remains in the laminar regime.

As shown in Figs. 7.22–7.24, most of the contour lines for $Ma = 10$ flows are well matched, with little difference at different Knundsen numbers. In the case $Kn = 0.01$, the contour lines before the cylinder are almost identical for the solution of DSMC and UGKS methods. As the Knudsen number increases, the shock in front of the cylinder becomes much more diffusive, and the density and temperature of UGKS solutions raise up little earlier than that of DSMC in front of the cylinder.

The flow fields for $Ma = 20$ are shown in Figs. 7.25–7.27. In order to

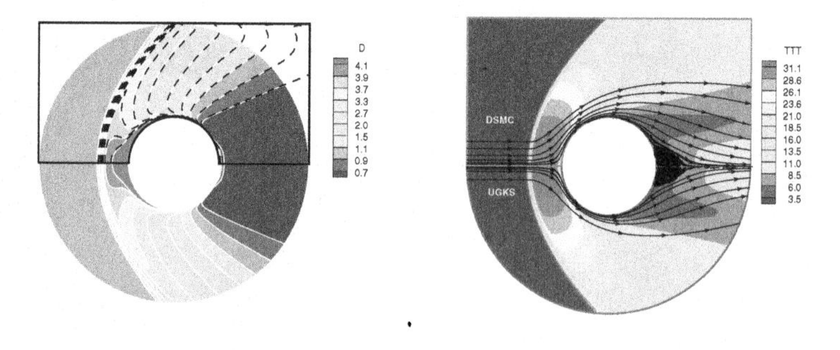

(a) density, dashed line: DSMC, solid Line: UGKS

(b) temperature and stream lines

Fig. 7.22 Flow distributions around a cylinder at $Ma = 10$ and $Kn = 0.01$ [Yu (2013)].

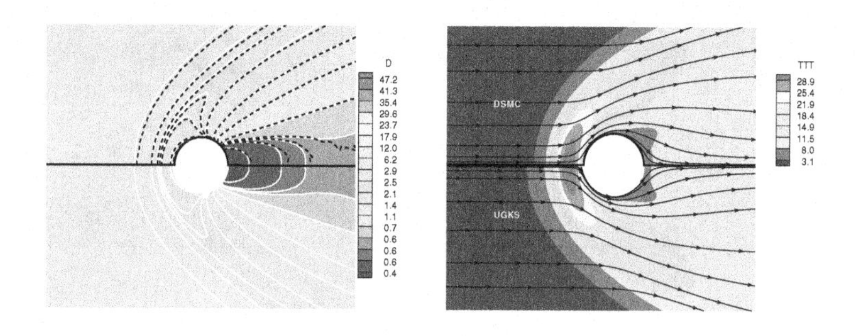

(a) density, dashed line: DSMC, solid Line: UGKS

(b) temperature and stream line

Fig. 7.23 Flow distributions around a cylinder at $Ma = 10$ and $Kn = 0.1$ [Yu (2013)].

satisfy the memory requirement of a personal computer (less than $10G$) at $Ma = 20$, the maximum level of local trees in the adaptive mesh method is limited to only 2-point Gauss-Legendre quadrature rule. So numerical quadrature error exists in the UGKS solutions. However, from the comparison with DSMC solutions, the contour lines are also well matched, and the difference can be acceptable in the practical applications.

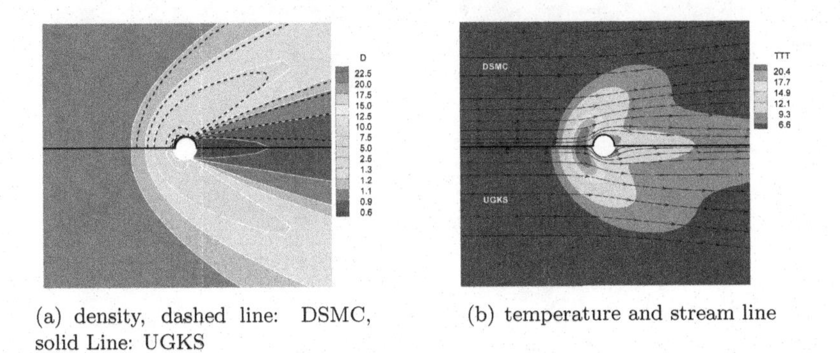

(a) density, dashed line: DSMC, solid Line: UGKS

(b) temperature and stream line

Fig. 7.24 Flow distributions around a cylinder at $Ma = 10$ and $Kn = 1.0$ [Yu (2013)].

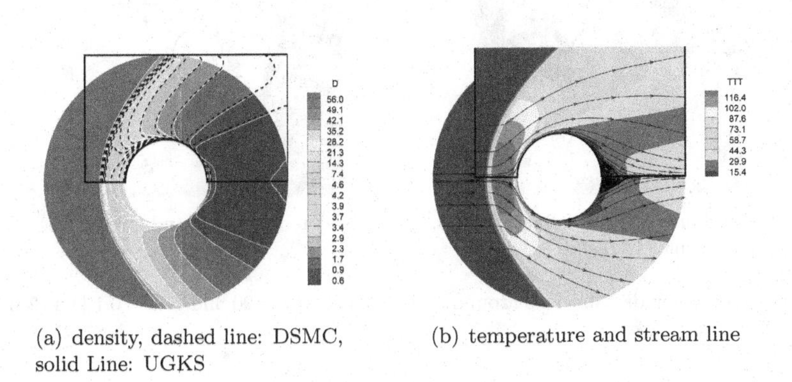

(a) density, dashed line: DSMC, solid Line: UGKS

(b) temperature and stream line

Fig. 7.25 Flow distributions around a cylinder at $Ma = 20$ and $Kn = 0.01$ [Yu (2013)].

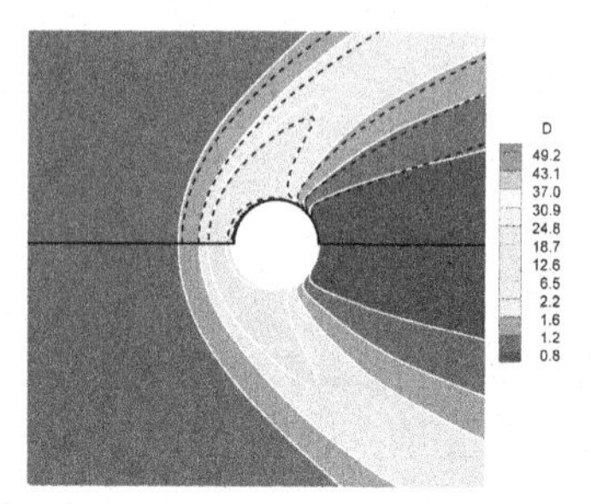

(a) density, dashed line: DSMC, solid Line: UGKS

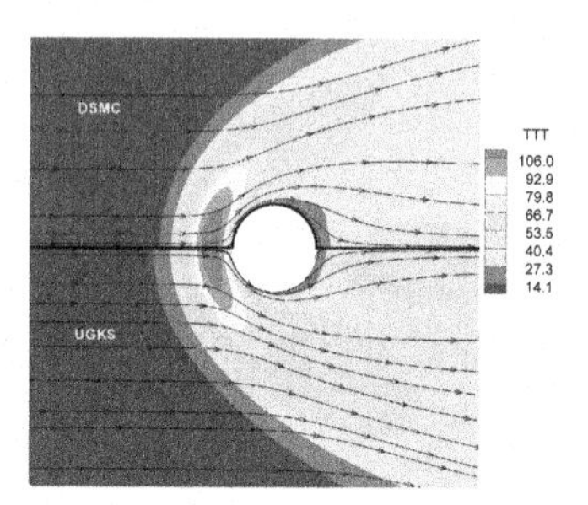

(b) temperature and stream line

Fig. 7.26 Flow distributions around a cylinder at $Ma = 20$ and $Kn = 0.1$ [Yu (2013)].

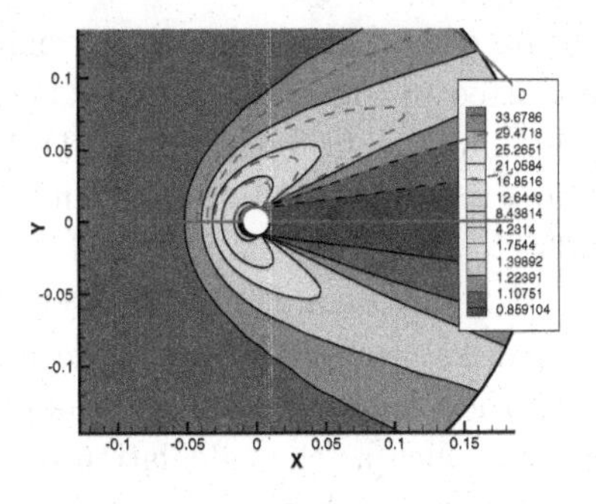

(a) density, dashed line: DSMC, solid Line: UGKS

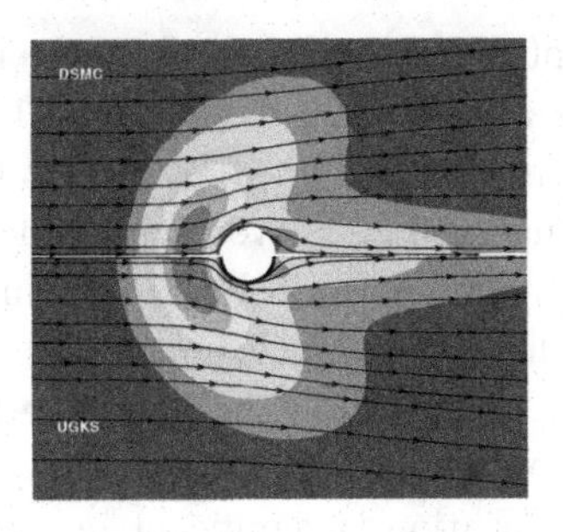

(b) temperature and stream line

Fig. 7.27 Flow distributions around a cylinder at $Ma = 20$ and $Kn = 1.0$ [Yu (2013)].

7.4.2 *Surface Properties*

The surface properties around the cylinder, such as the pressure, shear stress, and heat flux, are the most valuable quantities for the hypersonic flow study. The comparison between UGKS and DSMC solutions are presented in Figs. 7.28 to 7.33.

At $Kn = 0.01$, as shown in Figs. 7.28 and 7.31 for $Ma = 10$ and 20, the solutions of UGKS are not very satisfactory, especially in the case of $Ma = 20$. However, just as mentioned earlier, the main reason for the difference is due to the coarse mesh in physical space used in UGKS simulations, and the surface properties are sensitive to the mesh size in the near continuum regime, especially for the heat flux and shear stress. The minimum mesh size used in UGKS is about 8 times larger than the mean free path, while the mesh size used in the DSMC simulation is about 10% of a mean free path. To confirm the effect of mesh size, we recalculate the case of $Ma = 20$ and $Kn = 0.01$ in a domain of half cylinder and refined mesh around the cylinder surface. The new comparison is plotted in Fig. 7.34. Perfect agreements between UGKS and DSMC solutions are achieved near the cylinder surface.

As the Knudsen number increases, the mesh size effect gradually reduces due to the mixture of shock layer and the boundary layer. At $Kn = 0.1$, as shown in Figs. 7.29 and 7.32 for $Ma = 10$ and 20, respectively, the solutions of UGKS are well matched with DSMC solution and the differences are acceptable. In order to verify the mesh size effect, the results with a refined mesh at $Ma = 20$ and $Kn = 0.1$ are plotted in Fig. 7.35, where a better agreement is obtained. At $Kn = 1.0$, the solutions from UGKS and DSMC are matched well even with much coarse mesh used in UGKS, and the results are shown in Figs. 7.30 and 7.33.

The overall conclusion is that the UGKS can use a much larger cell size than the DSMC in the high-speed rarefied flow computations. However, if the non-equilibrium effect does appear in the mean free path scale, and the flow variation is significant in such a scale, such as the calculation of the heat flux and stress next to the boundary, the mesh size used there does need to resolve the flow physics. A large cell size there can smooth out the non-equilibrium solution due to the coarse grained averaging in the mesh size scale. However, if the non-equilibrium effect appears over a few mean free path, such as the shock layer, the UGKS can employ a mesh size which is a few times larger than the local particle mean free path. Even with the refined mesh in UGKS calculation, the mesh size used in UGKS next to the boundary is still much larger than that used in DSMC.

7.5 Summary

In this chapter, we present cases for the UGKS in the high speed flow computations with co-existence of multiple flow regimes. The UGKS can provide accurate solutions in all cases tested so far. In the rarefied regime, the validity of the UGKS has been tested in comparison with the benchmark solution of DSMC and experimental measurements. These tests clearly show that the UGKS can be faithfully used for rarefied high speed flow computation.

For the high speed and high temperature rarefied flow computation, currently there is no obvious advantage for the UGKS in comparison with DSMC method when fixed mesh points in the velocity space are used, especially in the high Mach number cases. The bottleneck for the kinetic solver for high speed flow is that a large number of grid points in the velocity space is needed to cover the whole spectrum of widespread particle velocity space. The use of adaptive mesh in velocity space is necessary and is implemented in the current study. As a result, all calculations presented here are calculated using personal computers. Many researchers working in the Boltzmann solvers fully realize the importance of the use of adaptive velocity space [Aristov (1977); Kolobov *et al.* (2011); Chen *et al.* (2012b);

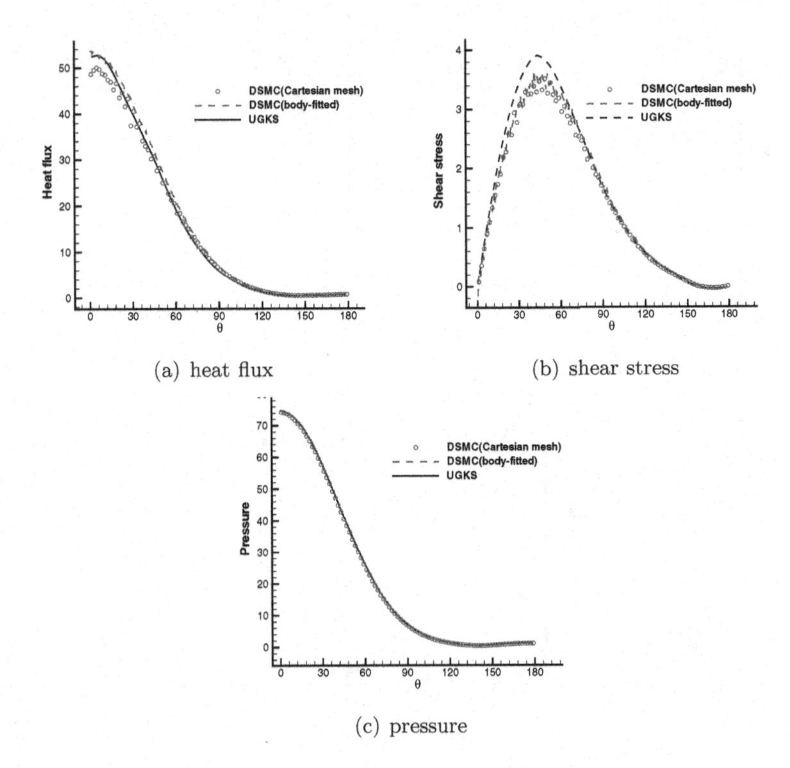

(a) heat flux

(b) shear stress

(c) pressure

Fig. 7.28 Heat flux, shear stress, and pressure along the surface of the cylinder at $Ma = 10$ and $Kn = 0.01$.

Kolobov and Arslanbekov (2012); Baranger *et al.* (2014)]. This will become main stream in the development of future solvers based on the kinetic equations.

Since the only free parameter in the UGKS is the collision time τ, which is related to the viscosity coefficient $\tau = \mu/p$. Different from the DSMC method, any complicated viscosity temperature relationship can be easily incorporated into the UGKS. There is no need to numerically validate this relationship before real computation. Due to multiscale modeling nature and high-order discretization, the UGKS can use much coarse mesh in physical space in comparison with direct Boltzmann solver and DSMC. The spatial cell size and time step used in the UGKS are not limited by the constraints of particle mean free path and collision time.

Based on the simulation results from the UGKS, it seems that the non-equilibrium flow behavior may not be so sensitive to the particle collision model, especially for the flow close to the boundary. With the

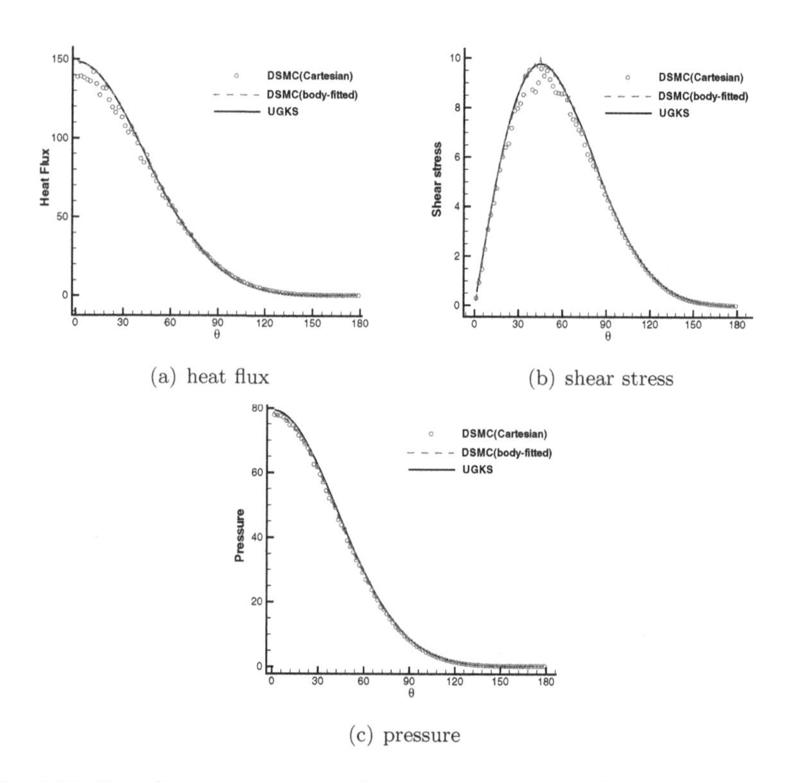

(a) heat flux

(b) shear stress

(c) pressure

Fig. 7.29 Heat flux, shear stress, and pressure along the surface of the cylinder at $Ma = 10$ and $Kn = 0.1$.

BGK-Shakhov model, even with the BGK model with additional heat flux modification [Xu and Huang (2010)], the rarefied flow behavior can be mostly and accurately captured by UGKS. Physically, the particle free transport deviates the gas distribution function to a non-equilibrium state, and the particle collision term drives the particle system back to "equilibrium" through the accumulating collision effect among the particles. The important mechanism in the particle collision process is the conservation laws and the rate to approach to the equilibrium. Since the macroscopic flow variables are the moments of the distribution function, such as pressure, stress, and heat flux, they will not be so sensitive to the detailed collision process at individual particle velocity, but to the evolution of the whole curve of the distribution function. The insensitivity to the particle collision model becomes more obvious when the mesh size is larger than the particle mean free path and the time step is a few times of the particle collision time. As tested in [Sun *et al.* (2014)], from an arbitrary initial

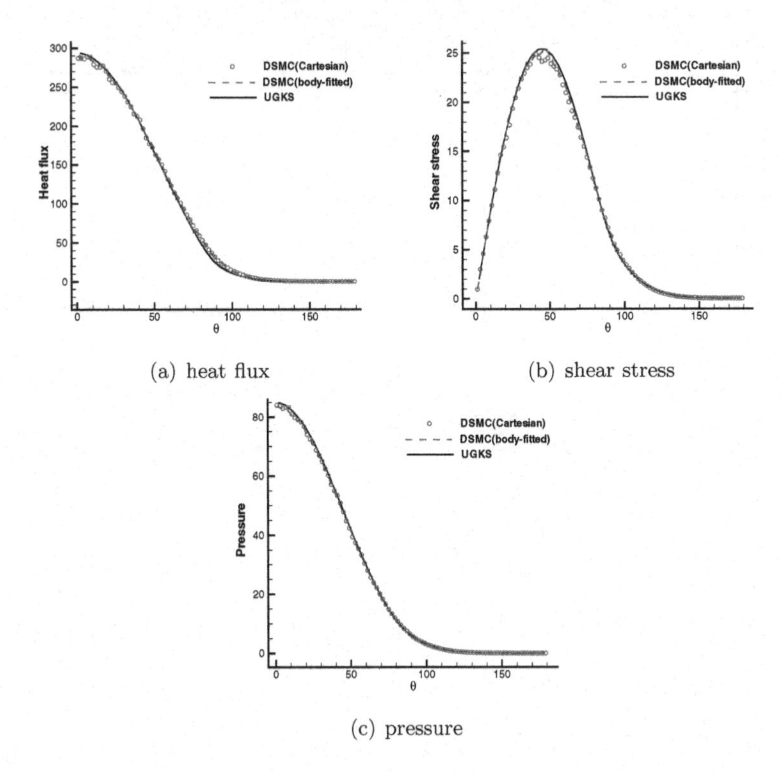

(a) heat flux

(b) shear stress

(c) pressure

Fig. 7.30 Heat flux, shear stress, and pressure along the surface of the cylinder at $Ma = 10$ and $Kn = 1.0$.

non-equilibrium state, the gas distribution function become indistinguishable from the full Boltzmann collision model and kinetic model equations after a few collision time, especially the values of pressure, stress, and heat flux. In the coarse graining process, the overall conservation will be important. In the continuum flow regime, the interface flux for the macroscopic flow variables becomes important, same as GKS, the NS solution can always be obtained from UGKS regardless the collision model used inside each cell. In the free molecule transport limit, the non-equilibrium state of the gas system is fully determined by the free transport part and the boundary condition, where the UGKS can present an exact solution as well. The merit of UGKS is the capability to present accurate solution from the low transition to the continuum flow regimes, where the full Boltzmann collision term plays a less dynamical role in this regime. In the highly non-equilibrium region with $\Delta t < \tau$, the full Boltzmann collision term can be used in UGKS if more accurate solution is necessary. Actually, the UGKS

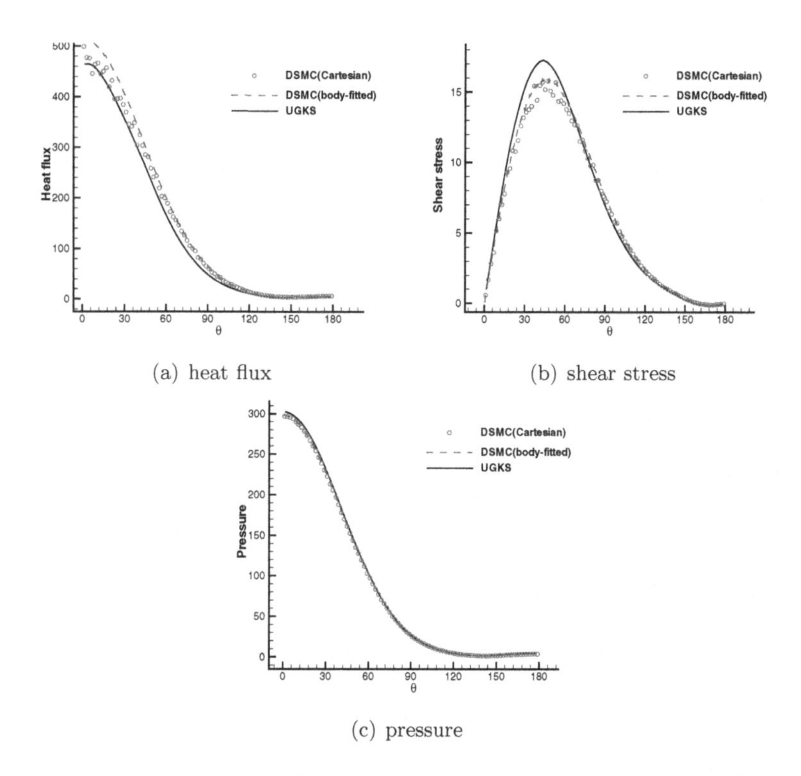

(a) heat flux

(b) shear stress

(c) pressure

Fig. 7.31 Heat flux, shear stress, and pressure along the surface of the cylinder at $Ma = 20$ and $Kn = 0.01$.

with the implementation of the full Boltzmann collision term for particle collision inside each cell in the sub-time step of $t \leq \tau$, and of kinetic model equation in $t > \tau$, has been developed [Liu *et al.* (2014)]. Based on the current study, the UGKS with the kinetic model equation only seems accurate enough for engineering applications.

For hypersonic flow, the shock layer temperature, such as in the Apollo reentry case, can reach to about $11,000K$. At such a high temperature, many thermo-chemical or real gas effects will be excited, which affect the properties of air and the aero-thermodynamic properties in a significant fashion. Typical thermal effects include excitation of internal energy modes in addition to translational and rotational ones, such as vibrational and electronic excitation. The chemical effects include molecular dissociation, recombination, and ionization. These effects have not been included in the current UGKS, and need to be studied in the future.

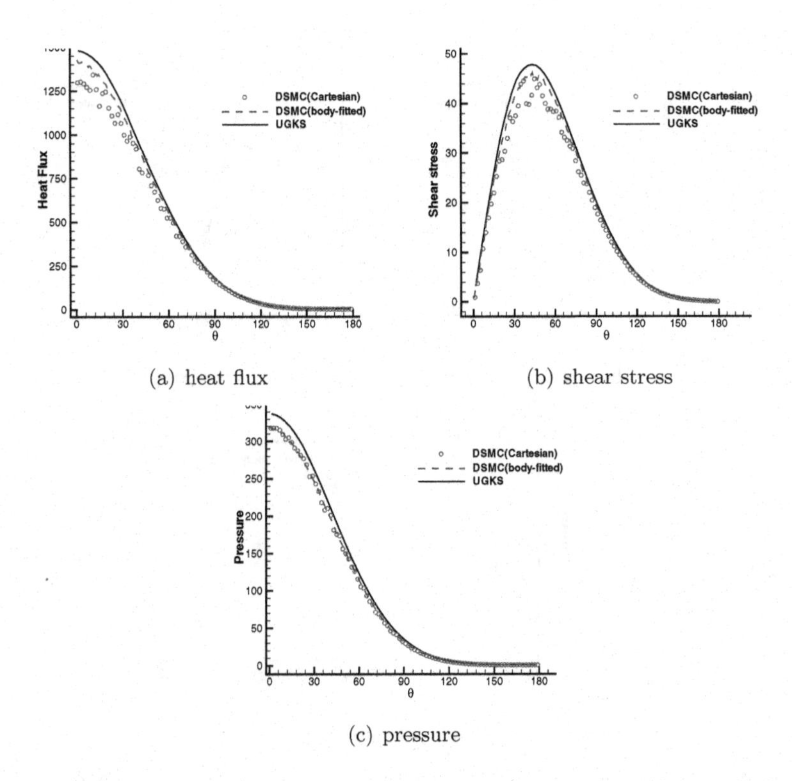

(a) heat flux

(b) shear stress

(c) pressure

Fig. 7.32 Heat flux, shear stress, and pressure along the surface of the cylinder at $Ma = 20$ and $Kn = 0.1$.

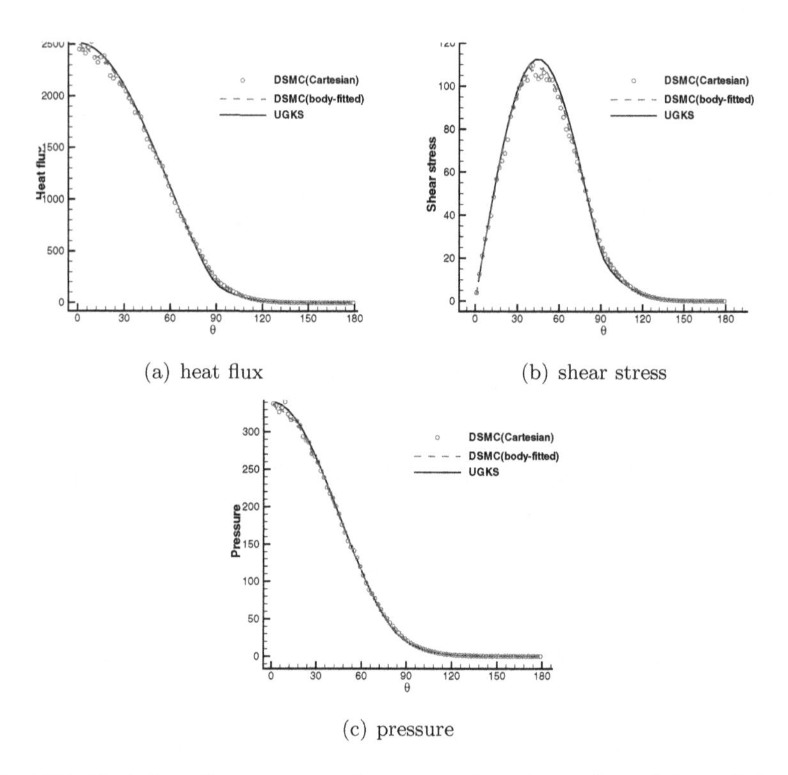

(a) heat flux (b) shear stress

(c) pressure

Fig. 7.33 Heat flux, shear stress, and pressure along the surface of the cylinder at $Ma = 20$ and $Kn = 1.0$.

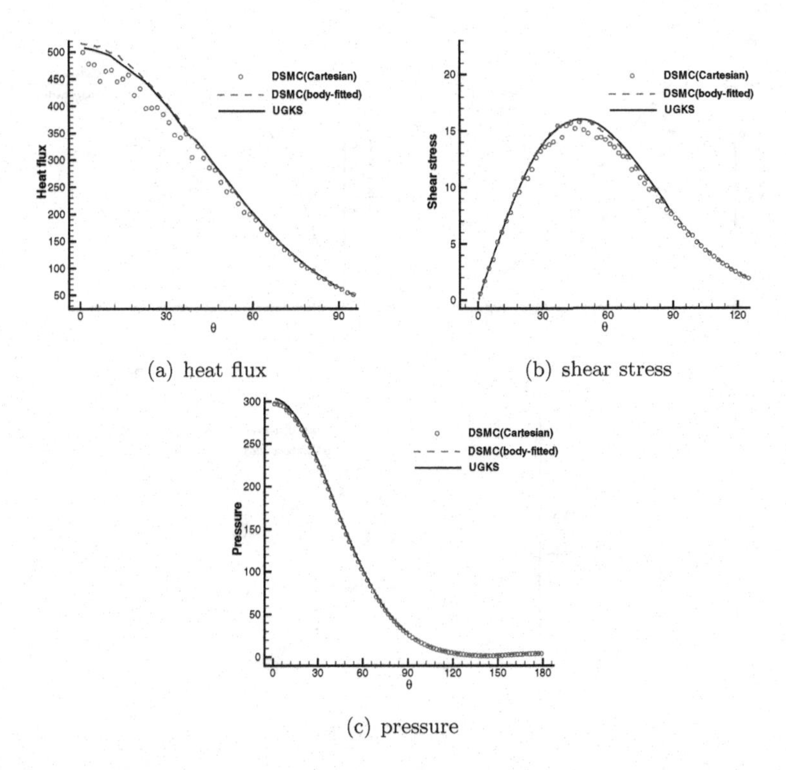

(a) heat flux

(b) shear stress

(c) pressure

Fig. 7.34 Flow properties along the surface of the cylinder from UGKS with a refined mesh at at $Ma = 20$ and $Kn = 0.01$. The comparison is Fig. 7.31.

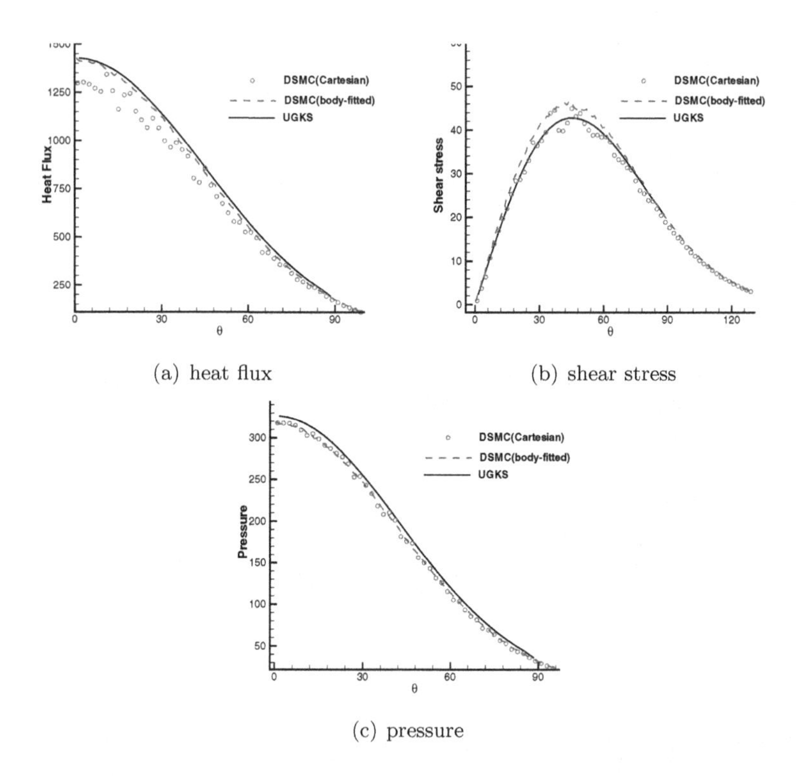

(a) heat flux

(b) shear stress

(c) pressure

Fig. 7.35 Flow properties along the surface of the cylinder from UGKS with a refined mesh at $Ma = 20$ and $Kn = 0.1$. The comparison is Fig. 7.32.

Chapter 8

Unified Gas Kinetic Scheme for Diatomic Gas

8.1 Diatomic Gas Model

The UGKS presented in last few chapters is for monatomic gas. This chapter concerns the extension of the UGKS to diatomic gases. In the diatomic gas modeling, the rotational degree of freedom is included and modeled through the energy exchange between translational and rotational ones. More specifically, the Rykov kinetic model is used to construct the time evolution solution of a gas distribution function [Rykov (1975)].

For diatomic gas, besides the translational degrees of freedom, the internal degrees of freedom have to be included as well in a gas distribution function, i.e., $f(t, \mathbf{x}, \mathbf{u}, \epsilon)$, where ϵ is the internal energy.

Here the Rykov model for diatomic gas will be used for the construction of UGKS [Rykov (1975)]. The Rykov model is an extension of the Shakhov model of monatomic gas. The Rykov model has the following form,

$$\frac{\partial f}{\partial t} + u\frac{\partial f}{\partial x} + v\frac{\partial f}{\partial y} + w\frac{\partial f}{\partial z} = \frac{g_{tran} - f}{\tau} + \frac{g_{rot} - g_{tran}}{Zr\tau}, \tag{8.1}$$

where the collision operator on the right hand of the equation is consisted of two terms corresponding to the elastic and inelastic collisions respectively. The translational kinetic energy is conserved in the elastic collisions, while, in the inelastic collisions, the energy exchange between the translational and rotational energy takes place. In Eq. (8.1), g_{tran} is the equilibrium state in the elastic collisions, and g_{rot} is the equilibrium state for inelastic collisions with the consideration of energy exchange and total energy conservation.

The equilibrium states are expressed as

$$
\begin{aligned}
g_{tran} = \rho\left(\frac{m}{2\pi kT_{tran}}\right)^{\frac{3}{2}} e^{-\frac{mc^2}{2kT_{tran}}} \left(\frac{1}{kT_{rot}}\right) e^{-\frac{\epsilon}{kT_{rot}}} \\
\left[1 - \frac{2m\mathbf{q}_{tran}\cdot\mathbf{c}}{15kT_{tran}p_{tran}}\left(\frac{5}{2} - \frac{mc^2}{2kT_{tran}}\right)\right. \\
\left. + (\sigma - 1)\frac{m\mathbf{q}_{rot}\cdot\mathbf{c}}{kT_{tran}}\frac{kT_{rot} - \epsilon}{\theta}\right], \\
g_{rot} = \rho\left(\frac{m}{2\pi kT}\right)^{\frac{3}{2}} e^{-\frac{mc^2}{2kT}} \left(\frac{1}{kT}\right) e^{-\frac{\epsilon}{kT}} \\
\left[1 - \omega_0\frac{2m\mathbf{q}_{tran}\cdot\mathbf{c}}{15kTp}\left(\frac{5}{2} - \frac{mc^2}{2kT}\right)\right. \\
\left. + (1 - \sigma)\omega_1\frac{m\mathbf{q}_{rot}\cdot\mathbf{c}}{kTp}\left(1 - \frac{\epsilon}{kT}\right)\right],
\end{aligned}
$$

where

$$p_{tran} = nkT_{tran} = \frac{1}{3}\int c^2 f dudvdwd\epsilon,$$

$$q_{tran,i} = \int c_i \frac{c^2}{2} f dudvdwd\epsilon,$$

$$q_{rot,i} = \int c_i \epsilon f dudvdwd\epsilon,$$

and $\mathbf{c} = (u - U, v - V, w - W)$. Here, $\tau = \mu(T_{tran})/p_{tran} = \mu(T_{tran})/nkT_{tran}$ is the mean collision time, $1/\tau$ is the mean collision frequency, $\mu(T_{tran})$ is the viscosity coefficient of a diatomic gas, and T is the equilibrium temperature for translational and rotational degrees of freedom. The mean collision time τ depends on translational temperature T_{tran}, because the translational temperature characterizes the mean relative velocity of colliding particles. Here Zr is the rotation collision number which is related to the ratio of elastic collision frequency to inelastic frequency. In the above model equations, $\Pr = 2/3$ has been assumed. The parameters used in the above models are $\omega_0 = 0.2354, \omega_1 = 0.3049$ and $1/\sigma = 1.55$ for nitrogen, and $\omega_0 = 0.5, \omega_1 = 0.286$ and $1/\sigma = 1.55$ for oxygen.

8.2 Unified Gas Kinetic Scheme

The UGKS is a scheme for the time evolution of a gas distribution function $f(t, \mathbf{x}, \mathbf{u}, \epsilon)$, where the particle velocity space $\mathbf{u}$ is discretized in the same way as that for the monatomic gas, and a continuous space is used for the rotational degree of freedom ϵ. In order to reduce the computational cost, reduced gas distribution functions can be defined,

$$G(t, \mathbf{x}, \mathbf{u}) = \int f(t, \mathbf{x}, \mathbf{u}, \epsilon)\, d\epsilon,$$

$$R(t, \mathbf{x}, \mathbf{u}) = \int \epsilon f(t, \mathbf{x}, \mathbf{u}, \epsilon)\, d\epsilon.$$

As a result, the relations between macroscopic flow variables and distribution functions can be written in terms of the moments of G and R,

$$\mathbf{W} = \begin{pmatrix} \rho \\ \rho U \\ \rho V \\ \rho W \\ \rho E \\ \rho E_{rot} \end{pmatrix} = \int \begin{pmatrix} G \\ uG \\ vG \\ wG \\ \frac{(u^2+v^2+w^2)}{2} G + R \\ R \end{pmatrix} d\Xi, \tag{8.2}$$

with $d\Xi = dudvdw$. Multiplying the Rykov equation by vector $(1, \epsilon)^T$ and integrating the equations, the following system can be obtained,

$$\begin{aligned} \frac{\partial G}{\partial t} + u\frac{\partial G}{\partial x} + v\frac{\partial G}{\partial y} + w\frac{\partial G}{\partial z} &= \frac{G_{tran} - G}{\tau} + \frac{G_{rot} - G_{tran}}{Zr\tau}, \\ \frac{\partial R}{\partial t} + u\frac{\partial R}{\partial x} + v\frac{\partial R}{\partial y} + w\frac{\partial R}{\partial z} &= \frac{R_{tran} - R}{\tau} + \frac{R_{rot} - R_{tran}}{Zr\tau}, \end{aligned} \tag{8.3}$$

where the evolution of the distribution function $f(t, \mathbf{x}, \mathbf{u}, \epsilon)$ is replaced by the time evolution of two new distribution functions G and R, which depend on $(t, \mathbf{x}, \mathbf{u})$ only. The above system (8.3) couples the time evolution of G and R through the collision terms. The equilibrium states in Eq. (8.3) depend on the macroscopic flow variables $\mathbf{W}$ in Eq. (8.2), which depend on G and R as well. These elastic and inelastic equilibrium states are denoted by G_{tran}, G_{rot}, R_{tran} and R_{rot}, which are given by

$$G_{tran} = \left[1 - \frac{2m\mathbf{q}_{tran} \cdot \mathbf{c}}{15kT_{tran}p_{tran}} \left(\frac{5}{2} - \frac{mc^2}{2kT_{tran}}\right)\right] G_{tran\_Guassian},$$

$$G_{rot} = \left[1 - \omega_0 \frac{2m\mathbf{q}_{tran} \cdot \mathbf{c}}{15kTp} \left(\frac{5}{2} - \frac{mc^2}{2kT}\right)\right] G_{rot\_Guassian},$$

$$R_{tran} = \left[1 - \frac{2m\mathbf{q}_{tran} \cdot \mathbf{c}}{15kT_{tran}p_{tran}}\left(\frac{5}{2} - \frac{mc^2}{2kT_{tran}}\right) + (1-\sigma)\frac{m\mathbf{q}_{rot} \cdot \mathbf{c}}{kT_{tran}p_{tran}}\right] R_{tran\_Guassian},$$

$$R_{rot} = \left[1 - \omega_0 \frac{2m\mathbf{q}_{tran} \cdot \mathbf{c}}{15kTp}\left(\frac{5}{2} - \frac{mc^2}{2kT}\right) + \omega_1 (1-\sigma)\frac{m\mathbf{q}_{rot} \cdot \mathbf{c}}{kTp}\right] R_{rot\_Guassian},$$

where $G_{tran\_Guassian}$, $G_{rot\_Guassian}$, $R_{tran\_Guassian}$, and $R_{rot\_Guassian}$ are Maxwellians defined by

$$G_{tran\_Guassian} = \rho\left(\frac{m}{2\pi kT_{tran}}\right)^{\frac{3}{2}} e^{-\frac{mc^2}{2kT_{tran}}},$$

$$G_{rot\_Guassian} = \rho\left(\frac{m}{2\pi kT}\right)^{\frac{3}{2}} e^{-\frac{mc^2}{2kT}},$$

$$R_{tran\_Guassian} = kT_{rot}G_{tran\_Guassian},$$

$$R_{rot\_Guassian} = kTG_{rot\_Guassian}.$$

In the above equations, the distributions $G_{tran\_Guassian}$, $G_{rot\_Guassian}$, $R_{tran\_Guassian}$, and $R_{rot\_Guassian}$ are fully determined by the macroscopic variables $\mathbf{W}$ in Eq. (8.2).

Defining

$$G_{eq} = \left(\frac{Zr-1}{Zr}G_{tran} + \frac{1}{Zr}G_{rot}\right),$$

and

$$R_{eq} = \left(\frac{Zr-1}{Zr}R_{tran} + \frac{1}{Zr}R_{rot}\right),$$

the system of model equations can be simplified as

$$\begin{aligned}
\frac{\partial G}{\partial t} + u\frac{\partial G}{\partial x} + v\frac{\partial G}{\partial y} + w\frac{\partial G}{\partial z} &= \frac{\left(\frac{Zr-1}{Zr}G_{tran} + \frac{1}{Zr}G_{rot}\right) - G}{\tau} = \frac{G_{eq} - G}{\tau}, \\
\frac{\partial R}{\partial t} + u\frac{\partial R}{\partial x} + v\frac{\partial R}{\partial y} + w\frac{\partial R}{\partial z} &= \frac{\left(\frac{Zr-1}{Zr}R_{tran} + \frac{1}{Zr}R_{rot}\right) - R}{\tau} = \frac{R_{eq} - R}{\tau}.
\end{aligned} \tag{8.4}$$

The construction of UGKS for diatomic gases will be based on the above two equations. The detailed procedures are the following.

(i). Integrating the model equations Eq. (8.4) over the control volume $\Omega_{i,j,k}$ in a physical space and in a time interval (t_n, t_{n+1}).

(ii). Discretizing the time integration of collision terms using a trapezoid rule.

(iii). Using the following equations for updating the gas distribution functions ($G_{i,j,k}$ and $R_{i,j,k}$) and macroscopic flow variables ($\mathbf{W}$) in the control volume $\Omega_{i,j,k}$,

$$
\begin{aligned}
G_{i,j,k}^{n+1} = G_{i,j,k}^{n} + \frac{1}{\Omega_{i,j,k}} \int_{t_n}^{t_{n+1}} \oint_{\partial\Omega_{i,j,k}} G_{cf}(u,t)\,\mathbf{u}\cdot d\mathbf{S}dt \\
+\Delta t \left(\frac{G_{eq,i,j,k}^{n} - G_{i,j,k}^{n}}{2\tau^{n}} + \frac{G_{eq,i,j,k}^{n+1} - G_{i,j,k}^{n+1}}{2\tau^{n+1}} \right),
\end{aligned}
\tag{8.5}
$$

$$
\begin{aligned}
R_{i,j,k}^{n+1} = R_{i,j,k}^{n} + \frac{1}{\Omega_{i,j,k}} \int_{t_n}^{t_{n+1}} \oint_{\partial\Omega_{i,j,k}} R_{cf}(u,t)\,\mathbf{u}\cdot d\mathbf{S}dt \\
+\Delta t \left(\frac{R_{eq,i,j,k}^{n} - R_{i,j,k}^{n}}{2\tau^{n}} + \frac{R_{eq,i,j,k}^{n+1} - R_{i,j,k}^{n+1}}{2\tau^{n+1}} \right),
\end{aligned}
\tag{8.6}
$$

$$
\begin{aligned}
\mathbf{W}_{ijk}^{n+1} = \begin{pmatrix} \rho \\ \rho U \\ \rho V \\ \rho W \\ \rho E \\ \rho E_{rot} \end{pmatrix}_{i,j,k}^{n+1} = \begin{pmatrix} \rho \\ \rho U \\ \rho V \\ \rho W \\ \rho E \\ \rho E_{rot} \end{pmatrix}_{i,j,k}^{n} \\
+\frac{1}{\Omega_{i,j,k}} \sum \int_{t_n}^{t_{n+1}} \oint_{\partial\Omega_{i,j,k}} \left[\begin{pmatrix} 1 \\ u \\ v \\ w \\ \left(\frac{u^2+v^2+w^2}{2}\right) \\ 0 \end{pmatrix} G_{cf}(u,t) + \begin{pmatrix} 0 \\ 0 \\ 0 \\ 0 \\ 1 \\ 1 \end{pmatrix} R_{cf}(u,t) \right] \mathbf{u}\cdot d\mathbf{S}dtd\Xi \\
+ \begin{pmatrix} 0 \\ 0 \\ 0 \\ 0 \\ 0 \\ \Delta t \left(\frac{(\rho E_{rot})^{n} - (\rho E_{rot,eq})^{n}}{2\tau^{n}} + \frac{(\rho E_{rot})^{n+1} - (\rho E_{rot,eq})^{n+1}}{2\tau^{n+1}} \right) \end{pmatrix},
\end{aligned}
\tag{8.7}
$$

where $(\rho E_{rot,eq})^{n+1}$ can be obtained from the updated total thermal and internal energy with equal partition to each degree of freedom.

In the above system, the equilibrium states ($G_{eq,i,j,k}^{n+1}$ and $R_{eq,i,j,k}^{n+1}$) at $(n+1)-th$ time step depend on the macroscopic flow variables at $(n+1)-th$ step, which can be provided through the solution of Eq. (8.7). The solutions in equations (8.5), (8.6), and (8.7) can be uniquely determined once the time dependent gas distribution functions $G_{cf}(\mathbf{u},t)$ and $R_{cf}(\mathbf{u},t)$ at a cell interface (cf refers to cell interface) can be obtained. Along the same line as UGKS, the cell interface gas distribution functions are obtained from the integral solutions of the Rykov models,

$$G_{cf}(t,\mathbf{u}) = G(t,\mathbf{u},\mathbf{x}_{cf}) = \frac{1}{\tau}\int_{t_n}^{t} G_{eq}(t',\mathbf{u},\mathbf{x}_{cf}-\mathbf{u}t+\mathbf{u}t')\, e^{\frac{t'-t}{\tau}}dt' + e^{\frac{t_n-t}{\tau}} G(t_n,\mathbf{u},\mathbf{x}_{cf}-\mathbf{u}(t-t_n)),$$

$$R_{cf}(t,\mathbf{u}) = R(t,\mathbf{u},\mathbf{x}_{cf}) = \frac{1}{\tau}\int_{t_n}^{t} R_{eq}(t',\mathbf{u},\mathbf{x}_{cf}-\mathbf{u}t+\mathbf{u}t')\, e^{\frac{t'-t}{\tau}}dt' + e^{\frac{t_n-t}{\tau}} R(t_n,\mathbf{u},\mathbf{x}_{cf}-\mathbf{u}(t-t_n)),$$

where $G(t_n,\mathbf{u},\mathbf{x}_{cf}-\mathbf{u}(t-t_n))$ and $R(t_n,\mathbf{u},\mathbf{x}_{cf}-\mathbf{u}(t-t_n))$ are the initial condition of the distribution functions. The method to obtain the integral solution is the same as the UGKS presented in Chapter 5.

The energy relaxation term of Rykov equation is modeled using a Landau-Teller-Jeans relaxation. The particle collision time multiplied by rotational collision number Z_r defines the relaxation rate for the rotational energy equilibrating with the translational energy. Currently, the value Z_r is defined as,

$$Z_r = \frac{{Z_r}^*}{1+(\pi^{3/2}/2)\sqrt{\tilde{T}/T_{tran}}+(\pi+\pi^2/4)\left(\tilde{T}/T_{tran}\right)}, \quad (8.8)$$

where the quantity $\tilde{T}$ is the characteristic temperature of intermolecular potential, and Z_r^* has a fixed value. Over a temperature range from $30K$ to $3000K$ for N_2, a value $Z_r^*=23.0$ and $\tilde{T}=91.5K$ are used. More complicated models for the energy relaxation can be found in [Parker (1959); Koura (1992); Boyd (1993); Ivanov and Gimelshein (1998)].

8.3 Diatomic Gas Studies

8.3.1 *Rotational Relaxation of a Homogenous Gas*

For a diatomic homogeneous gas with different initial rotational T_{rot} and translational temperature T_{tran}, the system will evolve into an equilibrium one with averaged temperature T, which is a constant. Intermolecular collision will relax T_{rot} to temperature T with a rate related to the collision frequency. Due to the homogeneous space distribution, the governing equations can be simplified as

$$\frac{\partial f}{\partial t} = \frac{g_{tran} - f}{\tau} + \frac{g_{rot} - g_{tran}}{Zr\tau}.$$

Multiplying the above equation with ϵ and integrating over the whole particle velocity and rotational energy space, the time evolution of rotational energy can be obtained,

$$\frac{\partial T_{rot}}{\partial t} = \frac{T - T_{rot}}{Zr\tau}. \tag{8.9}$$

In the case with constant rotational collision number Z_r and mean collision time τ, an analytical solution of Eq. (8.9) can be obtained,

$$T_{rot}(t) = T - (T - T_{rot}(0))\, e^{-\frac{t}{Zr\tau}}. \tag{8.10}$$

For diatomic gas, the mean collision time depends on the transitional temperature even in the homogenous case. For the VHS model, the collision time can be approximated as $\tau = \mu/p \sim T_{tran}^{\omega}/(\rho R T_{tran})$, which is a constant only for the Maxwell molecule with $\omega = 1$. For hard sphere ($\omega = 1/2$) and nitrogen ($\omega = 0.72$), τ depends on the temperature and there is no exact analytical solution. In this computation, only one cell in physical space is used with periodic boundary condition. The domain in velocity space $(u, v) \in [(-5, 5) \times (-5, 5)]$ is discretized with 50×50 mesh points. Figure 8.1 presents the UGKS solutions with $Zr = 3$ and 5 for Maxwell, hard sphere, and nitrogen molecules along with the analytic solution (Maxwell gas only), where the mean collision time (m.c.t.) is calculated using the averaged temperature T in all three cases to normalize the time. The UGKS solution for the Maxwell molecule matches completely with the theoretical prediction. The collision frequencies for the HS and VHS models are higher than the Maxwell one, and they present faster relaxation to the equilibrium state.

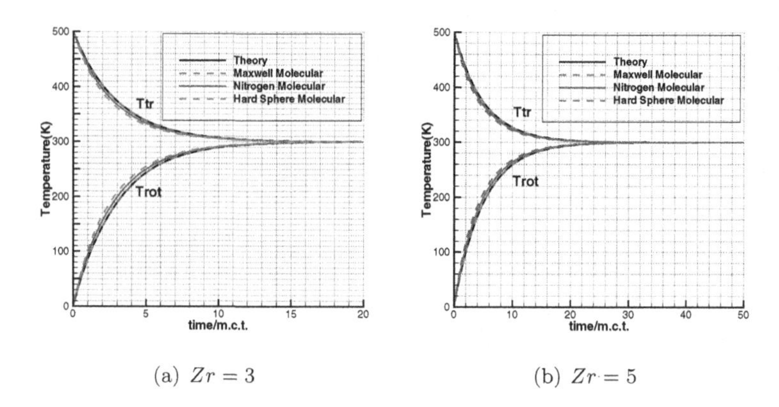

(a) $Zr = 3$ (b) $Zr = 5$

Fig. 8.1 Temperature relaxation in a homogenous gas [Liu *et al.* (2014)].

8.3.2 *Shock Structures*

For nitrogen gas, the collision time τ depends on the translational temperature only, and it is determined with the same formula as that of argon gas with $\omega = 0.72$. The collision rotation number Zr used in the UGKS has a value $Z_r = 2.4$. The comparison of density profiles at different Mach numbers between UGKS and experimental data [Alsmeyer (1976)] is illustrated in Fig. 8.2.

The nitrogen shock structures at Mach numbers 7 and 12.9 are obtained as well using UGKS, which are compared with the experimental measurements in [Robben and Talbot (1966)]. Figure 8.3 presents the comparison of density and rotational temperature between the UGKS solutions and experimental data.

8.3.3 *Flow around a Flat Plate*

Space vehicles, space stations, and planetary exploration systems fly in a rarefied gas environment. Due to the hypersonic velocity, their flight environment includes shock-shock interactions and shock-boundary interactions that cause high heat transfer and pressure on the body of the spacecraft. Strong thermal non-equilibrium is associated with these flows. It is important to study the physical flow around spacecraft in a hypersonic rarefied environment in order to understand flow phenomenon and to design a real size vehicle. Here following the experiment [Tsuboi and Matsumoto (2005)], the hypersonic rarefied gas flow over a flat plate is computed, and the results

Table 8.1 Flow condition of run34 in [Tsuboi and Matsumoto (2005)] and non-dimensional system.

Condition in run34 case		non-dimensional coefficients	
Nozzle exit Mach number	4.89	length L_{ref}	$1mm$
Nozzle exit temperature T_e	$116K$	density ρ_{ref}	ρ_e
Nozzle exit static pressure P_e	$2.12Pa$	velocity V_{ref}	$\sqrt{2R_{specical}T_e}$
Stagnation temperature T_0	$670K$	temperature T_{ref}	T_e
Stagnation pressure P_0	$983Pa$	energy e_{ref}	$\rho_e T_e$
Surface temperature T_w	$290K$		

are compared with the experimental measurements. The early study of this case has been done using the gas-kinetic scheme with multiple temperature model [Xu *et al.* (2008)].

The run 34 case in [Tsuboi and Matsumoto (2005)] is studied, where the condition is shown on Table 8.1. The flat plate is made of copper and is cooled by water to preserve a constant wall temperature $290K$. The mean free path λ_{mfp} and viscosity coefficient μ are defined as

$$\lambda_{mfp} = \frac{16}{5}\left(\frac{1}{2\pi RT}\right)^{1/2}\frac{\mu}{\rho},$$

and

$$\mu = \mu_0\left(\frac{T}{T_0}\right)^{\omega}.$$

The gas constant R and viscosity index ω of nitrogen gas are given by $297Jkg^{-1}K^{-1}$ and 0.75. Therefore, the density at nozzle exit can be determined, i.e., ρ_{ref} is $6.15 \times 10^{-5}kgm^{-3}$.

At inlet boundary, the inflow gas has a Maxwellian distribution. The outlet boundary uses a non-reflection boundary condition, where the flow distribution outside domain is extrapolated from the interior region. The interaction between the gas flow and the solid boundary is the Maxwell boundary condition with full accommodation, which has been widely used [Cercignani (2000); Patterson (1956)].

In this study, the shock wave and boundary layer at the sharp leading edge are merged. Highly non-equilibrium between translational and rotational temperatures appears above the flat plate. In the experiment, the non-equilibrium rotational temperatures were measured by an electron beam fluorescence technique [Tsuboi and Matsumoto (2005)]. The UGKS used in this case is the second order scheme and the cell size of the UGKS is not limited by the mean free path, thus a relatively coarse mesh is used to obtain the result. Here 3323 elements are used in physical space with

59×39 mesh points above the plate and 44×25 below the plate, which is shown in Fig. 8.4. The CFL number is set to be 0.5. The domain in velocity space $(u, v) \in [(-10, 10) \times (-8, 8)]$ is discretized with 80×60 mesh points. Fig. 8.5 presents the density, translational, rotational, and total temperature contours around the sharp edged flat plate. The temperature distributions along the vertical line above the flat plate at the locations of $x = 5mm$ and $x = 20mm$ from the leading edge are presented in Fig. 8.6. The numerical results match well with the experimental data.

8.3.4 *Flow around a Blunt Circular Cylinder*

In this case, we consider the hypersonic flow passes through a blunt cylinder at $Ma = 5.0$ and $Kn = 0.1$. Just as shown in Fig. 8.7, the cylinder radius has a value $0.01m$, and the computational domain is divided with $10,000$ cells for DSMC (half domain) and $4,900 \times 2$ quadrilateral cells for UGKS (whole domain), respectively.

For the DSMC method, the inflow nitrogen gas has a velocity $U_\infty = 1684.4835m/s$ with temperature $T_\infty = 273K$, molecule number density $n = 1.2944 \times 10^{21}/m^3$, and the viscosity coefficient at upstream condition $\mu_\infty = 1.65788 \times 10^{-5}Ns/m^2$. The cylinder has a cold surface with a constant temperature $T_w = 273K$, with diffusive reflection boundary condition. The rotational collision number Z_r is defined in Eq. (8.8), where $\tilde{T} = 91.5K$ and $(Z_r)_\infty = 18.1$.

For the UGKS method, the domain $(u, v) \in [-13.6, 13.6] \times [-13.6, 13.6]$ in velocity space is discretized with 93×93 mesh points based on Newton-Cotes rule, and the collision time is calculated with $\tau = \mu_\infty (T/T_\infty)^{0.78}/p_{tran}$.

The UGKS and DSMC solutions are plotted in Figs. 8.8 and 8.9. In Fig. 8.8, the contour lines from DSMC and UGKS for density, velocity, total average temperature, translational temperature, and rotational temperature are plotted together. Fig. 8.9 presents the flow variables along central symmetric line in front of the stagnation point, in which the density and U-velocity are matched very well, while for the temperature profiles, the temperature in UGKS solution upraise earlier than that in DSMC solution, and it is the same as the UGKS results from Shakhov model for monatomic gas. The UGKS with the inclusion of the full Boltzmann collision term in the kinetic sub-time step can totally remove this problem, as shown in the Mach 3 shock structure calculation for monatomic gas. The discrepancy in the temperature distribution at the upstream region of a shock wave does not deteriorate the quality of the UGKS. There basically have no effect on the accurate prediction of surface flow properties around the blunt cylinder.

8.4 Summary

In this chapter, the UGKS has been used for diatomic gas study. The current UGKS is based on the Rykov equation with a Landau-Teller-Jeans-type relaxation model for the transitional and rotational energy exchange of a diatomic gas. The new scheme has been tested in a few cases, which include relaxation test of homogeneous flow, shock structure calculations, hypersonic rarefied flow passing through a flat plate, and the hypersonic flow around a blunt cylinder. The numerical results are compared with the analytic solutions, experimental measurements, and DSMC computations. Reasonable agreements between the UGKS results and other validated solutions have been obtained. For rarefied flow computation, to capture non-equilibrium translational motion is the most difficult part. Once the monatomic gas behavior can be well captured, the additional rotational mode can be added to the kinetic solver in a straightforward way. Theoretically, there is no great difficulties to further develop UGKS by including additional physical effect, such as vibrational mode.

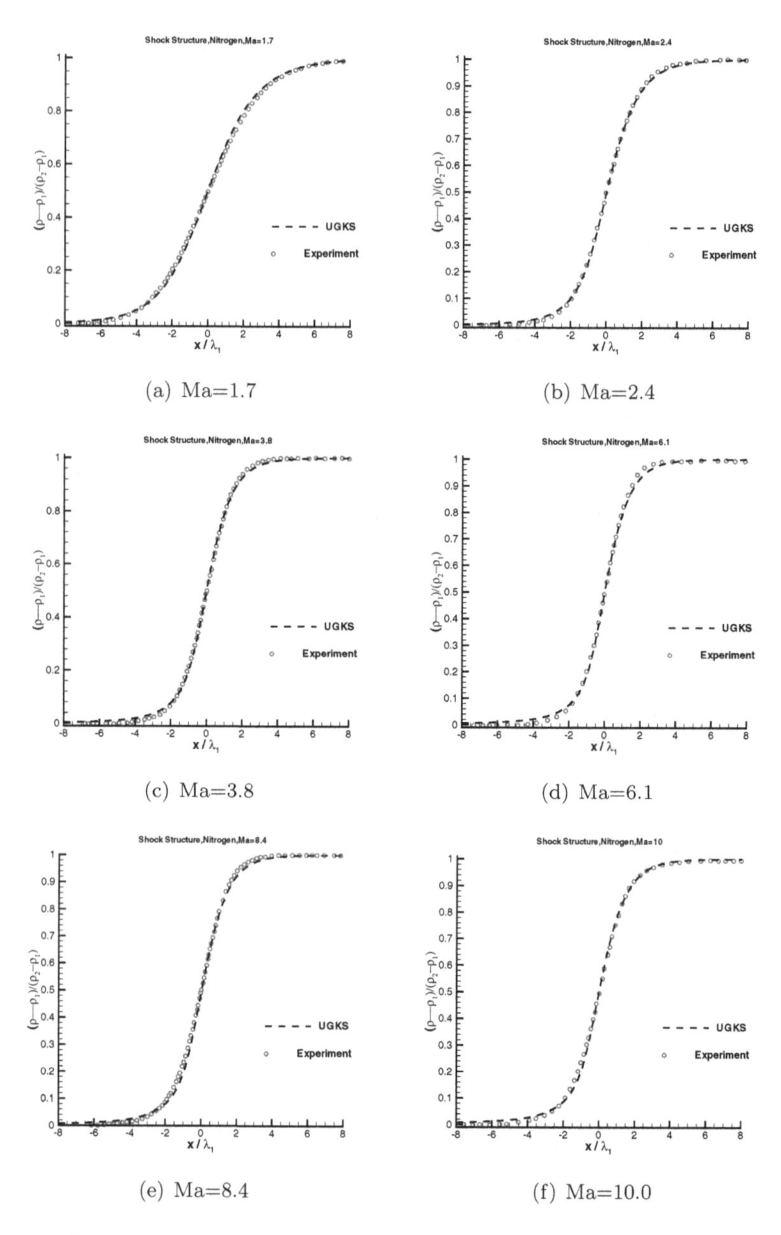

(a) Ma=1.7

(b) Ma=2.4

(c) Ma=3.8

(d) Ma=6.1

(e) Ma=8.4

(f) Ma=10.0

Fig. 8.2 Density profiles of diatomic shock structure at different Mach numbers. Circles: experiment measurements [Alsmeyer (1976)], dashed lines: UGKS solutions [Liu *et al.* (2014)].

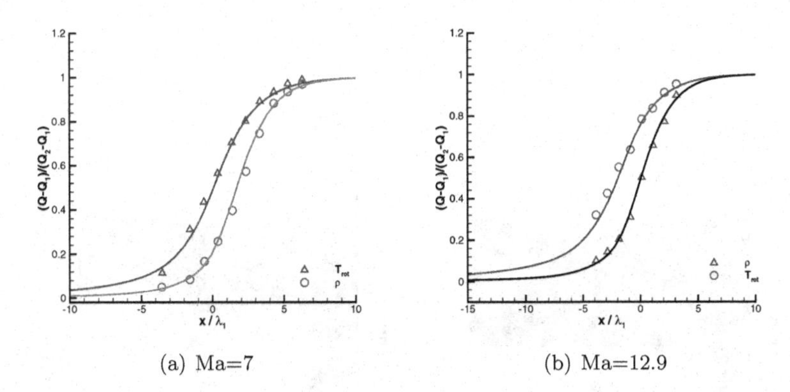

(a) Ma=7 (b) Ma=12.9

Fig. 8.3 Nitrogen shock structures at $M = 7$ and $M = 12.9$ from experiment measurements (symbols) [Robben and Talbot (1966)] and UGKS computations (lines).

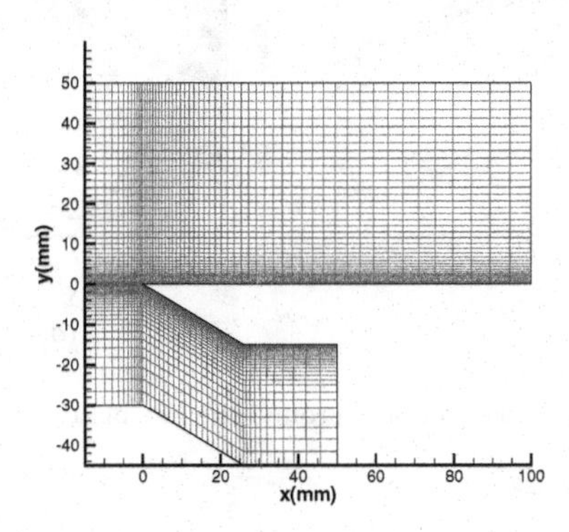

Fig. 8.4 Computational Mesh for the flat plate simulation [Liu *et al.* (2014)].

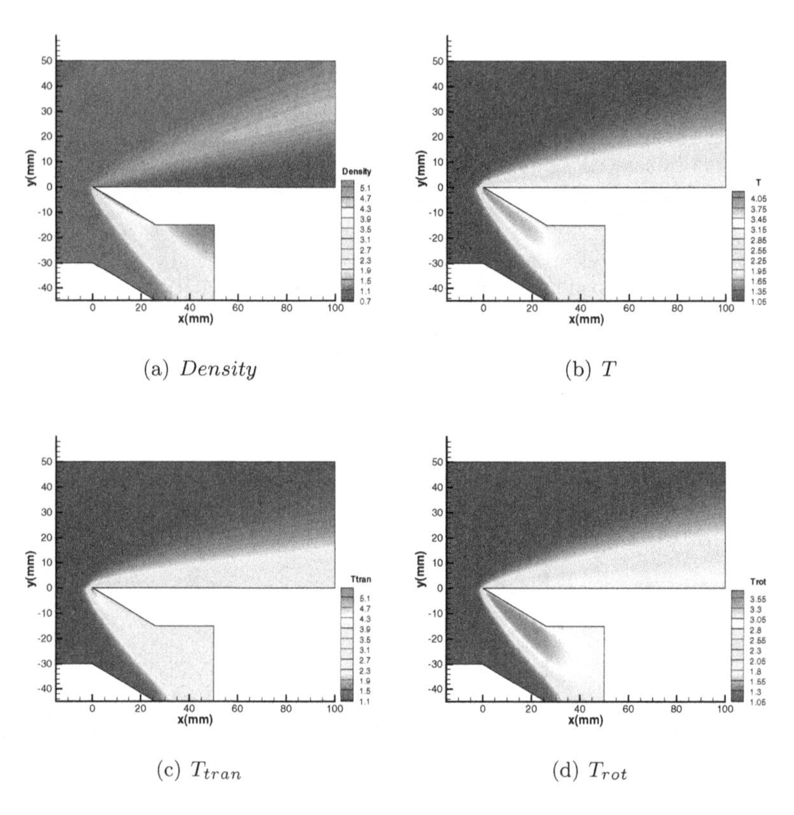

(a) *Density* (b) T

(c) T_{tran} (d) T_{rot}

Fig. 8.5 Flow variable contours around the flat plate [Liu *et al.* (2014)].

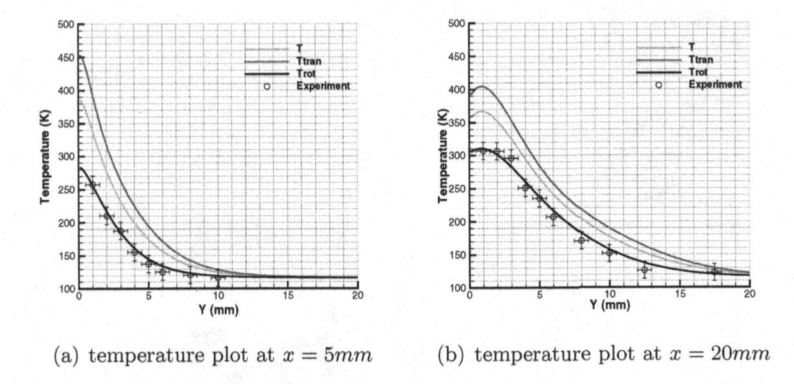

(a) temperature plot at $x = 5mm$ (b) temperature plot at $x = 20mm$

Fig. 8.6 Temperature plots along vertical lines at $x = 5mm$ and $x = 20mm$ [Liu *et al.* (2014)].

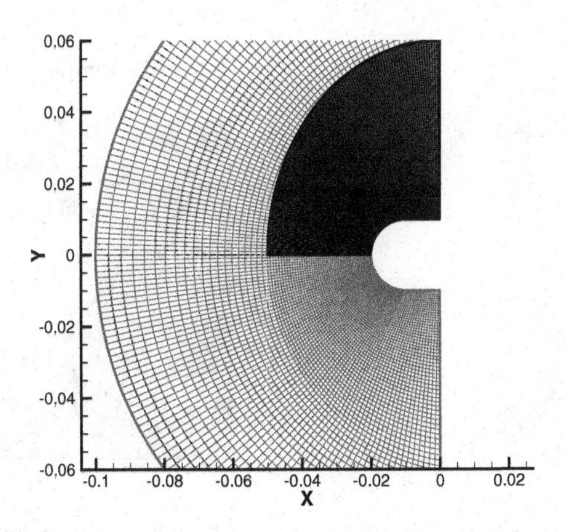

Fig. 8.7 Meshes for DSMC (black, half domain with 10, 000 points) and UGKS (red, whole domain with 9, 800 mesh points) computations.

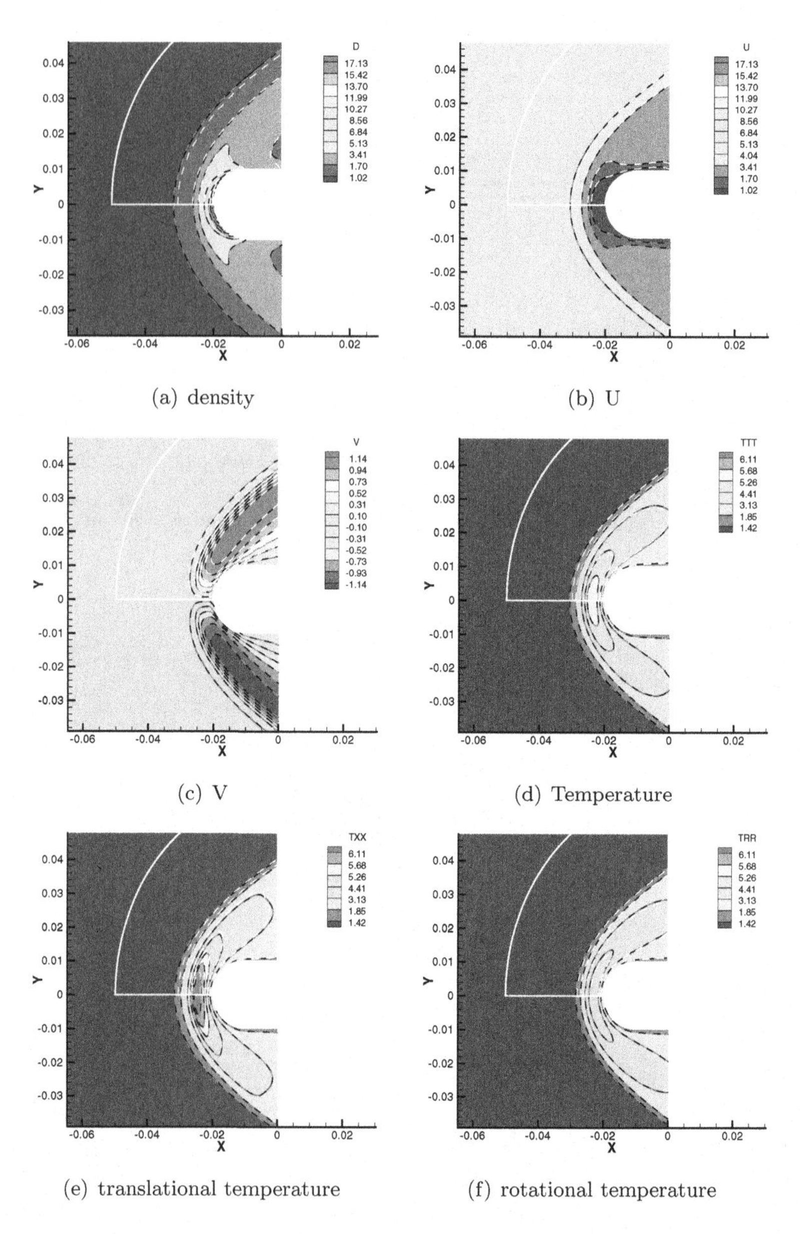

(a) density (b) U

(c) V (d) Temperature

(e) translational temperature (f) rotational temperature

Fig. 8.8 Flow contours for nitrogen flows around a blunt cylinder at Ma=5.0 and Kn=0.1. DSMC: white dashed lines, UGKS: background and black dashed lines [Liu *et al.* (2014)].

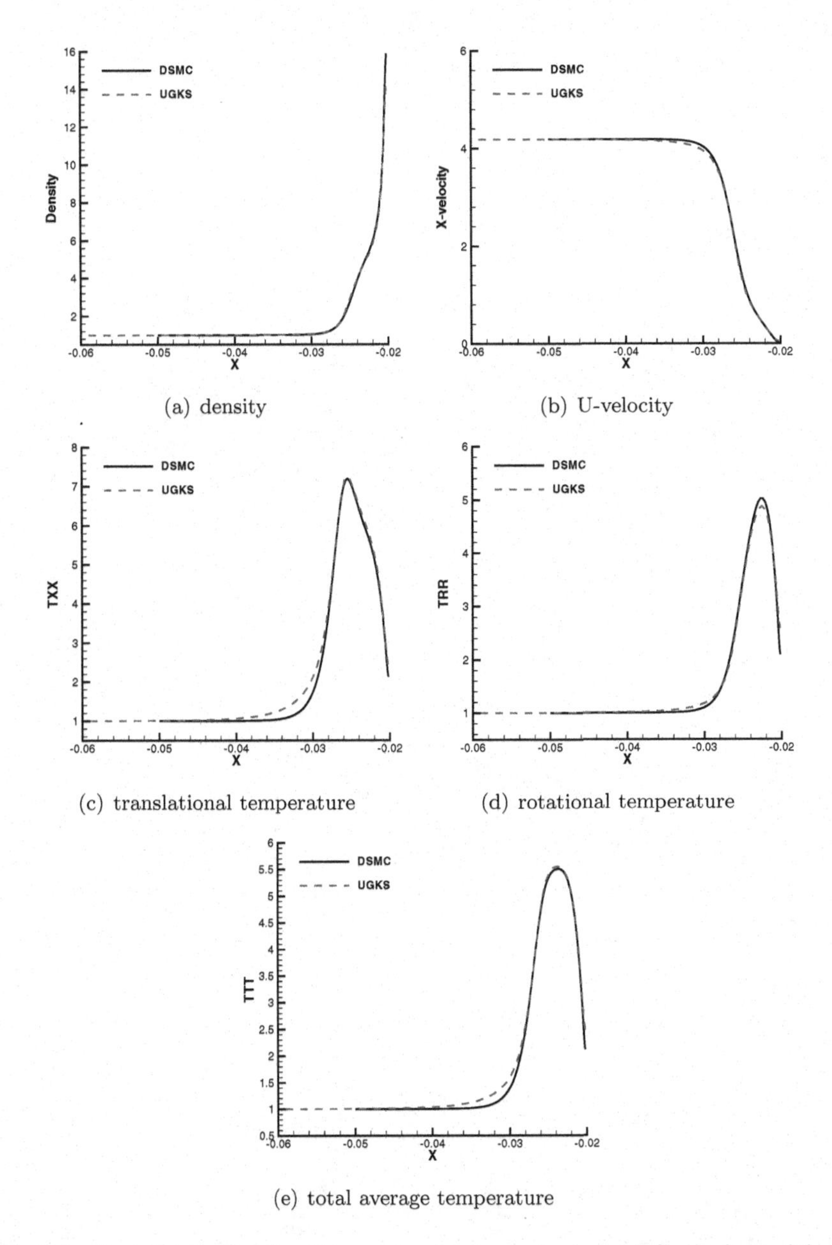

(a) density

(b) U-velocity

(c) translational temperature

(d) rotational temperature

(e) total average temperature

Fig. 8.9 Flow distributions for nitrogen gas along central symmetric line in front of the stagnation point at Ma=5.0 and Kn=0.1. Solid lines: DSMC, Dashed lines: UGKS [Liu *et al.* (2014)].

Chapter 9

Conclusion

This book presents the direct modeling as a general framework for computational fluid dynamics algorithm development. The principle underlying the direction modeling is that the mesh size and time step will actively participate in the gas evolution, and the aim of CFD is basically to describe and capture the flow physics in the mesh size scale. The unified gas kinetic scheme is constructed under such a principle and provides a multiple scale modeling for the flow simulations. Here the fluid dynamics refers to a general multiscale flow motion, which cannot be fully described by any special governing equation, such as the Euler, the NS, or the Boltzmann. The UGKS covers a continuous spectrum of governing equations from the rarefied to the continuum one. The cell size used in UGKS can be ranged from the particle mean free path to the hydrodynamic dissipative layer thickness for the capturing of flow physics when needed. This CFD principle is distinguishable from the current CFD methodology, where a direct discretization of the fluid dynamic equations is adopted. In this stereotype CFD approach, the cell size and time step don't play any dynamic role in the fluid modeling. The numerical mesh size seems introduce error only. The solution with diminishing mesh size is theoretically pursued. The numerical PDE CFD is laid on fundamental inconsistent basis due to the concealed dynamic difference between the PDE's modeling scale and the mesh size scale. This methodology also prevents the development of multiscale method once there is no valid governing equation in all scales. The dependence of the flow dynamics on the resolution of the discretized space in UGKS can be easily understood once we realize that all existing fluid dynamics equations are derived based on the physical modeling in their specific scales. Here we only change the modeling scale to the mesh size and time step, which are the resolution to describe the flow. Since the mesh size can be chosen freely according to dynamics requirement, such as the mean free path scale for resolving the shock structure and hundreds of mean free path in the hydrodynamic region, under such a direct modeling principle, the flow physics can be truthfully described and followed. Otherwise, based on numerical PDE we may need to use the physical modeling scale of the governing equations as the mesh size everywhere to resolve the corresponding flow physics. For example, if we are targeting to solve the Boltzmann equation numerically, we may use the particle mean free path as the cell size, which is not necessary and not practical at all in the low transition and continuum flow study. Also, for a gas flow without discernible scale separation, it is hard to imagine that the multiscale problem can be described correctly by a few distinguishable governing equations.

The key in the unified scheme is the construction of a time dependent crossing scale solution for the interface flux evaluation for the updates of both conservative macroscopic flow variables and the microscopic gas distribution function. This time evolution solution covers different flow regime from the kinetic to the hydrodynamic ones. The flow modeling used locally in the CFD algorithm depends on the ratio of the time step to the local particle collision time. In this book, we mainly present the cross scale modeling based on the evolution solution of the kinetic models, such as the BGK, Shakhov, or ES-BGK. The UGKS based on the full Boltzmann collision model has been developed recently as well [Liu *et al.* (2014)], which has been used to validate the UGKS based on the kinetic model equations. Even with the kinetic model equations for the collision term, extensive tests show that the UGKS can present accurate solutions in almost all cases in different flow regimes. The current study clearly indicates that the UGKS is a valuable and indispensable tool for the flow study, especially for the flow with the co-existence of both continuum and rarefied regions.

The CFD research has achieved great success in the past four decades, but has been on a plateau in the past years, and being saturated without new significant ideas recently. Due to the use of the nonlinear limiter and the implicit dissipation associated with it, the CFD made a quantum jump in 1970s-1980s in the construction of shock capturing schemes. But, with the further refinement of the computations, the CFD meets difficulties due to its underlying principle of direct numerical discretization of the Euler and Navier-Stokes equations, without considering the dynamic connection between the modeling scale of the PDEs and the mesh size scale. The problem facing the CFD is the absence of a fundamental physical rule for the algorithm construction. In the current CFD research, the mesh size and time step have no dynamic influence on the flow evolution, and the exact solution of specific PDEs is what the CFD aims to get. This is mainly a study of the equations, instead of fluid dynamics. Even though the Godunov method with the exact Riemann solver becomes the foundation of modern CFD method, progressive development of approximate Riemann solvers without fully sticking to the Euler equations, seems push the CFD algorithms to a more robust and reliable level. The inadequacy for the simulation of vortex dominant flows from the Godunov method is due to the dynamic deficiency in its direction-splitting modeling. The triggering of shock instability in Godunov method is due to the physical deficiency of equilibrium assumption in the description of the discontinuous shock layer. However, based on the numerical PDE approach, there is no much freedom to remedy the dynamic deficiency of the CFD algorithm, because the dynamics underlying the special governing equation is fixed beforehand. The numerical methodology through the direct discretization of PDE is to get the exact solution of the original governing equation with the continuous refinement of the mesh. This approach is clearly under the shadow of the calculus. The best situation is that accurate solution of the original equation can be obtained, but the solution is still limited by the modeling scale of the governing equation. But, the real case is that the governing equation underlying the numerical scheme can be never figured out due to the limited cell resolution. In the current methodology, the CFD aims ultimately to get rid of the mesh size and time step effect on the numerical solution, which is basically a mission of impossible. The current CFD ignores the possible dynamic connection between the mesh size scale and the physical modeling scale of the PDE. As a result, there is no distinguishable dynamics between resolved and unresolved flow structures, and the quality of the solution depends on the numerics only. All difficulties facing current CFD come from trying to identify the unresolved dynamics from the equations, which have no such a mechanism for the sub-scale structure construction, such as zero shock thickness of the Euler equations. The direct modeling principle for CFD emphasized in this book is mainly to bridge different scales in a consistent way. Instead of direct discretizing the fluid dynamic PDEs, the numerical governing equations or algorithm in the mesh size scale is directly constructed. In this case, even for the numerical shock structure with a few grid point thickness, there is corresponding non-equilibrium physical mechanism for its structure construction.

The unified gas-kinetic scheme has been successfully constructed and used in simulating flows in different regimes. The flow problems presented in this book cover the cases from the low speed microflow to the hypersonic aerodynamics, and from the free molecular motion to the dissipative Navier-Stokes solutions. Even for the inviscid flow computation, the direct modeling can provide some guidance about the construction of a physically consistent scheme inside the unresolved shock layer in order to avoid shock instability or carbuncle phenomenon.

Even with great success of UGKS, there is still space for its further development, which includes the following aspects.
(1) to merge the full Boltzmann collision term into the unified framework in the region where $\Delta t \leq \tau$, and connect it to other model equation to develop an efficient multiple scale numerical algorithm. The preliminary results have been obtained recently [Liu *et al.* (2014)].
(2) to use the UGKS as a tool to study and discover new physical phenomena in the transition regime; to provide the mechanism for turbulent modeling through further mesh size scale enlargement from the current resolved NS to unresolved LES; and to test the validity of the NS solution in the near continuum flow regime. For example, the validity of Fourier's law has been quantitatively studied in this book, but theoretical study needs to be followed. We believe that UGKS provides such a useful tool or experimental device for the physics study in the unexplored transition regime.
(3) to use the unified scheme to study non-equilibrium thermodynamics in complicated geometry, such as to check the validity of Onsager reciprocal relation in non-equilibrium regime; to study the engine efficiency in terms of the temperature difference in the transition regime; to validate the extended thermodynamic equations, and to figure out the dynamic structure in a highly non-equilibrium dissipative regime quantitatively.
(4) to extend UGKS framework to other transport phenomena, such as radiation and neutron transport. We believe that the multiscale UGKS will be more successful here because all these transport equations are relaxation types.
(5) to further improve the efficiency of UGKS in different ways, such as switch to GKS in the continuum flow regime to replace the discrete particle velocity space with a piecewise continuous one; to use adaptive velocity space, and to implement the current CFD techniques, such as local time stepping, multigrid, implicit treatment, and LU-SGS for fast convergence to steady state solution.
(6) to model and implement complicated real gas effect, such as vibration, ionization, and chemical reaction into UGKS.

Hopefully, the direct modeling principle for computational fluid dynamics can be gradually accepted by CFD community, and "when the theory of gases is again revived, not too much will have to be rediscovered" [Boltzmann (1898)].

Appendix A

Non-dimensionlizing Fluid Dynamic Variables

There are three fundamental units to describe physical quantities. These three units can be chosen as the units of length, time, and mass. Fluid dynamic equations can be represented in different unit system. Many text books present how to non-dimensionlize the fluid dynamic equations. But most times, the fundamental reasons for these choices are not given. For example, theoretically we can only use three fundamental units or independent flow variables to non-dimensionlize a physical system, but it seems that there are more than three independent quantities used in the non-dimensionlization of the fluid dynamic variables. Here, we will present how to non-dimensionlize fluid dynamic variables through a fundamental analysis.

For any fluid system, such as the NS or the Boltzmann equation, all physical quantities are measured in units. There are only three fundamental units as references to define a unit system. The equations can be considered as a physical law in such a unit frame. Different choices of units are equivalent to a transformation between different frames. Non-dimensionlization is not to make the physical quantity be dimensionless, but to take an appropriate unit system, where the equations related to the flow problem can be described conveniently.

Now let's consider a physical system described in two unit frames (l_0, t_0, m_0) and (l_1, t_1, m_1), such as $l_0 = 1m$ and $l_1 = 1mm$. Since in both unit frames, the same physical quantities are measured, where the measured length, time, and mass are given by

$$x(l_0) = x_*(l_1), t(t_0) = t_*(t_1), m(m_0) = m_*(m_1),$$

where the "times" is used between the value and unit. Here the values x, x_*, t, t_*, m and m_* are purely numbers, and the units are included in (l_0, t_0, m_0) and (l_1, t_1, m_1). Then, with the above connection, the specific values of length, time, and mass can be transformed from one unit frame to another one,

$$x = x_* \left(\frac{l_1}{l_0}\right), t = t_* \left(\frac{t_1}{t_0}\right), m = m_* \left(\frac{m_1}{m_0}\right).$$

Now let's study the transformation of flow variables between two unit frames. In the unit frame (0), the values of density, velocity and length can be expressed as ρ_∞, u_∞ and l_∞. Note that they are purely numbers in unit frame (0). In the unit frame (1), these numbers will be changed to $(\hat{\rho}, \hat{u}, \hat{l})$ with the connections,

$$\rho_\infty \left(\frac{m_0}{l_0^3}\right) = \hat{\rho} \left(\frac{m_1}{l_1^3}\right), u_\infty \left(\frac{l_0}{t_0}\right) = \hat{u} \left(\frac{l_1}{t_1}\right), l_\infty(l_0) = \hat{l}(l_1).$$

Obviously, for convenience, we can choose the units of unit frame (1) to have $\hat{\rho} = \hat{u} = \hat{l} = 1$. With the above requirement, i.e., any inflow condition around a flying vehicle becomes the same in frame (1), the connections between different unit frames are

$$l_1 = l_\infty l_0, t_1 = \frac{l_\infty}{u_\infty} t_0, m_1 = \rho_\infty l_\infty^3 m_0.$$

These are the basic rules to define the units for frame (1).

With the above choices, the values to describe the same length, time, and mass between two frames are connected by

$$x = l_\infty x_*, t = \frac{l_\infty}{u_\infty} t_*, m = \rho_\infty l_\infty^3 m_*.$$

Then, the values of flow variables can be changed from one unit system (0) to another one (1), such as

$$\rho \left(\frac{m_0}{l_0^3}\right) = \rho_* \left(\frac{m_1}{l_1^3}\right), p \left(\frac{m_0}{l_0 t_0^2}\right) = p_* \left(\frac{m_1}{l_1 t_1^2}\right), u \left(\frac{l_0}{t_0}\right) = u_* \left(\frac{l_1}{t_1}\right),$$

from which we have

$$\rho_* = \frac{\rho}{\rho_\infty}, p_* = \frac{p}{\rho_\infty u_\infty^2}, u_* = \frac{u}{u_\infty}.$$

Note that (ρ, p, u) and (ρ_*, p_*, u_*) are purely numbers in unit frames (0) and (1).

In terms of the dynamical viscosity coefficient, we have

$$\mu \left(\frac{m_0}{l_0 t_0}\right) = \mu_* \left(\frac{m_1}{l_1 t_1}\right),$$

and

$$\mu_* = \frac{\mu}{\rho_\infty l_\infty u_\infty}.$$

In unit frame (0), for a specific viscosity coefficient, such as the incoming flow with μ_∞, the value of corresponding viscosity coefficient in unit frame (1) becomes

$$\mu_{\infty,*} = \frac{\mu_\infty}{\rho_\infty l_\infty u_\infty} = \frac{1}{\text{Re}},$$

where $\text{Re} = \rho_\infty l_\infty u_\infty / \mu_\infty$. Then, for any other viscosity coefficient μ in unit frame (0), in frame (1) it becomes

$$\mu_* = \frac{1}{\text{Re}} \left(\frac{\mu}{\mu_\infty} \right).$$

Similarly, the particle collision time has the unit of time, the transformation of collision times between two frames are

$$\tau_* = \tau \left(\frac{u_\infty}{l_\infty} \right).$$

Theoretically, there are only three independent units for a fluid dynamic system. Under this unit system, the molecular energy, which is a variable in a microscopic scale, can be expressed as a value on the order of 10^{-23} in the units of meter, second, and kilogram. Therefore, for the convenience purpose, a new unit temperature is introduced. For example, in unit frame (0), the unit of energy can be expressed as $m_0 l_0^2 / t_0^2$, which is defined again as kT_0, and k is a constant. The constant is more or less an exchange rate if we define two money system, such as US dollar and Chinese Yuan. Then, with the new unit temperature T_0 we have $l_0^2/t_0^2 = (k/m_0)T_0$, and k/m_0 can be defined as a new constant R.

Now we have two unit frames (0) and (1) with the units (l_0, t_0, m_0, T_0) and (l_1, t_1, m_1, T_1), where T_0 and T_1 are temperatures in two frames. Besides previous choices for the units transformation from $\rho_\infty, u_\infty, l_\infty$ to $\hat{\rho} = \hat{u} = \hat{l} = 1$, we now have an additional one,

$$T_\infty(T_0) = \hat{T}(T_1).$$

Similar, we can define the relation between T_0 and T_1 to make sure $\hat{T} = 1$. As a result, the temperature transformation between unit frames is

$$T_1 = T_\infty T_0.$$

Then, for any temperature T is unit frame (0), it becomes T_* in system (1),

$$T_* = T \frac{T_0}{T_1} = \frac{T}{T_\infty}.$$

Besides the values of flow variable transformation, the constants of exchange rates, such as k and R in frame (0), will have different values in frame (1). For example, based on the velocity square we have

$$RT_\infty \left(\frac{l_0^2}{t_0^2}\right) = \hat{R}\hat{T}\left(\frac{l_1^2}{t_1^2}\right),$$

from which we have

$$\hat{R} = RT_\infty \left(\frac{l_\infty^2}{u_\infty^2}\frac{1}{l_\infty^2}\right) = \frac{RT_\infty}{u_\infty^2} = \frac{1}{\gamma M_\infty^2},$$

where the reference Mach number is defined as $M_\infty = u_\infty/\sqrt{\gamma RT_\infty}$. Even though the universal constants can be different in different unit frames, within the same unit frame, such as frame (1), the value of the universal constant should keep the same value, such as

$$R_* = \hat{R} = 1/(\gamma M_\infty^2).$$

If k is the Boltzmann constant in unit frame (0), then the corresponding value k_* in unit frame (1) can be derived from

$$R_* = \frac{k_*}{m_*} = \frac{1}{\gamma M_\infty^2}.$$

Since

$$m_* = \frac{m}{\rho_\infty l_\infty^3},$$

we get

$$k_* = \frac{1}{\gamma M_\infty^2}\frac{m}{\rho_\infty l_\infty^3} = \frac{T_\infty}{\rho_\infty l_\infty^3 u_\infty^2}k.$$

After the definition of temperature, we can talk about the heat flux $-\kappa\nabla T$ and the heat conduction coefficient κ. In unit frame (0), the heat conduction coefficient κ is defined as

$$\kappa = \frac{\mu C_p}{\mathrm{Pr}} = \frac{\mu}{\mathrm{Pr}}\frac{\gamma}{\gamma-1}R,$$

where the specific heat capacity at constant pressure is defined as $C_p = \gamma R/(\gamma-1)$. In unit frame (1), it becomes

$$\kappa_* = \frac{\mu_*}{\mathrm{Pr}}\frac{\gamma}{\gamma-1}R_* = \frac{\mu}{\mu_\infty}\frac{1}{(\gamma-1)\mathrm{Pr}\mathrm{Re}M_\infty^2}.$$

In many cases, the above expression can be written as

$$\kappa_* = \tilde{\mu}\frac{1}{(\gamma-1)\mathrm{Pr}\mathrm{Re}M_\infty^2},$$

where $\tilde{\mu} = \mu/\mu_\infty$. Then, the heat flux in unit frame (1) is

$$\vec{q}_* = -\kappa_* \left(\frac{\partial T_*}{\partial x_*} + \frac{\partial T_*}{\partial y_*} + \frac{\partial T_*}{\partial z_*} \right),$$

and the stress terms, such as τ_{xy}, become

$$\tau_{*,xy} = \frac{1}{\text{Re}} \left(\frac{\mu}{\mu_\infty} \right) \left(\frac{\partial u_*}{\partial y_*} + \frac{\partial v_*}{\partial x_*} \right).$$

All fluid dynamic equations, including the Boltzmann equation, can be expressed in the same way in unit frame (0) and unit frame (1), but with their own unit system. The connections between unit systems are based on the choices of $\rho_\infty, u_\infty, l_\infty, T_\infty$ in unit frame (0), which have the corresponding values $\hat{\rho} = \hat{u} = \hat{l} = \hat{T} = 1$ in unit frame (1). This brings great simplification in the computation in unit frame (1). This is only one of the choices, which is mainly used in applications of aerospace flow problems. Due to the freedom in choosing the units, sometimes it is much more convenient to use $k_B \to \hat{k} = 1, m \to \hat{m} = 1$, $u_\infty \to \hat{u} = 1$, and $l_\infty \to \hat{l} = 1$ to define a new unit system (1) in microflow applications. In the cosmology study, it may become easy to use the speed of light $\hat{c} = 1$, Planck constant $\hat{\hbar} = 1$, and the mass of the electron $\hat{m}_e = 1$ to define the unit frame (1).

Appendix B

Connection between BGK, Navier Stokes and Euler Equations

Derivation of the Navier-Stokes equations from the Boltzmann equation can be found in [Kogan (1969); Chapman and Cowling (1990)] and from the Bhatnagar-Gross-Krook equation in [Cercignani (1988)] and [Vincenti and Kruger (1965)] for the case of perfect monotonic gases. Here we reconsider the derivation of the Navier-Stokes and Euler equations from the BGK equation, right from the outset for polyatomic gases.

To derive the Navier-Stokes equations, let $\tau = \epsilon\widehat{\tau}$ where ϵ is a small dimensionless quantity, and suppose that g has a Taylor series expansion about some point x_i, t. Since τ depends on the local thermodynamic variables, and since these depend on the moments of g, we may assume that τ and consequently $\widehat{\tau}$ can be expanded about the point x_i, t. Now consider the formal solution of the BGK equation for f, supposing that g is known, and suppose that $t >> \tau$; *i.e.* that the initial condition were imposed many relaxation times ago. We can then ignore the initial value of f, and, with negligible error, the difference between $t' = 0$ and $t' = -\infty$ in the integral solution of the BGK model. It can be shown from the integral solution that the Taylor series expansion of τ and g about x_i, t may be written as power series in ϵ, and therefore f has an expansion in powers of ϵ. We can find the terms in this expansion from the formal solution for f, or, more easily, by putting

$$f = f_0 + \epsilon f_1 + \epsilon^2 f_2 + ...$$

and $\tau = \epsilon\widehat{\tau}$ into the BGK equation directly. Let

$$D_{\mathbf{u}} = \frac{\partial}{\partial t} + u_i \frac{\partial}{\partial x_i},$$

and write the BGK equation as $\epsilon\widehat{\tau}D_{\mathbf{u}}f + f - g = 0$. An expansion of this equation in powers of ϵ yields

$$f = g - \epsilon\widehat{\tau}D_{\mathbf{u}}g + \epsilon^2\widehat{\tau}D_{\mathbf{u}}(\widehat{\tau}D_{\mathbf{u}}g) + ...$$

and the compatibility condition, after dividing by $\epsilon\widehat{\tau}$, give

$$\int \psi_\alpha D_{\mathbf{u}} g d\Xi = \epsilon \int \psi_\alpha D_{\mathbf{u}}(\widehat{\tau} D_{\mathbf{u}} g) d\Xi + \mathcal{O}(\epsilon^2). \quad \text{(B.1)}$$

We define $\mathcal{L}_\alpha$ to be the integral on the left side of this equation, and $\mathcal{R}_\alpha$ to be the integral on the right, so that Eq. (B.1) can be written as

$$\mathcal{L}_\alpha = \epsilon \mathcal{R}_\alpha + \mathcal{O}(\epsilon^2). \quad \text{(B.2)}$$

We show that these equations give the Euler equations if we drop the term of $\mathcal{O}(\epsilon)$, and the Navier-Stokes equations if we drop terms of $\mathcal{O}(\epsilon^2)$. To simplify the notation, let

$$\langle \psi_\alpha(...) \rangle \equiv \int \psi_\alpha(...) g d\Xi,$$

and consider

$$\begin{aligned} \mathcal{L}_\alpha &\equiv \int \psi_\alpha D_{\mathbf{u}} g d\Xi \\ &= \int \psi_\alpha (g_{,t} + u_l g_{,l}) d\Xi \\ &= \langle \psi_\alpha \rangle_{,t} + \langle \psi_\alpha u_l \rangle_{,l}, \end{aligned}$$

since ψ_α is independent of x_i and t. Now Eq. (B.2) shows that

$$\langle \psi_\alpha \rangle_{,t} + \langle \psi_\alpha u_l \rangle_{,l} = \mathcal{O}(\epsilon) \quad \text{(B.3)}$$

for all α, and therefore, in reducing $\mathcal{R}_\alpha$ on the right side of Eq. (B.2), which is already $\mathcal{O}(\epsilon)$, we can drop $\mathcal{O}(\epsilon)$ quantities and their derivatives. Put differently, we first reduce the $\mathcal{L}_\alpha$ to find that $\mathcal{L}_\alpha = 0$ $(\alpha = 1, 2, ...5)$ is identical to the Euler equations; then we use the fact that $\mathcal{L}_\alpha$ is $\mathcal{O}(\epsilon)$ to simplify $\mathcal{R}_\alpha$ — the result is the Navier-Stokes equations.

The expression for $\mathcal{R}_\alpha$ contains time derivatives which must be eliminated. We have, from the definition of $\mathcal{R}_\alpha$,

$$\begin{aligned} \mathcal{R}_\alpha = &\ \widehat{\tau}[\langle \psi_\alpha \rangle_{,tt} + 2\langle \psi_\alpha u_k \rangle_{,tk} + \langle \psi_\alpha u_k u_l \rangle_{,lk}] \\ &+ \widehat{\tau}_{,t}[\langle \psi_\alpha \rangle_{,t} + \langle \psi_\alpha u_l \rangle_{,l}] + \widehat{\tau}_{,k}[\langle \psi_\alpha u_k \rangle_{,t} + \langle \psi_\alpha u_k u_l \rangle_{,l}]. \end{aligned} \quad \text{(B.4)}$$

According to Eq. (B.3) the coefficient of $\widehat{\tau}_{,t}$ in this expression is $\mathcal{O}(\epsilon)$, and can therefore be neglected. As for the first term, consider

$$\begin{aligned} \frac{\partial}{\partial t}[\langle \psi_\alpha \rangle_{,t} + \langle \psi_\alpha u_k \rangle_{,k}] &= \langle \psi_\alpha \rangle_{,tt} + \langle \psi_\alpha u_k \rangle_{,kt} \\ &= \mathcal{L}_{\alpha,t} = \mathcal{O}(\epsilon) \end{aligned}$$

Then the first term in Eq. (B.4) is

$$\hat{\tau}\frac{\partial}{\partial x_k}[\langle\psi_\alpha u_k\rangle_{,t} + \langle\psi_\alpha u_k u_l\rangle_{,l}] + \mathcal{O}(\epsilon),$$

which can be combined with the third term to give

$$\mathcal{R}_\alpha = \frac{\partial}{\partial x_k}\{\hat{\tau}[\langle\psi_\alpha u_k\rangle_{,t} + \langle\psi_\alpha u_k u_l\rangle_{,l}]\} + \mathcal{O}(\epsilon),$$

which eliminates the second time derivatives from $\mathcal{R}_\alpha$; the first time derivatives will be removed by using $\mathcal{L}_\alpha \simeq 0$.

The Euler equations follow from putting $\mathcal{L}_\alpha = 0$. To see this, consider

$$\mathcal{L}_1 = \langle\psi_1\rangle_{,t} + \langle\psi_1 u_k\rangle_{,k} = \rho_{,t} + (\rho U_k)_{,k},$$

since $\psi_1 = 1$; $\mathcal{L}_1 = \mathcal{O}(\epsilon)$ is the continuity equation if we neglect $\mathcal{O}(\epsilon)$. For $\alpha = 2, 3, 4$, it is convenient to define $\mathcal{L}_i$ and $\mathcal{R}_i$ such that $i = \alpha - 1$ and to let $w_i = u_i - U_i$. Then

$$\mathcal{L}_i = \langle u_i\rangle_{,t} + \langle u_i u_k\rangle_{,k} = (\rho U_i)_{,t} + [\rho U_i U_k + \langle w_i w_k\rangle]_{,k},$$

since all moments of g odd in w_l vanish. The pressure tensor is defined by

$$p_{ik} = \langle w_i w_k\rangle \equiv p\delta_{ik}.$$

(The diagonal form of p_{ik} is obvious from the fact that g can be written as a function of w_k^2.) Then

$$\mathcal{L}_i = (\rho U_i)_{,t} + (\rho U_i U_k + p\delta_{ik})_{,k} \tag{B.5}$$

and $\mathcal{L}_i = 0$ is the Euler equation for the conservation of momentum. For the energy equation we have

$$\mathcal{L}_5 = \frac{1}{2}\langle {u_n}^2 + \xi^2\rangle_{,t} + \frac{1}{2}\langle u_l({u_n}^2 + \xi^2)\rangle_{,l}$$

or

$$\mathcal{L}_5 = \left(\frac{1}{2}\rho {U_n}^2 + \frac{K+3}{2}p\right)_{,t} + \left(\frac{1}{2}\rho U_k {U_n}^2 + \frac{K+5}{2}pU_k\right)_{,k}.$$

Setting $\mathcal{L}_5 = 0$ gives the energy equation in the absence of dissipation.

We proceed to eliminate the time derivatives from $\mathcal{R}_\alpha$ using the fact that $\mathcal{L}_\alpha = \mathcal{O}(\epsilon)$. For $\alpha = 1$, we have

$$\mathcal{R}_1 = \{\hat{\tau}[\langle u_k\rangle_{,t} + \langle u_k u_l\rangle_{,l}]\}_{,k}$$

The quantity in square brackets is $\mathcal{L}_k$, which implies that $\mathcal{R}_1 = \mathcal{O}(\epsilon)$, and $\mathcal{L}_1 = \epsilon\mathcal{R}_1 = \mathcal{O}(\epsilon^2)$. Hence, to the order we have retained, $\mathcal{R}_1 = 0$ and $\mathcal{L}_1 = 0$, or

$$\rho_{,t} + (\rho U_k)_{,k} = 0, \tag{B.6}$$

which is the continuity equation. We can use the continuity equation to simplify the momentum equations and the energy equation. Multiplying the continuity equation by U_i and the subtracting the result from $\mathcal{L}_i$ gives, according to Eq. (B.5),

$$\mathcal{L}_i = \rho U_{i,t} + \rho U_k U_{i,k} + p_{,i} + \mathcal{O}(\epsilon^2). \tag{B.7}$$

For $\mathcal{L}_5$, we group the terms as follows:

$$\begin{aligned}\mathcal{L}_5 = & \frac{1}{2}{U_n}^2[\rho_{,t} + (\rho U_k)_{,k}] + \rho U_n U_{n,t} + \rho U_k U_n U_{n,k} + U_k p_{,k} \\ & + \frac{K+3}{2}[p_{,t} + U_k p_{,k}] + \frac{K+5}{2} p U_{k,k}\end{aligned}$$

The first term is $\frac{1}{2}{U_n}^2\mathcal{L}_1$ which is $\mathcal{O}(\epsilon^2)$, and the next three are $U_n\mathcal{L}_n$, and are therefore $\mathcal{O}(\epsilon)$. Then

$$\mathcal{L}_5 = \frac{K+3}{2}[p_{,t} + U_k p_{,k}] + \frac{K+5}{2} p U_{k,k} + U_n \mathcal{L}_n. \tag{B.8}$$

We can drop the last term in the reduction of $\mathcal{R}_\alpha$, but the term $U_n\mathcal{L}_n$ must be retained in the reduction of $\mathcal{L}_5$ when we finally write $\mathcal{L}_5 = \epsilon\mathcal{R}_5$ in detail.

For the right sides of the momentum equations, consider $\mathcal{R}_j = \left(\hat{\tau} F_{jk}\right)_{,k}$, where

$$F_{jk} \equiv \langle u_j u_k \rangle_{,t} + \langle u_j u_k u_l \rangle_{,l}$$

or

$$\begin{aligned}F_{jk} = & U_j[(\rho U_k)_{,t} + (\rho U_k U_l + p\delta_{kl})_{,l}] \\ & + \rho U_k U_{j,t} + (\rho U_k U_l + p\delta_{kl}) U_{j,l} \\ & + (p\delta_{jk})_{,t} + (U_l p \delta_{jk} + U_k p \delta_{jl})_{,l},\end{aligned}$$

using the fact that all moments odd in w_k vanish. The term in square brackets multiplying U_j is $\mathcal{L}_k$, *i.e.* it is $\mathcal{O}(\epsilon)$, and can therefore be ignored. Then, after gathering terms with coefficients U_k and p, we have

$$F_{jk} = U_k[\rho U_{j,t} + \rho U_l U_{j,l} + p_{,j}] + p[U_{k,j} + U_{j,k} + U_{l,l}\delta_{jk}] + \delta_{jk}[p_{,t} + U_l p_{,l}].$$

The coefficient of U_k is $\mathcal{L}_j$, according to Eq. (B.7), and can therefore be neglected. To eliminate $p_{,t}$ from the last term we use the Eq. (B.8) for $\mathcal{L}_5$; this gives

$$p_{,t} + U_k p_{,k} = -\frac{K+5}{K+3} p U_{k,k} + \mathcal{O}(\epsilon).$$

Finally, we decompose the tensor $U_{k,j}$ into its dilation and shear parts in the usual way, which gives

$$F_{jk} = p\left[U_{k,j} + U_{j,k} - \frac{2}{3}U_{l,l}\delta_{jk}\right] + \frac{2}{3}\left(\frac{K}{K+3}\right) p U_{l,l}\delta_{jk}. \tag{B.9}$$

The last term is due to bulk viscocity; it vanishes, as it should, for $K = 0$, since the physical mechanism for bulk viscosity involves energy sharing between translational and internal degrees of freedom of the molecules, and $K = 0$ corresponds to a monoatomic ($\gamma = \frac{5}{3}$) gas.

For $\alpha = 5$, we write

$$\mathcal{R}_5 = (\widehat{\tau} N_k)_{,k} \tag{B.10}$$

where

$$N_k \equiv \left\langle u_k \frac{(u_n{}^2 + \xi^2)}{2} \right\rangle_{,t} + \left\langle u_k u_l \frac{(u_n{}^2 + \xi^2)}{2} \right\rangle_{,l}$$

which can be written as $N_k = N_k{}^{(1)} + N_k{}^{(2)}$, where

$$N_k{}^{(1)} = \left[U_k \frac{\langle u_n{}^2 + \xi^2 \rangle}{2} \right]_{,t} + \left[U_k \left\langle u_l \frac{(u_n{}^2 + \xi^2)}{2} \right\rangle \right]_{,l}$$

and

$$N_k{}^{(2)} = \langle w_k \frac{u_n{}^2 + \xi^2}{2} \rangle_{,t} + \left\langle w_k u_l \frac{(u_n{}^2 + \xi^2)}{2} \right\rangle_{,l}$$

For $N_k{}^{(1)}$ we have

$$\begin{aligned} N_k{}^{(1)} &= U_k \left[\frac{\langle u_n{}^2 + \xi^2 \rangle_{,t}}{2} + \frac{\langle u_l (u_n{}^2 + \xi^2) \rangle_{,l}}{2} \right] \\ &+ \left[\frac{1}{2} \rho U_n{}^2 + \frac{K+3}{2} p \right] U_{k,t} + \frac{1}{2} \rho U_l \left[U_n{}^2 + \frac{(K+5)p}{\rho} \right] U_{k,l}. \end{aligned}$$

The coefficient of U_k in the equation above is $\mathcal{L}_5$, and can therefore be neglected, and the remaining terms can be rewritten as

$$\left[\frac{1}{2} \rho U_n{}^2 + \frac{K+3}{2} p \right] [U_{k,t} + U_l U_{k,l}] + p U_l U_{k,l},$$

or, using the fact that $\mathcal{L}_k = \mathcal{O}(\epsilon)$,

$$N_k{}^{(1)} = - \left[\frac{1}{2} U_n{}^2 + \frac{K+3}{2} \frac{p}{\rho} \right] p_{,k} + p U_l U_{k,l}.$$

For $N_k{}^{(2)}$, remembering that moments odd in w_k vanish, we have

$$\begin{aligned} N_k{}^{(2)} &= \langle U_n w_n w_k \rangle_{,t} + \langle U_l U_n w_n w_k \rangle_{,l} \\ &+ \frac{1}{2} \langle U_n{}^2 w_k w_l \rangle_{,l} + \frac{1}{2} \langle w_k w_l (w_n{}^2 + \xi^2) \rangle_{,l}, \end{aligned}$$

or

$$N_k{}^{(2)} = (p U_k)_{,t} + (p U_k U_l)_{,l} + \frac{1}{2} (U_n{}^2 p)_{,k} + \frac{K+5}{2} \left(\frac{p^2}{\rho} \right)_{,k}.$$

This result can rewritten as

$$N_k{}^{(2)} = p[U_{k,t} + U_l U_{k,l} + U_k U_{l,l} + U_l U_{l,k}]$$
$$+U_k(p_{,t} + U_l p_{,l}) + \frac{1}{2}U_l{}^2 p_{,k} + \frac{K+5}{2}\left(\frac{p^2}{\rho}\right)_{,k},$$

and the time derivatives can be removed by using $\mathcal{L}_k = \mathcal{O}(\epsilon)$, and $\mathcal{L}_5 = \mathcal{O}(\epsilon)$, neglecting $\mathcal{O}(\epsilon)$, since we are evaluating $\mathcal{R}_5$. Finally, $N_k{}^{(1)} + N_k{}^{(2)}$ can be combined to give (after some algebra)

$$N_k = \frac{K+5}{2}p\left(\frac{p}{\rho}\right)_{,k} + p\left[-\frac{2}{K+3}U_k U_{l,l} + U_l(U_{k,l} + U_{l,k})\right]. \qquad \text{(B.11)}$$

All time derivatives have now been removed from $\mathcal{R}_\alpha$ (for all α). The remaining steps in the derivation of the Navier-Stokes equations may be summarized briefly as follows:

1). Drop $\mathcal{O}(\epsilon^2)$ in Eq. (B.2).

2). Combine ϵ and $\widehat{\tau}$ to recover $\tau = \epsilon\widehat{\tau}$.

3). Define the stress tensor

$$\sigma'_{jk} = \eta\left[U_{j,k} + U_{k,j} - \frac{2}{3}U_{l,l}\delta_{jk}\right] + \varsigma U_{l,l}\delta_{jk},$$

where

$$\eta = \tau p$$

and

$$\varsigma = \frac{2}{3}\frac{K}{K+3}\tau p$$

are the dynamic viscosity and second viscosity coefficients respectively.

4). From Eq. (B.7) for $\mathcal{L}_j$ and Eq. (B.9) for F_{jk}, it follows that $\mathcal{L}_j = \epsilon\mathcal{R}_j$ may now be written as

$$\rho U_{j,t} + \rho U_k U_{j,k} + p_{,j} = \sigma'_{jk,k},$$

which is the Navier-Stokes equation.

5). The energy equation follows from $\mathcal{L}_5 = \epsilon\mathcal{R}_5$ by using Eqs. (B.8), (B.7) and (B.6) to write $\mathcal{L}_5$ in detail, and using Eqs. (B.10) and (B.11) for $\mathcal{R}_5$. The result is

$$\frac{K+3}{2}(p_{,t} + U_k p_{,k}) - \frac{K+5}{2}p(\rho_{,t} + U_k\rho_{,k}) = (\kappa T_{,k})_{,k} + (U_l\sigma'_{lk})_{,k},$$

where

$$\kappa = \frac{K+5}{2}\frac{k}{m}\tau p$$

is the thermal conductivity, k is the Boltzmann constant, m is the mass of a molecule and T is the temperature. The equations can be written in terms of γ instead of K by using $K = (5-3\gamma)/(\gamma-1)$ for 3-Dimensional gas flow.

Appendix C

Moments of Maxwellian Distribution Function and Expansion Coefficients

Moments of a Maxwellian Distribution Function

In the gas-kinetic scheme, we need to evaluate moments of a Maxwellian distribution function with bounded and unbounded integration limits. Here, we list the formula for moment evaluation.

The Maxwellian distribution in 3D is

$$g = \rho\left(\frac{\lambda}{\pi}\right)^{\frac{K+3}{2}} e^{-\lambda((u-U)^2+(v-V)^2+(w-W)^2+\xi^2)},$$

where ξ has K degrees of freedom, such as $\xi^2 = \xi_1^2 + \xi_2^2 + ... + \xi_K^2$, $K = 0$ for monatomic gas, and $K = 2$ for diatomic gas with two rotational degree of freedom. By introducing the following notation for the moments of g,

$$\rho\langle ... \rangle = \int (...) g du dv dw d\xi,$$

where $d\xi = d\xi_1 d\xi_2 ... d\xi_K$, the general moment formula becomes

$$\langle u^n v^m w^k \xi^l \rangle = \langle u^n \rangle \langle v^m \rangle \langle w^k \rangle \langle \xi^l \rangle,$$

where n, m, and k are integers, and l is an even integer (owing to the symmetrical property of ξ). The moments of $\langle \xi^l \rangle$ are:

$$\langle \xi^0 \rangle = 1,$$

$$\langle \xi^2 \rangle = \left(\frac{K}{2\lambda}\right),$$

$$\langle \xi^4 \rangle = \left(\frac{3K}{4\lambda^2} + \frac{K(K-1)}{4\lambda^2}\right),$$

and

$$\langle \xi^{2l} \rangle = \frac{K + 2(l-1)}{2\lambda} \langle \xi^{2(l-1)} \rangle,$$

for $l = 1, 2, 3, ...$

The values of $\langle u^n \rangle$ depend on the integration limits. If the limits are $-\infty$ to $+\infty$, we have

$$\langle u^0 \rangle = 1,$$

$$\langle u \rangle = U,$$

and

$$\langle u^{n+2} \rangle = U \langle u^{n+1} \rangle + \frac{n+1}{2\lambda} \langle u^n \rangle,$$

for $n = 0, 1, 2, 3, ...$ When the integral is from 0 to $+\infty$ as $\langle ... \rangle_{>0}$ or from $-\infty$ to 0 as $\langle ... \rangle_{<0}$, the error function and the complementary error function, appear in the formulation. Thus, the moments for u^n in the half space are,

$$\langle u^0 \rangle_{>0} = \frac{1}{2} \text{erfc}(-\sqrt{\lambda} U),$$

$$\langle u \rangle_{>0} = U \langle u^0 \rangle_{>0} + \frac{1}{2} \frac{e^{-\lambda U^2}}{\sqrt{\pi \lambda}},$$

and

$$\langle u^{n+2} \rangle_{>0} = U \langle u^{n+1} \rangle_{>0} + \frac{n+1}{2\lambda} \langle u^n \rangle_{>0},$$

for $n = 0, 1, 2, 3, ...$ And

$$\langle u^0 \rangle_{<0} = \frac{1}{2} \text{erfc}(\sqrt{\lambda} U),$$

$$\langle u \rangle_{<0} = U \langle u^0 \rangle_{<0} - \frac{1}{2} \frac{e^{-\lambda U^2}}{\sqrt{\pi \lambda}},$$

and

$$\langle u^{n+2} \rangle_{<0} = U \langle u^{n+1} \rangle_{<0} + \frac{n+1}{2\lambda} \langle u^n \rangle_{<0},$$

for $n = 0, 1, 2, 3, ...$ Similar formulation can be obtained for $\langle v^m \rangle$ and $\langle w^k \rangle$ by changing U to V and W in the above moments of $\langle u^n \rangle$.

Derivatives of Macroscopic Variables and Expansion Coefficients of a Maxwellian

In both GKS and UGKS, based on derivatives of macroscopic flow variables, such as $\partial \mathbf{W}/\partial x$ or $\partial \mathbf{W}/\partial t$, we need to evaluate the corresponding coefficients of the expansion of a Maxwellian distribution function, such as a and $\bar{A}$ in $\partial g/\partial x = ga$ and $\partial g/\partial t = g\bar{A}$. The Maxwellian distribution functions can be defined as

$$g = \rho\left(\frac{\lambda}{\pi}\right)^{\frac{K+1}{2}} e^{-\lambda((u-U)^2+\xi^2)}$$

in 1D case,

$$g = \rho\left(\frac{\lambda}{\pi}\right)^{\frac{K+2}{2}} e^{-\lambda((u-U)^2+(v-V)^2+\xi^2)}$$

in 2D case, and

$$g = \rho\left(\frac{\lambda}{\pi}\right)^{\frac{K+3}{2}} e^{-\lambda((u-U)^2+(v-V)^2+(w-W)^2+\xi^2)}$$

in 3D case.

With the definition of moments and integrands:

1D case:

$$\psi_\alpha = \left(1, u, \frac{1}{2}(u^2+\xi^2)\right)^T \quad \text{and} \quad d\Xi = dud\xi,$$

2D case:

$$\psi_\alpha = \left(1, u, v, \frac{1}{2}(u^2+v^2+\xi^2)\right)^T \quad \text{and} \quad d\Xi = dudvd\xi,$$

3D case:

$$\psi_\alpha = \left(1, u, v, w, \frac{1}{2}(u^2+v^2+w^2+\xi^2)\right)^T \quad \text{and} \quad d\Xi = dudvdwd\xi,$$

where $\xi^2 = \xi_1^2+\xi_2^2+...+\xi_K^2$ and $d\xi = d\xi_1 d\xi_2 ... d\xi_K$, the connections between the derivatives of macroscopic flow variables, such as mass, momentum, energy, and the expansion of the equilibrium distribution function becomes

$$\frac{\partial W_\alpha}{\partial x} = \int ga\psi_\alpha d\Xi,$$

where $a = a_\beta \psi_\beta$, such that

1D case:

$$\mathbf{W} = (\rho, \rho U, \rho E)^T, \quad a = a_1 + a_2 u + a_3 \frac{1}{2}(u^2 + \xi^2),$$

2D case:

$$\mathbf{W} = (\rho, \rho U, \rho V, \rho E)^T, \quad a = a_1 + a_2 u + a_3 v + a_4 \frac{1}{2}(u^2 + v^2 + \xi^2),$$

3D case:

$$\mathbf{W} = (\rho, \rho U, \rho V, \rho W, \rho E)^T, \quad a = a_1 + a_2 u + a_2 v + a_4 w + a_5 \frac{1}{2}(u^2 + v^2 + w^2 + \xi^2).$$

Then, the above connections become

$$\frac{\partial W_\alpha}{\partial x} = \int g a \psi_\alpha d\Xi = \int \psi_\alpha \psi_\beta a_\beta g d\Xi,$$

which can be written as

$$\frac{1}{\rho}\frac{\partial W_\alpha}{\partial x} = M_{\alpha,\beta} a_\beta,$$

where

$$M_{\alpha,\beta} = (1/\rho) \int \psi_\alpha \psi_\beta g d\Xi,$$

and

$$a_\beta = M_{\beta\alpha}^{-1} \frac{1}{\rho} \left(\frac{\partial W_\alpha}{\partial x} \right).$$

The solution for coefficients a is the following.

1D case:

Define

$$A = 2\frac{\partial(\rho E)}{\partial x} - \left(U^2 + \frac{K+1}{2\lambda} \frac{\partial \rho}{\partial x} \right),$$

$$B = \frac{\partial(\rho U)}{\partial x} - U \frac{\partial \rho}{\partial x},$$

we have

$$a_3 = \frac{4\lambda^2}{K+1}(A - 2UB),$$

$$a_2 = 2\lambda B - U a_3,$$

and

$$a_1 = \frac{\partial \rho}{\partial x} - U a_2 - \frac{1}{2}\left(U^2 + \frac{K+1}{2\lambda} \right) a_3.$$

2D case:

Define

$$A = 2\frac{\partial(\rho E)}{\partial x} - \left(U^2 + V^2 + \frac{K+2}{2\lambda}\frac{\partial \rho}{\partial x}\right),$$

$$B = \frac{\partial(\rho U)}{\partial x} - U\frac{\partial \rho}{\partial x},$$

$$C = \frac{\partial(\rho V)}{\partial x} - V\frac{\partial \rho}{\partial x},$$

we have

$$a_4 = \frac{4\lambda^2}{K+2}(A - 2UB - 2VC),$$

$$a_3 = 2\lambda C - Va_4,$$

$$a_2 = 2\lambda B - Ua_4,$$

and

$$a_1 = \frac{\partial \rho}{\partial x} - Ua_2 - Va_3 - \frac{1}{2}\left(U^2 + V^2 + \frac{K+2}{2\lambda}\right)a_4.$$

3D case:

Define

$$A = 2\frac{\partial(\rho E)}{\partial x} - \left(U^2 + V^2 + W^2 + \frac{K+3}{2\lambda}\frac{\partial \rho}{\partial x}\right),$$

$$B = \frac{\partial(\rho U)}{\partial x} - U\frac{\partial \rho}{\partial x},$$

$$C = \frac{\partial(\rho V)}{\partial x} - V\frac{\partial \rho}{\partial x},$$

$$D = \frac{\partial(\rho W)}{\partial x} - W\frac{\partial \rho}{\partial x},$$

we have

$$a_5 = \frac{4\lambda^2}{K+3}(A - 2UB - 2VC - 2WD),$$

$$a_4 = 2\lambda D - Wa_5,$$

$$a_3 = 2\lambda C - Va_5,$$

$$a_2 = 2\lambda B - Ua_5,$$

and

$$a_1 = \frac{\partial \rho}{\partial x} - Ua_2 - Va_3 - Wa_4 - \frac{1}{2}\left(U^2 + V^2 + W^2 + \frac{K+3}{2\lambda}\right)a_5.$$

Similar formulations can be obtained for the time derivative of a Maxwellian from the corresponding derivatives of macroscopic flow variables.

Appendix D

Flux Evaluation through Stationary and Moving Cell Interfaces

Stationary cell interface

In the inertia reference of frame with (x, y) coordinate system, for any shaped control volume $S_{i,j}$, the update of the conservative flow variables $\mathbf{W}_{i,j}$ inside this control volume is

$$\mathbf{W}_{i,j}^{n+1} = \mathbf{W}_{i,j}^{n} + \frac{1}{S_{i,j}} \sum_{k} \int_{t^n}^{t^{n+1}} \vec{\mathbf{F}}_k \cdot \vec{l}_k dt, \tag{D.1}$$

where $\vec{l}_k$ is the k_{th} section of the interface of the control volume $S_{i,j}$ with the normal direction $\vec{n}$ and tangential direction $\vec{t}$, which is shown in Fig. D.1.

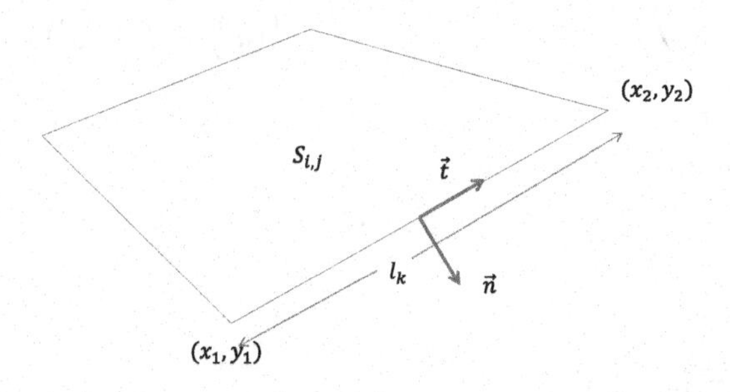

Fig. D.1 Schematic control volume and boundary surfaces.

First, let's consider the flux function $\vec{F}$ across the cell interface l_k. With the definition of vertex locations (x_1, y_1) and (x_2, y_2), the length of the cell interface k can be evaluated as $l = \sqrt{(x_2 - x_1)^2 + (y_2 - y_1)^2}$, and the normal and tangential directions become

$$\vec{n} = \left(\frac{y_2 - y_1}{l}, -\frac{x_2 - x_1}{l} \right),$$

$$\vec{t} = \left(\frac{x_2 - x_1}{l}, \frac{y_2 - y_1}{l} \right).$$

In the x-y coordinate system, the unit directions are defined by $\vec{i} = (1,0)$ and $\vec{j} = (0,1)$. Then, all dot products of

$$\vec{n} \cdot \vec{i}, \vec{n} \cdot \vec{j}, \vec{t} \cdot \vec{i}, \vec{t} \cdot \vec{j},$$

can be obtained.

The conservative flow variables $\mathbf{W}$ in the $x-y$ coordinate system can be transformed to the local $n-t$ coordinate system with the values $\mathbf{W}'$, such as $\mathbf{W} = (\rho, \rho U, \rho V, \rho E)^T$ and $\mathbf{W}' = (\rho, \rho U', \rho V', \rho E)^T$, where $(U', V') = ((U,V) \cdot \vec{n}, (U,V) \cdot \vec{t})$. At the same time, the particle velocity changes from (u, v) in $x-y$ system to (u', v') in local $n-t$ system.

So, in the local $n-t$ coordinate system, with the initial flow distribution $\mathbf{W}'$, we can use the gas-kinetic scheme to evaluate the gas distribution function f at the center of the cell interface. The fluxes across the cell interface in the normal direction become

$$\vec{F}' = \begin{pmatrix} F'_\rho \\ F'_{\rho U'} \\ F'_{\rho V'} \\ F'_{\rho E} \end{pmatrix} = \int u' f(0,t,u',v',\xi) \begin{pmatrix} 1 \\ u' \\ v' \\ \frac{1}{2}(u'^2 + v'^2 + \xi^2) \end{pmatrix} du'dv'd\xi. \tag{D.2}$$

The fluxes $\vec{\mathbf{F}}$ in Eq. (D.1) for the update of conservative flow variables in the $x-y$ coordinate system is

$$\vec{\mathbf{F}} = \vec{F}\vec{n}_k,$$

and

$$\vec{F} = \begin{pmatrix} F_\rho \\ F_{\rho U} \\ F_{\rho V} \\ F_{\rho E} \end{pmatrix} = \int u' f(0,t,u',v',\xi) \begin{pmatrix} 1 \\ u \\ v \\ \frac{1}{2}(u^2 + v^2 + \xi^2) \end{pmatrix} du'dv'd\xi, \tag{D.3}$$

where the physical meaning is that the particles passing through the cell interface $u'f$ will carry the mass, momentum, and energy measured in the

$x-y$ coordinate system. Here (u, v) in the above equation can be expressed in terms of (u', v'), such as

$$u = u'(\vec{n} \cdot \vec{i}) + v'(\vec{t} \cdot \vec{i}),$$

$$v = u'(\vec{n} \cdot \vec{j}) + v'(\vec{t} \cdot \vec{j}),$$

and

$$u'^2 + v'^2 = u^2 + v^2.$$

As a result, the fluxes $\vec{F}$ in Eq. (D.3) become a linear combination of fluxes $\vec{F}'$ in Eq. (D.2), such that

$$\vec{F} = \begin{pmatrix} F'_{\rho} \\ F'_{\rho U'}(\vec{n} \cdot \vec{i}) + F'_{\rho V'}(\vec{t} \cdot \vec{i}) \\ F'_{\rho U'}(\vec{n} \cdot \vec{j}) + F'_{\rho V'}(\vec{t} \cdot \vec{j}) \\ F'_{\rho E} \end{pmatrix}. \tag{D.4}$$

Moving mesh method

If the cell interface is moving with grid velocity (U_g, V_g), we can similarly evaluate the fluxes $\vec{F}'$ standing on the moving cell interface. Note in the moving $n-t$ coordinate system, the initial transformation from $\mathbf{W}$ to $\mathbf{W}'$ needs to take into account the moving interface velocity. Based on the flow variables in the stationary $x-y$ coordinate system,

$$\mathbf{W} = \left(\rho, \rho U, \rho V, \frac{1}{2}\rho(U^2 + V^2) + \rho e\right)^T,$$

the corresponding flow variables in the local $n-t$ coordinate system are

$$\mathbf{W}' = \begin{pmatrix} \rho \\ (\rho(U - U_g), \rho(V - V_g)) \cdot \vec{n} \\ (\rho(U - U_g), \rho(V - V_g)) \cdot \vec{t} \\ \frac{1}{2}\rho((U - U_g)^2 + (V - V_g)^2) + \rho e \end{pmatrix}, \tag{D.5}$$

where ρe is the thermal energy which has the same value in both coordinate systems. Based on initial condition $\mathbf{W}'$, the gas-kinetic scheme can be used to evaluate the fluxes across the cell interface in the local $n-t$ coordinate system, such as

$$\vec{F}' = (F'_{\rho}, F'_{\rho U'}, F'_{\rho V'}, F'_{\rho E'})^T,$$

which are

$$\vec{F}' = \begin{pmatrix} F'_{\rho} \\ F'_{\rho U'} \\ F'_{\rho V'} \\ F'_{\rho E} \end{pmatrix} = \int u' f(0, t, u', v', \xi) \begin{pmatrix} 1 \\ u' \\ v' \\ \frac{1}{2}(u'^2 + v'^2 + \xi^2) \end{pmatrix} du' dv' d\xi. \tag{D.6}$$

Then, the fluxes across the moving interface in the $x - y$ coordinate system, which will be used for the updating of conservative flow variables of a deformed control volume are

$$\vec{F} = \begin{pmatrix} F_{\rho} \\ F_{\rho U} \\ F_{\rho V} \\ F_{\rho E} \end{pmatrix} = \int u' f(0, t, u', v', \xi) \begin{pmatrix} 1 \\ u \\ v \\ \frac{1}{2}(u^2 + v^2 + \xi^2) \end{pmatrix} du' dv' d\xi, \tag{D.7}$$

which become

$$\vec{F} = \begin{pmatrix} F'_{\rho} \\ F'_{\rho U'}(\vec{n} \cdot \vec{i}) + F'_{\rho V'}(\vec{t} \cdot \vec{i}) + F'_{\rho} U_g \\ F'_{\rho U'}(\vec{n} \cdot \vec{j}) + F'_{\rho V'}(\vec{t} \cdot \vec{j}) + F'_{\rho} V_g \\ F'_{\rho E'} + \alpha F'_{\rho U'} + \beta F'_{\rho V'} + \gamma F'_{\rho} \end{pmatrix}, \tag{D.8}$$

where $\alpha = (\vec{n} \cdot \vec{i})U_g + (\vec{n} \cdot \vec{j})V_g$, $\beta = \vec{t} \cdot \vec{i} U_g + \vec{t} \cdot \vec{j} V_g$, and $\gamma = (U_g^2 + V_g^2)/2$. With the above fluxes, the conservative flow variables inside each deforming control volume $S_{i,j}$ can be updated as

$$S_{i,j}^{n+1} \mathbf{W}_{i,j}^{n+1} = S_{i,j}^{n} \mathbf{W}_{i,j}^{n} + \sum_k \int_{t^n}^{t^{n+1}} \vec{\mathbf{F}}_k \cdot \vec{l}_k dt, \tag{D.9}$$

where $\vec{\mathbf{F}} = \vec{F} \vec{n}_k$ is the flux given in Eq. (D.8), and the length of the the cell interface $\vec{l}_k$ should be an averaged length between time step n and $n + 1$.

Bibliography

ALEXANDER, F.J., GARCIA, A.L. AND ALDER, B.J. (1994). *Direct simulation Monte Carlo for thin-film bearings*, Physics of Fluids, **6**, No. 12, pp. 3854–3860.

ALEXANDER, F.J., GARCIA, A.L. AND ALDER, B.J. (1998). *Cell size dependence of transport coefficients in stochastic particle algorithm*, Physics of Fluids A, **10**, pp. 1540-1542.

ALEXEENKO, A.A., GIMELSHEIN, S.F., MUNTZ, E.P. AND KETSDEVER, A.D. (2005). *Modeling of thermal transpiration flows for Knudsen compressor optimization meeting*, 43rd Aerospace Sciences Meeting, Reno, NV, AIAA Paper 2005-963.

ALSMEYER, H. (1976). *Density profiles in argon and nitrogen shock waves measured by the absorption of an eletron beam*, J. Fluid Mech., **74**, pp. 497.

ANDRIES, P., TALLEC, P.L., PERLAT, J.P. AND PERTHAME, B. (2000). *The Gaussian-BGK model of Boltzmann equation with small Prandtl number*, European Journal of Mechanics - B/Fluids, **16**, pp. 813-830.

ANIKIN, Y.A. (2011a). *Numerical study of radiometric forces via the direct solution of the Boltzmann kinetic equation*, Comput. Math. Math. Phys. **51**, pp. 1251-1266.

ANIKIN, Y.A. (2011b). *Numerical study of the radiometric phenomenon exhibited by a rotating Crookes radiometer*, Comput. Math. Math. Phys. **51**, pp. 1923-1932.

ANNIS, B.K. (1972). *Thermal creep in gases*, Journal of Chemical Physics **57**, pp. 2898-2905.

AOKI, K., SONE, Y. AND YANO, T. (1989). *Numerical analysis of a flow induced in a rarefied gas between noncoaxial circular cylinders with different temperatures for the entire range of the Knudsen number*, Physics of Fluids A **1** (2), pp. 409-419.

AOKI, K., TAKATA, S., HIDEFUMI, A. AND GOLSE, F. (2001). *A rarefied gas flow caused by a discontinuous wall temperature*, Physics of Fluids **13**, pp. 2645-2661.

AOKI, K., TAKATA, S. AND NAKANISHI T. (2002). *Poiseuille-type flow of a rarefied gas between two parallel plates driven by a uniform external force*,

Phys. Rev. E **65**, 026315:1.

AOKI, K., DEGOND, P. AND MIEUSSENS, L. (2009). *Numerical simulations of rarefied gases in curved channels: thermal creep, circulating flow, and pumping effect*, Commun. Comput. Phys. **6**, No. 5, pp. 919-954.

ARISTOV, V.V. (1977). *Method of adaptative meshes in velocity space for the intense shock wave problem*, USSR J. Comput. Math. Math. Phys., **17**, pp. 261-267.

ARISTOV, V.V. (2001). *Direct methods for solving the Boltzmann equation and study of nonequilibrium flows*, Kluwer Academic Publishers.

ARORA, M. AND ROE, P.L. (1997). *On postshock oscillations due to shock capturing schemes in unsteady flows*, J. Comput. Phys., **130**, pp. 25-40.

BARANGER, C., CLAUDEL, J., HEROUARD, N. AND MIEUSSENS, L. (2014). *Locally refined discrete velocity grids for stationary rarefied flow simulations*, J.Comput. Physics, **257**, pp. 572-593.

BARISIK, M. AND BESKOK, A. (2011). *Molecular dynamics simulations of shear-driven gas flows in nano-channels*, Microfluid Nanofluid **11**, pp. 611-622.

BELOTSERKOVSKII, O.M. AND KHLOPKOV, Y.I. (2010). *Monte Carlo methods in mechanics of fluid and gas*, World Scientific.

BEN-ARTZI, M. AND FALCOVITZ, J. (1984). *A second-order Godunov-type scheme for compressible fluid dynamics*, J. Comput. Phys., **55**, pp. 1-32.

BEN-ARTZI, M. AND FALCOVITZ, J. (2003). *Generalized Riemann problems in computational fluid dynamics*, Cambridge University Press.

BEN-ARTZI, M. AND LI, J. (2007). *Hyperbolic balance laws: Riemann invariants and the generalized Riemann problem*, Numer. Math. **106**, pp. 369-425.

BEN-ARTZI, M., LI, J. AND WARNECKE, G. (2006). *A direct Eulerian GRP scheme for compressible fluid flows*, J. Comput. Phys., **218**, pp. 19-34.

BENNOUNE, M., LEMOU, M. AND MIEUSSENS, L. (2008). *Uniformly stable numerical schemes for the Boltzmann equation preserving the compressible Navier-Stokes asymptotics*, J. Comput. Phys. **227**, pp. 3781-3803.

BEYLICH, A.E. (2000). *Solving the kinetic equation for all Knudsen numbers*, Physics of fluids **12**, pp. 444.

BHATNAGAR, P.L., GROSS, E.P. AND KROOK, M. (1954). *A model for collision processes in gases I: small amplitude processes in charged and neutral one-component systems*, Phys. Rev. **94**, pp. 511-525.

BIELENBERG, J.R. AND BRENNER, H. (2006). *A continuum model of thermal transpiration*, Journal of Fluid Mechanics **546**, pp. 123.

BINNIG, G., QUATE, C.F. AND GERBER, C. (1986). *Atomic force microscope*, Phys. Rev. Lett. **56**, pp. 930-933

BIRD, G.A. (1970). *Aspects of the structure of strong shock waves*, Phys. Fluids, **13**, pp. 1172-1177.

BIRD, G.A. (1994). *Molecular gas dynamics and the direct simulation of gas flows*, Oxford Science Publications.

BOLTZMANN, L. (1898). *Lectures on gas theory*, Dover Publications in 1995, INC, New York.

BORIS, J.P. AND BOOK, D.L. (1973). *Flux-corrected transport, I. SHASTA. a fluid transport algorithm that works*, J. Comput. Phys. **11**, pp. 38.

Botella, O. and Peyret, R. (1998). *Benchmark spectral results on the lid-driven cavity flow*, Computers Fluids **27**, pp. 421-433.

Bouchut, F. (2004). *Nonlinear stability of finite volume methods for hyperbolic conservation laws, and well-balanced schemes for sources*, Frontiers in Mathematics, Birkhauser.

Bourgat, J.F., Le Tallec, P. and Tidriri, M.D. (1996). *Coupling Boltzmann and Navier-Stokes equations by friction*, J. Comput. Phys. **127**, pp. 227-245.

Boyd, I. (1993). *Relaxation of discrete rotational energy distributions using a Monte Carlo method*, Phys. Fluids A **5**, pp. 2278-2286.

Buckner, J.K. and Ferziger, J.H. (1966). *Linearized boundary value problem for a gas and sound propagation*, Physics of Fluids, **9**, no. 12, pp. 2315-2322.

Burt, J.M., Josyula, E., Deschenes, T.R. and Boyd, I. (2011). *Evaluation of a hybrid Boltzmann-continuum method for high-speed nonequilibrium flows*, J. of Thermophysics and Heat Transfer **25**, pp. 500-515.

Caflisch, R.E., Jin, S. and Russo, G. (1997). *Uniformly accurate schemes for hyperbolic systems with relaxation*, SIAM J. Numer. Anal. **34**, pp. 246-281.

Cai, Z. and Li, R. (2010). *Numerical regularized moment method of arbitrary order for Boltzmann-BGK equation*, SIAM J. Sci. Comput. **32**, No. 5, pp. 2875-2907.

Cai, Z., Li, R. and Qiao Z.H. (2012). *NRxx simulation of microflows with Shakhov model*, SIAM J. Sci. Comput. **34**, No. 1, pp. A339-A369.

Candler, G.V., Nompelis, I. and Druguet, M.C. (2001). *Navier-Stokes predictions of hypersonic double-cone and cylinder-flare fields*, AIAA paper 2001-1024, 39th AIAA Aerospace Sciences Meeting and Exhibit, January 8-11.

Cardy, J., Falkovich, G. and Gawedzki, K.(2008). *Non-equilibrium statistical mechanics and turbulence*, Cambridge University Press.

Carrillo, J.A., Goudon, T., Lafitte, P. and Vecil, F. (2008). *Numerical schemes of diffusion asymptotics and moment closures for kinetic equations*, J. Sci. Comput. **36**(2008), pp. 113-149.

Cercignani, C. (1988). *The Boltzmann equation and its applications*, Springer-Verlag.

Cercignani, C. (2000). *Rarefied gas dynamics: from basic concepts to actual calculations*, Cambridge texts in applied mathematics, Cambridge University Press.

Chae, D., Kim, C. and Rho, O. (2000). *Development of an improved gas-kinetic BGK scheme for inviscid and viscous flows*, J. Comput. Phys. **158**, pp. 1.

Chapman, S. and Cowling, T.G. (1990). *The mathematical theory of non-uniform gases*, Cambridge University Press.

Chen, S.Y. and Doolen, G.D. (1998). *Lattice Boltzmann method for fluid flows*, Annual Review of Fluid Mechanics **30**, pp. 329-364.

Chen, H., Kandasamy, S., Orszag, S., Shock, R., Succi, S., and Yakhot, V. (2003). *Extended Boltzmann kinetic equation for turbulent flows*, Science **301**, pp. 633-636.

Chen, S.Z. and Xu, K. (2013). *A comparative study of an asymptotic preserving scheme and unified gas-kinetic scheme in continuum flow limit*, arXiv:1307.4961v1, physics-fluid-dynamics, July.

Chen, S.Z., Xu, K. and Cai, Q.D. (2013). *A comparison and unification of ellipsoidal statistical and Shakhov BGK*, arXiv:1304.0865v2, physics-fluid-dynamics, April.

Chen, S.Z., Xu, K. and Lee, C.B. (2012a). *The dynamic mechanism of a moving Crookes radiometer*, Physics of Fluids **24**, 111701.

Chen, S.Z., Xu, K., Lee, C.B. and Cai, Q.D. (2012b). *A unified gas kinetic scheme with moving mesh and velocity space adaptation*, J. Comput. Phys. **231**, pp. 6643-6664.

Chou, S.Y. and Baganoff, D. (1997). *Kinetic flux-vector splitting for the Navier-Stokes equations*, J. Comput. Phys. **130**, 217-230.

Chu, C.K. (1965). *Kinetic-theoretic description of the formation of a shock wave*, Phys. Fluids **8**, pp. 12-22.

Coron, F. and Perthame, B. (1991). *Numerical passage from kinetic to fluid equations*, SIAM J. Numer. Anal. **28**, pp. 26-42.

Crookes, W. (1874). *On attraction and repulsion resulting from radiation*, Philos. Trans. R. Soc. London **164**, pp. 501-527.

Daru, V. and Tenaud, C. (2000). *Evaluation of TVD high resolution schemes for unsteady viscous shocked flows*, Comput. Fluids **30**, pp. 89-113.

Daru, V. and Tenaud, C. (2009). *Numerical simulation of the viscous shock tube problem by using a high resolution monotonicity-preserving scheme*, Comput. Fluids **38**, pp. 664-676.

Degond, P., Liu, J.G. and Mieussens, L. (2006). *Macroscopic fluid models with localized kinetic upscale effects*, Multiscale Model. Simul. **5**, pp. 940-979.

Deshpande, S.M. (1986). *A second order accurate, kinetic-theory based method for inviscid compressible flows*, NASA Langley Tech. paper No. 2613.

Dongari, N. (2012). *Micro gas flows: modeling the dynamics of Knudsen layers*, PhD Thesis, University of Strathclyde.

Drikakis, D. and Tsangaris, S. (1993). *On the solution of the compressible Navier-Stokes equations using improved flux vector splitting methods*, Appl. Math. Modeling, **17**, pp. 282-297.

E, W.N. (2012). *Principles of multiscale modeling*, Science Press.

Einstein, A. (1924). *Zur theorie der radiometerkrafte*, Z. Phys. **27**, pp. 1-6.

Elling, V. (2009). *The carbuncle phenomenon is incurable*, Acta Math. Sci. Ser. *B* **29**, pp. 1647-1656.

Erwin, D.A., Pham-Van-Diep, G.C. and Muntz, E.P. (1991). *Nonequilibrium gas flows. I: a detailed validation of Monte Carlo simulation for monatomic gases*, Phys. Fluids *A* **3**, pp. 697-705.

Estivalezes, J.L. and Villedieu, P. (1996). *High-order positivity-preserving kinetic schemes for the compressible Euler equations*, SIAM J. Numer. Anal. **33**, pp. 2050-2067.

Fan, J. and Shen, C. (2001). *Statistical simulation of low-speed rarefied gas flows*, J. Comput. Phys. **167**, pp. 393-412.

Filbet, F., Mouhot, C. and Pareschi, L. (2006). *Solving the Boltzmann equation in Nlog2N.* , SIAM J. Sci. Comput. **28**, pp. 1029-1053.

FILBET, F. AND JIN, S. (2010). *A class of asymptotic preserving schemes for kinetic equations and related problems with stiff sources*, J. Comput. Phys. **229**, pp. 7625-7648.

GABETTA, E., PARESCHI, L. AND TOSCANI, G. (1997). *Relaxation schemes for nonlinear kinetic equations*, *SIAM J. Numer. Anal.* **34**, pp. 2168-2194.

GAITONDE, D.V. AND CANUPP, P.W. (2002). *Heat transfer prediction in a laminar hypersonic viscous/inviscid interaction*, J. of Thermophysics and Heat Transfer, **16**, pp. 481-489.

GARCIA, R. AND SIEWERT, C.E. (2005). *The linearized Boltzmann equation: Sound-wave propagation in a rarefied gas*, Zeitschrift für angewandte Mathematik und *Physik*, **57**, pp. 94-122.

GHIA, U., GHIA, K.N. AND SHIN, C.T. (1982). *High-Re solutions for incompressible flow using the Navier-Stokes equations and a multigrid method*, J. *Comput. Phys.* **48**, pp. 387-411.

GNOFFO, P.A. (2001). *Computational aerothermodynamics in aeroassist applications*, AIAA paper 2001-2632.

GODUNOV, S.K. (1959). *A difference scheme fornumerical computation of discontinuous solutions of equations of fluid dynamics*, Mat. Sb. **47** (89), pp. 271-306.

GOTSMANN, B. AND DURIG, U. (2005). *Experimental observation of attractive and repulsive thermal forces on microcantilevers*, Applied Physics Letters **87**(19), 194102.

GRAD, H. (1949). *On the kinetic theory of rarefied gases*, Commun. Pure Appl. *Math.* **2**, pp. 325.

GREENBERG, J.M. AND LEROUX, A.Y. (1996). *A well-balanced scheme for the numerical processing of source terms in hyperbolic equations*, *SIAM J. Numer. Anal.* **33**, pp. 1-16.

GREENSPAN, M. (1956). *Propagation of Sound in Five Monatomic Gases*, The *Journal of the Acoustical Society of America* **28**, pp. 644-648.

GORTH, C. AND MCDONALD J. (2009). *Towards physically realizable and hyperbolic moment closures for kinetic theory*, Continuum Mech. Thermodyn. **21**, pp. 467-493.

GU, X.J. AND EMERSON, D.R. (2009). *A high-order moment approach for capturing non-equilibrium phenomena in the transition regime*, Journal of Fluid *Mechanics* **636**, pp. 177-216.

GUO, Z.L. AND SHU, C.(2013). *Lattice Boltzmann method and its applications in engineering*, *World Scientific publishing*.

GUO, Z.L. AND ZHENG, C.G.(2008). *Analysis of Lattice Boltzmann equation for miscroscale gas flows: relaxation times, boundary conditions, and the Knudsen layers*, *Int. J. Comput. Fluid Dynamics* **22**, pp. 465-473.

GUO, Z.L., LIU, H.W., LUO, L.S. AND XU, K. (2008). *A comparative study of the LBM and GKS methods for 2D near incompressible flows*, J. Comput. *Phys.* **227**, pp. 4955-4976.

GUO, Z.L., XU, K. AND WANG, R.J. (2013). *Discrete unified gas kinetic scheme for all Knudsen number flows: Low-speed isothermal case*, Physical Review *E* **88**, 033305.

Guo, Z.L., Qin, J. and Zheng, C. (2014). *Generalized second-order slip boundary condition for nonequilibrium gas flows*, Physical Review E **89**, 013021.

Gutnic, M, Haefele, M., Paun, I. and Sonnendrcker, E. (2004). *Vlasov simulations on an adaptive phase-space grid*, Comput. Phys. Commun. **164**, pp. 214-219.

Hadjiconstantinou, N.G. and Garcia, A.L. (2001). *Molecular simulations of sound wave propagation in simple gases*, Physics of Fluids, **13**, pp. 1040-1046.

Han, Y.L. (2010). *Working gas temperature and pressure change for microscale thermal creep-driven flow caused by discontinuous wall temperature*, Fluid Dyn. Res. **42**, pp. 1-23.

Han, L.H., Wu, S.M., Condit, J.C., Kemp, N.J., Milner, T.E., Feldman, M.D. and Chen, S.C. (2010). *Light-powered micromotor driven by geometry-assisted, asymmetric photon-heating and subsequent gas convection*, Appl. Phys. Lett. **96**, 213509.

Harten, A. (1983). *High resolution schemes for hyperbolic conservation laws*, J. Comput. Phy. **49**, pp. 357-393.

Harten, A., Engquist, B., Osher S. and Chakravarthy, S. (1987). *Uniformly high order essentially non-oscillatory schemes, III*, J. Comput. Phy. **71**, pp. 231-303.

Hilbert, D. (1912). *Grundzuge einer allgemeinen theorie der linearen integralgleichungen* (Teubner, Leipzig).

Hirsch, C. (2007). *Numerical computation of internal and external flows: the fundamentals of computational fluid dynamics* (Butterworth-Heinemann).

Holden, M.S. and Wadhams, T.P. (2003). *A review of experimental studies for DSMC and Navier-Stokes code validation in laminar regions of shock/shock and shock boundary layer interaction including real gas effects in hypervelocity flows*, AIAA 2003-3641.

Holway, H. (1966). *New statistical models for kinetic theory: methods of construction*, Phys. Fluids **9**, pp. 1658.

Huang, J.C., Xu, K. and Yu, P.B. (2012). *A unified gas-kinetic scheme for continuum and rarefied flows II: multi-dimensional cases*, Communications in Computational Physics **12**, no. 3, pp. 662-690.

Huang, J.C., Xu, K. and Yu, P.B. (2013). *A unified gas-kinetic scheme for continuum and rarefied flows III: microflow simulations*, Commun. Comput. Phys. **13**, no. 5, pp. 1147-1173.

Hui, W.H., and Xu, K. (2012). *Computational fluid dynamics based on unified coordinates*, Science Press.

Huynh, H.T. (2007). *A flux reconstruction approach to high-order schemes including discontinuous Galerkin methods*, AIAA Paper 2007-4079.

Huynh, H.T. (2009). *A reconstruction approach to high-order schemes including Discontinuous Galerkin for diffusion*, AIAA Paper 2009-403.

Ilgaz, M. and Tuncer, I.H. (2009). *A parallel gas-kinetic BhatnagarCGross-CKrook method for the solution of viscous flows on two-dimensional hybrid grids*, Int. J. Comput. Flu. Dyn. **23**, No. 10, pp. 699-711.

Ivanov, M. and Gimelshein, S. (1998). *Computational hypersonic rarefied flows*, Annual Review of Fluid Mech. **30**, pp. 469-505.

JIANG, G.S. AND SHU, C.W. (1996). *Efficient implementation of Weighted ENO schemes*, J. Comput. Phys. **126**, pp. 202-228.

JIANG, J. AND QIAN, Y.H. (2012). *Implicit gas-kinetic BGK scheme with multigrid for 3D stationary transonic high-Reynolds number flows*, Computers and Fluids **66**, pp. 21-28.

JIN, S. (1995). *Runge-Kutta methods for hyperbolic conservation laws with stiff relaxation terms*, J. Comput. Phys. **122**, pp. 51-67.

JIN, S. (2012). *Asymptotic preserving (AP) schemes for multiscale kinetic and hyperbolic equations: a review*, Riv. Mat. Univ. Parma **3**, pp. 177-216.

JIN, S. AND LEVERMORE, C.D. (1991). *The discrete-ordinate method in diffusive regimes*, Transport Theory Statist. Phys. **20**, pp. 413-439.

JIN, S. AND LEVERMORE, C.D. (1993). *Fully discrete numerical transfer in diffusive regimes*, Transport Theory Statist. Phys. **22**, pp. 739-791.

JIN, S. AND LEVERMORE, C.D. (1996). *Numerical schemes for hyperbolic conservation laws with stiff relaxation terms*, J. Comput. Phys. **126**, pp. 449-467.

JIN, S., PARESCHI, L. AND TOSCANI, G. (2000). *Uniformly accurate diffusive relaxation schemes for multiscale transport equations*, SIAM J. Numer. Anal. **38**, pp. 913-936.

JIN, C.Q. AND XU, K. (2007). *A unified moving grid gas-kinetic method in Eulerian space for viscous flow computation*, J. Comput. Phys. **222**, pp. 155-175.

JOHN, B., GU, X.J. AND EMERSON, D.R. (2011). *Effects of incomplete surface accommodation on non-equilibrium heat transfer in cavity flow: A parallel DSMC study*, Comput. and Fluids **45**, pp. 197-201.

JOHN, E.B.(1996). *Grid fins for missile applications in supersonic flow*, AIAA 96-0194.

JOU, D., GASAS-VAZQUEZ, J. AND LEBON, G. (1996). *Extended irreversible therodynamics*, Springer.

KARNIADAKIS, G., BESKOK, A. AND ALURU, N. (2005). *Microflows and nanoflows*, Springer Science+Business Media, Inc.

KERIMO, J. AND GIRIMAJI, S. (2007). *Boltzmann-BGK approach to simulating weakly compressible 3D turbulence: comparison between lattice Boltzmann and gas kinetic methods*, J. of Turbulence **8**, No. 46, DOI:10.1080/14685240701528551.

KIM, S.S., KIM, C., RHO, O.H. AND HONG, S.K. (2001). *Methods for the Accurate Computations of Hypersonic Flows I. AUSMPW+ scheme*, J. Comput. Phys. **174**, pp. 38-80.

KITAMURA, K. AND SHIMA, E. (2013). *Towards shock-stable and accurate hypersonic heating computations: A new pressure flux for AUSM-family schemes*, J. Comput. Phys. **245**, pp. 62-83.

KLAR, A. (1998). *An asymptotic-induced scheme for nonstationary transport equations in the diffusive limit*, SIAM J. Numer. Anal. **35**, pp. 1073-1094.

KOGAN, M.N. (1969). *Rarefied gas dynamics*, Plenum Press, New York.

KOLOBOV, V.I., ARSLANBEKOV, R.R., ARISTOV, V.V., FROLOVA, A.A. AND ZABELOK, S.A. (2007). *Unified solver for rarefied and continuum flows with adaptive mesh and algorithm refinement*, J. Comput. Phys. **223**, pp. 589-608.

KOLOBOV, V.I., ARSLANBEKOV, R.R. AND FROLOVA, A.A. (2010). *Boltzmann solver with adaptive mesh in velocity space, 27th International Symposium on Rarefied Gas*, Dynamics **133**, pp. 928-933.

KOLOBOV, V.I. AND ARSLANBEKOV, R.R. (2012). *Towards adaptive kinetic-fluid simulations of weakly ionized plasmas*, J. Comput. Phys. **231**, pp. 839-869.

KOLOBOV, V.I., ARSLANBEKOV, R.R. AND FROLOVA, A.A. (2011). *Boltzmann solver with adaptive mesh in velocity space*, In 27th International Symposium on Rarefied Gas Dynamics **133** of AIP Conf. Proc., pp. 928-933.

KNUDSEN, M. (1950). *The kinetic theory of gases*, third ed., Wiley, New York.

KOURA, K. (1992). *Statistical inelastic crosssection model for the Monte Carlo simulation of molecules with discrete internal energy*, Phys. Fluids A **4**, pp. 1782-1788.

KROOK, M. AND WU, T.T. (1977). *Exact solutions of the Boltzmann equation*, Phys. Fluids **20**, pp. 1589.

KUMAR, G., GIRIMAJI, S.S. AND KERIMO, J. (2013). *WENO-enhanced gas-kinetic scheme for direct simulations of compressible transition and turbulence*, J. Comput. Phys. **234**, pp. 499-523.

KUZMIN, D., LOHNER, R. AND TUREK, S. (2012). *Flux-corrected transport: principles, algorithms, and applications*, Springer.

LALLEMAND, P. AND LUO, L.S. (2000). *Theory of the lattice Boltzmann method: Dispersion, dissipation, isotropy, Galilean invariance, and stability*, Phys. Rev. E **61**, 6546.

LANDAU, L.D. AND E.M. LIFSHITZ (1959). *Fluid mechanics*, London: Pergamon Press.

LARSEN, A.W. AND MOREL, J.E. (1989). *Asymptotic solutions of numerical transport problems in optically thick, diffusive regimes. II*, J. Comput. Phys. **83**, pp. 212-236.

LARSEN, A.W., MOREL, J.E. AND MILLER, W.F. (1987). *Asymptotic solutions of numerical transport problems in optically thick, diffusive regimes*, J. Comput. Phys. **69**, pp. 283-324.

LEE, C. B. AND WANG, S. (1995). *Study of the shock motion in a hypersonic shock system/turbulent boundary layer interaction*, Experiments in fluids **19**, pp. 143-149.

LEE, C. B. AND WU, J.Z. (2008). *Transition in wall-bounded flows*, Applied Mechanics Reviews **61** (3), 030802.

LEMOU, M. AND MIEUSSENS, L. (2008). *A new asymptotic preserving scheme based on micro-macro formulation for linear kinetic equations in the diffusion limit*, SIAM J. Sci. Comput. **31**, pp. 334-368.

LEREU, A.L., PASSIAN, A., WARMACK, R.J., FERRELL, T.L. AND THUNDAT, T.(2004). *Effect of thermal variations on the Knudsen forces in the transitional regime*, Appl. Phys. Lett. **84**, pp. 1013-1015.

LEVEQUE, R.J.(2002). *Finite volume methods for hyperbolic problems*, Cambridge university press.

LEVERMORE, C.D. (1996). *Moment closure hierarchies for kinetic theories, J. Statistical Phys.* **83**, pp. 1021.

LI, J.Q., LI, Q.B. AND XU, K. (2011). *Comparison of the generalized Riemann solver and the gas-kinetic scheme for inviscid compressible flow simulations, J. Comput. Phys.* **230**, pp. 5080-5099.

LI, Q.B. AND FU, S. (2003). *Numerical simulation of high-speed planar mixing layer, Comput. Fluids* **32**, pp. 1357-1377.

LI, Q.B. AND FU, S. (2006). *On the multidimensional gas-kinetic BGK scheme, J. Comput. Phys.* **220**, pp. 532-548.

LI, Q.B., XU, K. AND FU, S. (2010). *A High-order gas-kinetic Navier-Stokes solver, J. Comput. Phys.* **229**, pp. 6715-6731.

LI, J. (2012). *Gas-kinetic BGK scheme and its application in continuum and transition regimes*, Master Thesis (in Chinese), China Aerodynamics Research and Development Center.

LI, J., JIANG, D.W., MAO, M.L. AND DENG, X.G.(2013). *Research on the application of BGK-NS method to complex flows*, Acta Aerodynamics Sinica (in Chinese), **31** (4), pp. 449-454.

LI, Z.H. AND ZHANG, H.X. (2009). *Gas-kinetic numerical studies of three-dimensional complex flows on spacecraft re-Entry, J. Comput. Phys.* **228**, pp. 1116-1138.

LIAO, W., PENG, Y. AND LUO, L.S. (2009). *Gas-kinetic schemes for direct numerical simulations of compressible homogeneous turbulence, Physical Review E* **80**, 046702.

LIAO, W., PENG, Y. AND LUO, L.S. (2010). *Effects of multitemperature nonequilibrium on compressible homogeneous turbulence, Physical Review E* **81**, 046704.

LIOU, M.S.(2006). *A sequel to AUSM, part II: AUSM+-up for all speeds, J. Comput. Phys.* **214**, pp. 137-170.

LIU, C., LU, P., CHEN, L. AND YAN, Y.(2012). *New theories on boundary layer transition and turbulence formation, J.* Modeling and Simulation in Engineering **2012**, ID 649419.

LIU, G.(1990). *A method for constructing a model form for the Boltzmann equation, Phys. Fluids* **A2**, pp. 277.

LIU, C., XU, K., SUN, Q.H., AND CAI, Q.D. (2014). *A unified gas-kinetic scheme for continuum and rarefied flows, direct modeling, and full Boltzmann collision term, preprint.* arXiv:1405.4479V1.

LIU, N. AND TANG, H.Z.(2014). *A high-order accurate gas-kinetic scheme for one- and two-dimensional flow simulation*, Communications in Comput. Phys. **15**, pp. 911-943.

LIU, S., YU, P.B., XU, K. AND ZHONG, C.W.(2014). *Unified Gas Kinetic Scheme for Diatomic Molecular Simulations in All Flow Regimes, J. Comput. Phys.* **259**, pp. 96-113.

LIU, S. AND ZHONG, C.W.(2014). *Investigation of the kinetic model equations, Physical Review E* **89**, 033306.

LOCKERBY, D.A., REESE, J.M., EMERSON, D.R. AND BARBER, R.W. (2004). *Velocity boundary condition at solid walls in rarefied gas calculations,*

Physical Review E **70**, 017303.

LOCKERBY, D.A., REESE, J.M. AND GALLIS, M.A. (2005). *Capturing the Knudsen layer in continuum-fluid models of non-equilibrium gas flows, AIAA Journal* **43**, pp. 1391-1993.

LOH, C.Y. AND JORGENSON, P.C.E. (2009). *Multi-dimensional Dissipation for Cure of Pathological Behaviors of Upwind Scheme*, J. Comput. Phys. **228**, pp. 1343-1346.

LOYALKA, S.K. AND CHENG, T.C. (1979). *Sound-wave propagation in a rarefied gas, Physics of Fluids*, **22**, pp. 830-836.

LUI, S.H. AND XU, K. (2001). *Entropy analysis of gas-kinetic schemes for the compressible Euler equations*, Z. angew. Math. Phys. **52**, pp. 62-78.

LUO, H., LUO, L. AND XU, K. (2009). *A Discontinuous Galerkin method based on BGK scheme for the Navier-Stokes equations on arbitrary grids*, Advances in Applied Mathematics and Mechanics **1**, pp. 301-318.

LUO, J. (2012). *A high-order Navier-Stokes flow solver and gravitational system modeling based on gas-kinetic equation, PhD Thesis*, Hong Kong University of Science and Technology.

LUO, J. AND XU, K. (2013). *A high-order multidimensional gas-kinetic scheme for hydrodynamic equations*, Science China, Technological Sciences **56**, No. 10, pp. 2370-2384.

LUO, J. AND XU, K. (2013). *A compact high-order multidimensional gas-kinetic scheme for inviscid and viscous flow simulations*, preprint.

LUO, J., XU, K. AND LIU, N. (2011). *A well-balanced symplecticity-preserving gas-kinetic scheme for hydrodynamic equations under gravitational field, SIAM J. Sci. Comput.* **33**, No. 5, pp. 2356-2381.

LUO, J., XUAN, L.J. AND XU, K. (2013). *Comparison of fifth-order WENO scheme and WENO-gas-kinetic scheme for inviscid and viscous flow simulation*, Commun. Comput. Phys. **14**, No. 3, pp. 599-620.

MALEK,M.M., BARAS, F. AND GARCIA, A.L. (1997). *On the validity of hydrodynamics in plane Poiseuille flows*, Physica A **240**, pp. 255.

MANDAL, J.C. AND DESHPANDE, S.M. (1994). *Kinetic flux vector splitting for Euler equations, Computers & Fluids* **23**, pp. 447.

MASLACH, G.J. AND SCHAAF, S.A. (1963). *Cylinder drag in the transition from continuum to free-molecule flow*, Phys. Fluids **16**, pp. 315.

MASTERS, N.D. AND YE, W. (2007). *Octant flux splitting information preserving DSMC method for thermally driven flows*, J Comput. Phys. **226**, pp. 2044-2062.

MAXWELL, J.C. (1879). *On stresses in rarefied gases arising from inequalities of temperature*, Philos. Trans. R. Soc. London **170**, pp. 231-256.

MAY, G., SRINIVASAN, B. AND JAMESON, A. (2007). *An improved gas-kinetic BGK finite-volume method for three-dimesnional transonic flow*, J. Comput. Phys. **220**, pp. 856-878.

MEHRENBERGER, M., VIOLARD, E., HOENEN, O., PINTO, M.C. AND SONNENDRCKER, E. (2006). *A parallel adaptive Vlasov solver based on hierarchical finite element interpolation*, Nuclear Instrum. Methods Phys. Res. A **558**, pp. 188-191.

MENG, J., ZHANG, Y., HADJICONSTANTINOU, N. G., RADTKE, G. A. AND SHAN, X. (2013a). *Lattice ellipsoidal statistical BGK model for thermal non-equilibrium flows*, J. Fluid Mech. **718**, pp. 347-370.

MENG, J., ZHANG, Y. AND REESE, J. M. (2013b). *Assessment of the ellipsoidal-statistical Bhatnagar-Gross-Krook model for force-driven poiseuille flows.*, J. Comput. Phys. **251**, pp. 383-395.

METCALF, S.C., BERRY, C.J. AND DAVIS, B.M. (1965). *An investigation of the flow about circular cylinders placed normal to a low-density, supersonic stream*, Aeronautical Research Council, Reports and Memoranda No. *3416*, Her Majesty's Stationery Office, London.

MEYER, E. AND SESSLER, G. (1957). *Schallausbreitung in gasen bei hohen frequenzen und sehr niedrigen Drucken*, Zeitschrift für Physik A Hadrons and Nuclei, **149**, pp. 15-39.

MIEUSSENS, L. (2000). *Discrete-velocity models and numerical schemes for the Boltzmann-BGK equtaion in plane and axisymmetric geometries*, J. Comput. Phys. **162**, pp. 429-466.

MIEUSSENS, L. (2013). *On the Asymptotic preserving property of the Unified Gas Kinetic Scheme for the diffusion limit of linear kinetic models*, J. Comput. Phys. **253**, pp. 138-156.

MIEUSSENS, L. AND STRUCHTRUP, H. (2004). *Numerical Comparison of Bhatnagar-Gross-Krook models with Proper Prandtl Number*, Phys Fluids **16**, pp. 2797-2813.

MOHAMMADZADEH, A., ROOHI, E., NIAZMAND, H., STEFANOV, S. AND MYONG, R.S. (2012). *Thermal and second-law analysis of a micro- or nanocavity using direct-simulation Monte Carlo*, Phys. Rev. E **85**, 056310.

MORINISHI, K. (2006). *Numerical simulation for gas microflows using Boltzmann equation*, Computers Fluids **35**, pp. 978-985.

MOTT-SMITH, H.M. (1951). *The solution of the Boltzmann equation for a shock wave*, Phys. Rev., **82**, pp. 885-892.

MULLER, I. AND RUGGERI, T. (1998). *Rational extended thermodynamics, Springer, New York.*

MYSONG, R.S. (1999). *Thermodynamically consistent hydrodynamic computational models for high-Knudsen-number gas flows*, Phys. Fluids **11**, 2788.

MYSONG, R.S. (2001). *A computational method for Eu's generalized hydrodynamic equations of rarefied and microscale gasdynamics*, J. Comput. Phys. **168**, No. 1, pp. 47-72.

MYSONG, R.S., REESE, J.M., BARBER, R.W. AND EMERSON, D.R. (2005). *Velocity slip in microscale cylindrical Couette flow: The Langmuir model*, Phys. Fluids **17**, 087105.

NALDI, G. AND PARESCHI, L. (1998). *Numerical schemes for kinetic equations in diffusive regimes*, Appl. Math. Lett. **11**, pp. 29-35.

NABETH, J., CHIGULLAPALLI, S. AND ALEXEENKO, A.A. (2011). *Quantifying the Knudsen force on heated microbeams: A compact model and direct comparison with measurements*, Phys. Rev. E **83**, 066306.

NOMPELIS, I., CANDLER, G.V. AND HOLDEN, M.S. (2003). *Effect of vibrational nonequilibrium on hypersonic double-cone experiments*, AIAA J. **41**, 2162-2169.

OHWADA, T. (1993). *Structure of normal shock waves: direct numerical analysis of the Boltzmann eqaution for hard-sphere molecules*, Phys. Fluids A, **5**, pp. 217.

OHWADA, T. (2002). *On the construction of kinetic schemes*, J. Comput. Phys. **177**, pp. 156-175.

OHWADA, T., ADACHI, R., XU, K. AND LUO, J. (2013). *On the remedy against shock anomalies in kinetic schemes*, J. Comput. Phys. **255**, pp. 106-129.

OHWADA, T. AND KOBAYASHI, S. (2004). *Management of discontinuous reconstruction in kinetic schemes*, J. Comput. Phys. **197**, pp. 116-138.

OHWADA, T., SONE, Y. AND AOKI, K. (1989). *Numerical analysis of the Poiseuille and thermal transpiration flows between two parallel plates on the basis of the Boltzmann equation for hard-sphere molecules*, Phys. Fluids A **1** (12), pp. 2024-2049.

OHWADA, T. AND XU, K. (2004). *The kinetic scheme for full Burnett equations*, J. Comput. Phys. **201**, pp. 315-332.

OTA, M., NAKAO, T. AND SAKAMOTO, M. (2001). *Numerical simulation of molecular motion around laser microengine blades*, Math. Comput. Simul. **55**, pp. 223-230.

PARKER, J. (1959). *Rotational and vibrational relaxation in diatomic gases*, Phys. Fluids **2**, pp. 449-462.

PASSIAN, A., WIG, A., MERIAUDEAU, F., FERRELL, T.L. AND THUNDAT, T. (2002). *Knudsen forces on microcantilevers*, J. Appl. Phys. **92**, pp. 6326-6333.

PASSIAN, A., WARMACK, R.J., WIG, A., FARAHI, R.H., MERIAUDEAU, F., FERRELL, T.L. AND THUNDAT, T.(2003). *Observation of Knudsen effect with microcantilevers*, Ultramicroscopy **97**, pp. 401-406.

PASSIAN, A., WARMACK, R.J., FERRELL, T.L. AND THUNDAT, T. (2003). *Thermal transpiration at the microscale: A Crookes cantilever*, Phys. Rev. Lett. **90**, 124503.

PATTERSON, G.N. (1956). *Molecular flow of gases, John Wiley & Sons, INC.*.

PEKERIS, C.L., ALTERMAN, Z., FINKELSTEIN, L. AND FRANKOWSKI, K. (1962). *Propagation of sound in a gas of rigid spheres*, Physics of Fluids, **5**, pp. 1608-1610.

PERTHAME, B. (1992). *Second-order Boltzmann schemes for compressible Euler equation in one and two space dimensions*, SIAM J. Numer. Anal., **29**, pp. 1-19.

PIERACCINI, S. AND PUPPO, G. (2007). *Implicit-explicit schemes for BGK kinetic equations*, J. Scientific Computing **32**, pp. 1-28.

PHAM-VAN-DIEP, G.C., ERWIN, D.A. AND MUNTZ, E.P. (1989). *Nonequilibrium molecular motion in a hypersonic shock wave*, Science **245**, pp. 624.

PRENDERGAST, K.H. AND XU, K. (1993). *Numerical hydrodynamics from gas-kinetic theory*, J. Comput. Phys. **109**, pp. 53-66.

PULLIN, D.I. (1980). *Direct simulation methods for compressible inviscid ideal gas flow*, J. Comput. Phys., **34**, pp. 231-244.

Qian, J.Z., Li, J.Q. and Wang, S.H. (2013). *The generalized Riemann problems for hyperbolic balance laws: A unified formulation towards high order, arXiv:1303.2941.*

Qian, Y.H., D'Humieres, D., and Lallemand, P. (1992). *Lattice BGK models for Navier-Stokes equation*, Europhys. Lett. **17**, pp. 479-484.

Quirk, J. (1994). *A contribution to the great Riemann solver debate*, Int. J. Num. Met. in Fluids **18**, no.6, pp. 555-574.

Radar, D.J., Gallis, M.A., Torczynski, J.R. and Wagner, W. (2006). *Direct simulation Monte Carlo convergence behavior of the hard-sphere-gas thermal conductivity for Fourier heat flow*, Physics of Fluids **18**, 077102.

Radtke, G.A., Hadjiconstantinou, N.G. and Wagner, W. (2011). *Low-noise Monte Carlo simulation of the variable hard sphere gas*, Physics of Fluids **23**, 030606.

Reitz, R.D. (1981). *One-dimensional compressible gas dynamics calculations using the Boltzmann equations*, J. Comput. Phys., **42**, pp. 108-123.

Reynolds, O. (1876). *On the forces caused by the communication of heat between a surface and a gas; and on a new photometer*, Philos. Trans. R. Soc. London **166**, pp. 725-735.

Reynolds, O. (1879). *On certain dimensional properties of matter in the gaseous state*, Philos. Trans. R. Soc. London **170**, pp. 727-845.

Righi, M. (2014). *A modified gas-kinetic scheme for turbulent flow*, Commun.Comput. Phys., to appear.

Roe, P.L. (1981). *Approximare Riemann solvers, parameter vector, and difference schemes*, J. Comput. Phys. **43**, pp. 357-372.

Roe, P.L. (1986). *Characteristic-based schemes for the Euler equations*, Ann. Rev. Fluid Mech. **18**, pp. 337.

Rovenskaya, O.I., Polikarpov, A.P. and Graur, I.A. (2013). *Comparison of the numerical solutions of the full Boltzmann and S-model kinetic equations for gas flow through a slit*, Computers & Fluids **80**, pp. 71-78.

Robben, F. and Talbot, L. (1966). *Experimental study of the rotational distribution function of nitrogen in a shock wave*, Phys. Fluids **9**, pp. 653-662.

Rykov, V. (1975). *A model kinetic equation for a gas with rotational degrees of freedom*, Translated from Izvestiya Akademii Nauk SSSR, iVlekhanika Zhidkostii Gaza, Moscow **1**, pp. 107-115.

Sanders, R.H. and Prendergast, K.H. (1974). *The possible relation of the three-kiloparsec arm to explosions in the galactic nucleus*, Astrophysical Journal, **188**, pp. 489-500.

Santos, A., Brey, J.J., Kim, C.S. and Dufty, J.W. (1989). *Velocity distribution for a gas with steady heat flow*, Phys. Rev. A **39**, pp. 320.

Scandurra, M., Iacopetti, F. and Colona, P. (2007). *Gas kinetic forces on thin plates in the presence of thermal gradients*, Phys. Rev. E **75**, 026308.

Schotter, R. (1974). *Rarefied gas acoustics in the noble gases*, Physics of Fluids, **17**, pp. 1163-1168.

Schwartzentruber, T.E. and Boyd, I.D. (2006). *A hybrid particle-continuum method applied to shock waves*, J. Comput. Phys. **215**, pp. 402-416.

Schwartzentruber, T.E., Scalabrin, L.C. and Boyd, I.D. (2007).

A molecular particle-continuum method for hypersonic non-equilibrium gas flows, J. Comput. Phys. **225**, pp. 1159-1174.

Schwartzentruber, T.E., Scalabrin, L.C. and Boyd, I.D. (2008). *Multi-scale particle-continuum simulations of hypersonic flow over a planetary probe*, J. Spacecraft Rockets **45**, pp. 1196-1206.

Selden, N., Ngalande, C., Gimelshein, S., Muntz, E.P., Alexeenko, A. and Ketsdever, A. (2009). *Area and edge effects in radiometric forces*, Phys. Rev. E **79**, 041201.

Selden, N., Ngalande, C., Gimelshein, N., Gimelshein, S. and Ketsdever, A. (2009). *Origins of radiometric forces on a circular vane with a temperature gradient*, J. Fluid Mech. **634**, pp. 419-431.

Shakhov, E.M. (1968). *Generalization of the Krook kinetic Equation*, Fluid Dyn.**3**, pp. 95.

Shan, X.W., Yuan, X.F. and Chen, H.D. (2006). *Kinetic theory representation of hydrodynamics: a way beyond the Navier-Stokes equations*, J. Fluid Mech. **550**, pp. 413-441.

Sharipov, F. (1999). *Non-isothermal gas flow through rectangular microchannels*, Journal of Micromechanics and Microengineering**9**, pp. 394-401.

Sharipov, F. (2002). *Free molecular sound propagation*, The Journal of the Acoustical, **112**, pp. 395-401.

Shizgal, B. (1981). *A Gaussian quadrature procedure for use in the solution of the Boltzmann equation and related problems*, J. Comput. Phys. **41**, pp. 309-328.

Shu, C.W. (1998). *Essentially non-oscillatory and weighted essentially non-oscillatory schemes for hyperbolic conservation laws*, Lecture Notes in Mathematics, Springer.

Shyy, W. (2006). *Computational modeling for fluid flow and interfacial transport*, Dover Books on Engineering.

Sirovich L. and Thurber, J.K. (1965). *Propagation of forced sound waves in rarefied gasdynamics*, The Journal of the Acoustical Society of America, **37**, pp. 329-339.

Slyz, A. and Prendergast, K.H. (1999). *Time-independent gravitational fields in the BGK scheme for hydrodynamics*, Astron. Astrophys. Suppl. Ser. **139**, pp. 199-217.

Sone, Y. (2007). *Molecular gas dynamics: theory, techniques, and applications*, Birkhauser Basel.

Sone, Y., Takata, S. and Ohwada, T. (1990). *Numerical analysis of the plane Couette flow of a rarefied gas on the basis of the linearized Boltzmann equation for hard-sphere molecules*, Eur. J. Mech., B/Fluids **9**, pp. 273.

Steen, N.M., Byrne, G.D. and Gelbard, E.M. (1969). *Gaussian quadratures for the integrals* $\int_0^\infty exp(-x^2)f(x)dx$, Math. Comp. **23**, pp. 661-671.

Steger, J.L. and Warming, R.F. (1981). *Flux vector splitting of the inviscid gas-dynamic equations with applications to finite difference methods*, J. Comput. Phys. **40**, pp. 263-293.

Steinhilper, E.A. (1972). *Electron beam measurements of the shock wave structure: Part 1, the inference of intermolecular potential from shock structure experiments, Ph.D. Thesis*, California Institute of Technology.

STRUCHTRUP, H. (2005). *Macroscopic transport equations for rarefied gas flows: approximation methods in kinetic theory*, Interaction of Mechanics and Mathematics Series, Springer, Heidelberg.

STRUCHTRUP, H. (2012). *Resonance in rarefied gases, Continuum Mechanics and Thermodynamics*, **24**, pp. 361-376.

STRUCHTRUP, H. AND TORRIHON, M. (2003). *Regularization of Grads 13 Moment Equations: Derivation and Linear Analysis*, *Phys. Fluids* **15**, pp. 2668.

SU, M.D., XU, K. AND GHIDAOUI, M.S. (1999). *Low-speed flow simulation by the gas-kinetic scheme*, *J. Comput. Phys.* **150**, pp. 17-39.

SUMAN, S., AND GIRIMAJI, S. (2013). *Velocity gradient dynamics in compressible turbulence: Characterization of pressure-Hessian tensor*, *Phy. Fluids* **25**, 125103.

SUN, M., SAITO, T., JACOBS, P.A., TIMOFEEV, E.V., OHTANI, K. AND TAKAYAMA K. (2005). *Axisymmetric shock wave interaction with a cone: a benchmark test* , *Shock Waves* **14**, pp. 313-331.

SUN, Q. (2003). *Information preservation methods for modeling micro-scale gas flows*, *Ph.D. thesis*, The University of Michigan.

SUN, Q. AND BOYD, I.D. (2002). *A direct simulation method for subsonic microscale gas flows*, *J. Comput. Phys.* **179**, pp. 400-425.

SUN, Q., CAI, C.P. AND GAO, W. (2014). *On the validity of the Boltzmann-BGK model through relaxation evaluation*, Acta Mech. *Sin.* **30**, pp. 133-143.

SUN, W.J., JIANG, S. AND XU, K. (2014). *Asymptotic preserving of the unified gas kinetic scheme for grey radiative transfer equations*, preprint.

TAGUCHI, S. AND AOKI, K. (2011). *Numerical analysis of rarefied gas flow induced around a flat plate with a single heated side*, *AIP Conf. Proc.* **1333**, pp. 790-795.

TANG, H.Z. AND XU, K. (2000). *A High-order Gas-kinetic Method for Multidimensional Ideal Magnetohydrodynamics*, *J. Comput. Phys.* **165**, pp. 69-88.

TANG, H.Z. AND XU, K. (2001). *Pseudo-particle Representation and Positivity Analysis of Explicit and Implicit Steger-Warming FVS Schemes*, *Z.* angew. *Math. Phys.* **52**, pp. 847-858.

TANG, L. (2012). *Prpgress in gas-kinetic upwind schemes for the solution of Euler/Navier-Stokes equations-I: overview*, Comput. Fluids **56**, pp. 39-48.

TANG, T. AND XU, K. (1999). *Gas-kinetic Schemes fro the Compressible Euler Equations I: Positivity-Preserving Analysis*, *Z.* angew. *Math. Phys.* **50**, pp. 258-281.

TCHEREMISSINE, F.G. (2008). *Solution of the Boltzmann kinetic equation for low speed flows*, Transport Theory and Statistical Physics, **37**, pp. 564-575.

THOMAS, J.R. AND SIEWERT, C.E. (1979). *Sound-wave propagation in a rarefied gas*, Transport Theory and Statistical Physics, **8**, pp. 219-240.

TIAN, C.L., XU, K., CHAN, K.L., AND DENG, L.C. (2007). *A three-dimensional multidimensional gas kinetic scheme for the Navier-Stokes equations under gravitational fields*, *J.* Comput. *Phys.* **226**, pp. 2003-2027.

TIWARI, S. (1998). *Coupling of the Boltzmann and Euler Equations with Automatic Domain Decomposition*, *J.* Comput. *Phys.* **144**, pp. 710-726.

TORO, E. (2009). *Riemann Solvers and Numerical Methods for Fluid Dynamics*, Springer.

TORRILHON, M. AND XU, K. (2006). *Stability and consistency of kinetic upwinding for advection-diffusion equations*, IMA Journal of Numerical Analysis **26**, pp. 686-722.

TSUBOI, N. AND MATSUMOTO, Y., *Experimental and numerical study of hypersonic rarefied gas flow over flat plates*, *AIAA J* **43**, pp. 1243-1255.

URIBE, F.J. AND GARCIA, A.L. (1999). *Burnett description for plane Poiseuille flow*, *Physical Review E* **60**, pp. 4063-4078.

VAN LEER, B. (1977). *Towards the ultimate conservative difference scheme IV, a new approach to numerical convection*, J. Comput. Phys., **23**, pp. 276-299.

VAN LEER, B. (1979). *Towards the ultimate conservative difference scheme V, A Second Order Sequel to Godunov's Method*, *J. Comput. Phys.* **32**, pp. 101-136.

VAN. LEER, B. (1982). *Flux-vector splitting for the Euler equations*, ICASE report, NO. 82-30.

VAN LEER, B. (2006). *Upwind and high-resolution methods for compressible flow: from donor cell to residual distribution schemes*, *Commun. Comput. Phys.* **1**, No. 2, pp. 192-206.

VAN. LEER, B., THOMAS, J.L., ROE, P.L. AND NEWSOME, R.W. (1987). *A comparison of numerical flux formulas for the Euler and Navier-Stokes equations*, AIAA paper 87-1104.

VARGO, S.E., MUNTZ, E.P., SHIFLETT, G.R. AND TANG, W.C. (1999). *Knudsen compressor as a micro- and macroscale vacuum pump without moving parts or fluids*, Journal of Vacuum Science and Technology A**7**, pp. 2308-2313.

VENUGOPAL, V. AND GIRIMAJI, S.S. (2014). *Unified gas kinetic scheme and Direct Simulation Monte Carlo Computations of high-speed lid-driven microcavity flows*, *Commun. Comput. Phys.* to appear.

VIDES, J., BRACONNIER, B., AUDIT, E., BERTHON, C. AND NKONGA, B. (2014). *A Godunov-Type Solver for the Numerical Approximation of Gravitational Flows*, *Commun. Comput. Phys.***15**, pp. 46-75.

VINCENTI, W.G. AND KRUGER, C.H. (1965). *Introduction to Physical Gas Dynamics*, Krieger.

VOGENITZ, F.W., BIRD, G.A., BROADWELL, J.E. AND RUNGALDIER H. (1968). *Theoretical and experimental study of rarefied supersonic flows about several simple shapes*, *AIAA Journal* **6**, pp. 2388.

WAGNER, W. (1992). *A convergence proof for Bird's Direct Simulation Monte Carlo Method for the Boltzmann equation*, *J. Statistical Phys.* **66**, pp. 1011-1044.

WANG CHANG, C.S. AND UHLENBECK, G.E. (1970). *On the propagation of sound in monatomic gases*, Studies in statistical mechanics, **5**, pp. 43-75, J. D. Boer and G. E. Uhlenbeck, Eds. Amsterdam: North Holland.

WANG, P., ZHU, L.H., GUO, Z.L. AND XU, K. (2014). *A comparative study of LBE and DUGKS for near incompressible flows*, preprint.

WANG, R.J. AND XU, K. (2012). *The study of sound wave propagation in rarefied gases using unified gas-kinetic scheme*, Acta Mech. Sinica, **28**, pp. 1022-1029.

WANG, R.J. AND XU, K. (2014). *Unified gas-kinetic simulation of slider air bearing*, Theoretical and Applied Mechanics Letters, **4**, 022001.

WANG, Z.J. (2011). *Adaptive High-order methods in computational fluid dynamics*, Advances in Computational Fluid Dynamics Vol. 2, World Scientific Publishing Co..

WANG, Z.J. AND GAO, H. (2009). *A unifying lifting collocation penalty formulation including the discontinuous Galerkin, spectral volume/ difference methods for conservation laws on mixed grids*, J. Comput. Phys. **228**, pp. 8161-8186.

WESTPHAL, H. (1920). *Messungen am radiometer*, Z. Phys. **1**, pp. 92-100.

WIJESINGHE, H.S., HORNUNG, R.D., GARCIA, A.L. AND HADJICONSTANTINOU, N.G. (2004). *Three-dimensional hybrid continuum-atomistic simulations for multiscale hydrodynamics*, J. Fluid Engineering **126**, pp. 768-777.

WOODS, L.C. (1993). *An introduction to the kinetic theory of gases and magnetoplasmas*, Oxford University Press.

WOODWARD, P. AND COLELLA, P. (1984). *Numerical simulations of two-dimensional fluid flow with strong shocks*, J. Comput. Phys. **54**, pp. 115-173.

WRIGHT, M.J., SINHA, K., OLEJNICZAK, J., CANDLER, G.V., MAGRUDER, T.D. AND SMITS A.J. (2000). *Numerical and experimental investigation of double-cone shock interactions*, AIAA Journal **38**, pp. 2268-2276.

WU, K.L., YANG, Z.C. AND TANG, H.Z. (2013). *A third-order accurate direct Eulerian GRP scheme for the Euler equations in gas dynamics*, J. Comput. Phys. submitted.

WU, L. (2013). *Deterministic Numerical Simulation of the Boltzmann and Kinetic Model Equations for Classical and Quantum Dilute Gases*, Ph.D. thesis, University of Strathclyde.

WU, L., REESE, J.M. AND ZHANG, Y.H. (2014). *Solving the Boltzmann equation deterministically by the fast spectral method: application to gas microflows*, J. Fluid Mech. **746**, pp. 53-84.

WU, L., WHITE, C., SCANLON, T.J., REESE, J.M. AND ZHANG, Y.H. (2013). *Deterministic numerical solutions of the Boltzmann equation using the fast spectral method*, J. Comput. Phys. **250**, pp. 27-52.

XING, Y. AND SHU, C.W. (2013). *High order well-balanced WENO scheme for the gas dynamics equations under gravitational fields*, ournal of Scientific Computing **54**, pp. 645-662.

XU, K. (1993). *Numerical hydrodynamics from gas-kinetic theory*, Ph.D. thesis, Columbia University.

XU, K. (1998). *Gas-kinetic schemes for unsteady compressible flow simulations*, von Karman Institute for Fluid Dynamics Lecture Series 1998-03.

XU, K. (1999). *Does perfect Riemann solver exist ?*, AIAA 99-3344.

XU, K. (1999). *Gas-kinetic theory based flux splitting method for Ideal MHD equations*, J. Comput. Phys. **153**, pp. 334-352.

XU, K. (2001). *A gas-kinetic BGK scheme for the Navier-Stokes equations and its connection with artificial dissipation and Godunov method*, J. Comput.

Phys. **171**, pp. 289-335.

XU, K. (2002). *Regularization of the Chapman-Enskog expansion and its description of shock structure*, Physics of Fluids **14**, pp. L17-L20.

XU, K. (2002). *A Slope-update Scheme for Compressible Flow Simulation*, J. Comput. Phys. **178** (2002), pp. 252-259.

XU, K. (2002). *A well-balanced gas-kinteic scheme for the shallow water equations with source terms*, J. Comput. Phys. **178** (2002), pp. 533-562.

XU, K. AND GUO, Z.L. (2011). *Multiple temperature gas dynamic equations for non-equilibrium flows*, J. Comput. Math. **29**, pp. 639-660.

XU, K. AND HUANG, J.C. (2010). *A unified gas-kinetic scheme for continuum and rarefied flows* J. Comput. Phys. **229**, pp. 7747-7764.

XU, K. AND HUANG, J.C. (2011). *An improved unified gas-kinetic scheme and the study of shock structures*, IMA J. of Appl. Math., **76**, pp. 698-711.

XU, K., HE, X. AND CAI, C. (2008). *Multiple Temperature Kinetic Model and Gas-Kinetic Method for Hypersonic Nonequilibrium Flow Computations*, J. Comput. Phys. **227**, pp. 6779-6794.

XU, K. AND HE, X.Y. (2003). *Lattice Boltzmann method and gas-kinetic BGK scheme in the low-Mach number viscous flow simulations*, J. Comput. Phys. **190**, pp. 100-117.

XU, K. AND JOSYULA, E. (2006). *Continuum formulation for non-equilibrium shock structure calculation*, Communications in Computational Physics **1**, pp. 425-450.

XU, K., KIM, C., MARTINELLI, L. AND JAMESON, A. (1996). *BGK-based schemes for the simulation of compressible flow*, International Journal of Computational Fluid Dynamics **7**, pp. 213-234.

XU, K. AND LI, Z.H. (2004). *Microchannel flows in slip flow regime: BGK-Burnett solutions*, J. Fluid Mech., **513**, pp. 87-110.

XU, K. AND LI, Z.W. (2001). *Dissipative Mechanism in Godunov-Type Schemes*, Int. J. Numer. Methods in Fluids, **37**, pp. 1-22.

XU, K., LIU, H. AND JIANG, J.Z. (2007). *Multiple temperature kinetic model for continuum and near continuum flows*, Physics of Fluids **19**, 016101.

XU, K. AND LIU, H. (2008). *A Multiple temperature kinetic model and its application to near continuum flows*, Commun. Comput. Phys. **4**, pp. 1069-1085.

XU, K. AND LUO, L.S. (1998). *Connection between Lattice Boltzmann Equation and Beam Scheme*, Int. J. Modern Physics C, **9**, pp. 1177-1188.

XU, K., MAO, M.L. AND TANG, L., *A Multidimensional Gas-Kinetic BGK scheme for hypersonic viscous flow*, J. Comput. Phys. **203**, pp. 405-421.

XU, K., MARTINELLI, L. AND JAMESON, A. (1995). *Gas-kinetic finite volume methods, flux-vector splitting and artificial diffusion*, J. Comput. Phys. **120**, pp. 48-65.

XU, K. AND PRENDERGAST, K.H. (1994). *Numerical Navier-Stokes solutions from gas-kinetic theory*, J. Comput. Phys. **114**, pp. 9-17.

XUAN, L.J. AND XU, K. (2013). *A new gas-kinetic scheme based on analytical solutions of the BGK equation*, J. Comput. Phy. **234**, pp. 524-539.

YANG, J.Y. AND HUANG, J.C. (1995). *Rarefied flow computations using nonlinear model Boltzmann equations*, J. Comput. Phys. **120**, pp. 323-339.

YANG, L.M., SHU, C., WU, J., ZHAO, N. AND LU, Z.L. (2013). *Circular function-based gas-kinetic scheme for simulation of inviscid compressible flows*, J. Comput. Phys. **255**, pp. 540-557.

YANG, L.M., SHU, C. AND WU, J. (2014). *Circular function-based gas-kinetic scheme for simulation of viscous incompressible and compressible flows*, to appear in J. Comput. Phys.

SJOGREEN, B. AND YEE, H.C. (2003). *Grid convergence of high order methods for multiscale complex unsteady viscous compressible flows*, J. Comput. Phys. **185**, pp. 1-26.

YU, P.B. (2013). *A Unified Gas Kinetic Scheme For All Knudsen Number Flows*, *PhD Thesis*, Hong Kong University of Science and Technology.

ZHANG, C.H., TANG, Q. AND LEE, C.B. (2013). *Hypersonic boundary-layer transition on a flared cone*, Acta Mechanica Sinica **29** (1), pp. 48-53.

ZHANG, J., FAN, J. AND JIANG, J.Z. (2011). *Multiple Temperature Model for the Information Preservation Method and Its Application to Nonequilibrium Gas Flows*, J. Comput. Phys. **230**, pp. 7250-7265.

ZHENG, Y., GARCIA, A.L. AND ALDER, B.J. (2002). *Comparison of kinetic theory and hydrodynamics for Poiseuille flow*, J. Statistical Phys. **109**, pp. 495-505.

ZHENG, Y., GARCIA, A.L. AND ALDER, B.J. (2002). *Comparison of kinetic theory and hydrodynamics for Poiseuille flow*, Rarefied Gas Dynamics **23**, Whistler, Canada.

ZHONG, X., MACCORMACK, R.W. AND CHAPMAN, D.R. (1993). *Stabilization of the Burnett equations and application to hypersonic flows*, AIAA J. **31**, pp. 1036.

ZHOU, J.G., CAUSON, D.M., MINGHAM, C.G. AND INGRAM, D.M. (2001). *The surface gradient method for the treatment of source terms in the shallow-water equations*, J. Comput. Phys. **168**, pp. 1-25.

ZHU, T.S. AND YE, W.J. (2010). *Origin of Knudsen forces on heated microbeams*, Phys. Rev. E **82**, 036308.

ZHUK, V.I., RYKOV, V.A. AND SHAKHOV, E.M. (1973). *Kinetic models and the shock structure problem*, Fluid Dynamics **8**(4), pp. 620-625.

Index

Printed in the United States
By Bookmasters

Printed in the United States
By Bookmasters